Stochastic problems in dynamics

Symposium on Stochastic Problems in Dynamics...

Stochastic problems in dynamics

B L Clarkson (Editor)

University of Southampton

J K Hammond, P J Holmes and A Kistner (Associate Editors)

Portsmouth Polytechnic, Cornell University and Universität Stuttgart

Pitman

LONDON · SAN FRANCISCO · MELBOURNE

PITMAN PUBLISHING LIMITED
39 Parker Street, London WC2B 5PB

FEARON–PITMAN PUBLISHERS INC.
6 Davis Drive, Belmont, California 94002, USA

Associated Companies
Copp Clark Ltd, Toronto
Pitman Publishing Co. SA (Pty) Ltd, Johannesburg
Pitman Publishing New Zealand Ltd, Wellington
Pitman Publishing Pty Ltd, Melbourne

First published 1977

AMS Subject Classifications: (main) 70L05, 93E10, 93E15

British Library Cataloguing in Publication Data

Stochastic problems in dynamics.
1. Dynamics – Congresses 2. Stochastic processes – Congresses
1. International Union of Theoretical and Applied Mechanics II. Clarkson, Brian Leonard III. Series
531'.11'015192 QA845

ISBN 0-273-01104-9

Reproduced and printed by photolithography
in Great Britain at Biddles of Guildford

Foreword

Everyone knows that mathematical methods play an essential part in modern engineering, and most mathematicians realise that engineering problems raise interesting questions of mathematical theory, particularly concerning differential equations. But it is easy for the applied mathematician to think of real systems as evolving smoothly and deterministically according to some differential equation, and it is easy too for the engineer to regard the mathematician as incapable of contemplating them otherwise.

Systems involving random fluctuations are certainly more difficult to analyse, but there are contributions which mathematical theory can make to their understanding. Notably the theory of stochastic differential equations and the Itô integral has a good deal to offer if tempered with a dose of engineering scepticism. But there remains a barrier of language which can only be overcome when mathematicians and engineers work closely together and subject themselves to the discipline of real problems.

Such a synthesis was the purpose of the IUTAM symposium of which this book is the expression, and its success can be judged by the reader. Many problems remain unsolved, but enough has been done to show how fruitful can be the collaboration of workers with different backgrounds, and to inspire others in the future to attack this difficult, fascinating and undeniably important area.

J.F.C. Kingman

Oxford, July 1977

Contents

PAGE

Introduction 1

Decision criteria for stability of stochastic systems from observed data 8

F. Kozin and S. Sugimoto

On the moments of linear systems excited by a coloured noise process 36

A. Kistner

Stochastic boundary and eigenvalue problems 54

W. Wedig

Moment stability of linear white noise and coloured noise systems 67

J.L. Willems

Moment stability of coupled linear systems under combined harmonic and stochastic excitation 90

S.T. Ariaratnam and D.S.F. Tam

Homogenization in deterministic and stochastic problems 106

A. Bensoussan, J.L. Lions and G.C. Papanicolaou

On stochastic singular problem of linear dynamical system 116

T. Nakamizo and M. Oshiro

On the asymptotic behaviour of nonlinear stochastic dynamical systems considering the initial states 138

Y. Sunahara, T. Asakura and Y. Morita

Influence of random disturbances on determined nonlinear vibrations 176

K. Piszczek

Probability densities of parametrically excited random vibrations 197

G. Schmidt

Identification of aerodynamic characteristics of a suspension bridge based on field data 214

M. Shinozuka, H. Imai, Y. Enami and K. Takemura

On a class of non-robust problems in stochastic dynamics 237

R.F. Drenick

Probability of first passage failure for lightly damped oscillators 256

J.B. Roberts

On the application of various crossing probabilities in the structural aseismic reliability problem 283

R. Grossmayer

Procedures for frequency decomposition of multiple input/output relationships 308

J.S. Bendat

Normal coordinates and residual spectra in the analysis of random vibration response 323

J.D. Robson and C.J. Dodds

Spectrum estimation through parametric model fitting 348

H. Akaike

Structured response patterns due to wide-band random excitation 366

S.H. Crandall

Flutter and random vibrations in plates 390

I. Elishakoff

Non-stationary random vibrations of systems travelling with variable velocity 412

K. Sobczyk and D.B. Macvean

Response of bridges to moving random loads 435

L. Frýba

Mean number of loads and acceleration roll of an airplane flying in turbulence 449

G. Coupry

Response of periodic beam to supersonic boundary-layer pressure fluctuations 468

Y.K. Lin, S. Maekawa, H. Nijim and L. Maestrello

Optimum structural design of sheet-stringer panels subjected to jet noise excitation 487

S. Narayanan and N.C. Nigam

Method of spectral moments to estimate structural damping 515

E.H. Vanmarcke

Random vibrations of magnetically levitated vehicles on flexible guideways 525

W. Schiehlen

Design problems of systems subject to random loads 528

J. Murzewski

Nonlinear response of ships on the sea 540

Y. Yamanouchi

The stochastic control of ship's course keeping motion 551

K. Otzu and G. Kitagawa

On the identification of ship's steering dynamics 559

G. Kitagawa

Introduction

The Symposium on Stochastic Problems in Dynamics was held in the University of Southampton, England, during the period July 19 to 23 1976. It was sponsored by the International Union of Theoretical and Applied Mechanics (IUTAM) and organised by the Institute of Sound and Vibration Research of the University. Participation was by invitation only and following IUTAM policy younger workers were encouraged to attend and participate in the meeting. The presentation of prepared contributions from about half of the participants took about half of the time available. The remaining time was devoted to extensive discussion of the points raised.

The objective of the symposium was to bring together those mathematicians working on stochastic processes and those engineers who are trying to solve dynamical problems which are of a stochastic nature. The papers fall into two broad categories which reflect these starting points. The value of the symposium was that it presented to the mathematicians a wide range of applications for their work from across the many engineering disciplines. Equally importantly it showed the engineers the latest developments which were being made in the theory and methods of handling stochastic problems. Sixty of the world's leading workers in these general areas gathered together under the sponsorship of the International Union of Theoretical and Applied Mechanics. Lively discussion took place after each paper and an edited version of this discussion is contained in this book.

The excitation of linear systems by stationary Gaussian coloured noise processes is considered by Kistner. Closed form expressions for the first and second moments of the state variables are derived. These are used to show the evolution of the moments for finite time intervals. The stability of these and higher moments is discussed in the paper by Willems. Criteria for almost sure stability are derived and examples given.

In a contribution from the field of control theory Nakamizo and Oshiro consider an optimal stochastic control problem for a linear dynamical system and quadratic performance index, in which the matrix weighting the control and the observation noise intensity matrix are singular.

Wedig shows how the time variable of the Wiener process can be replaced by a space co-ordinate to obtain a model for randomly distributed loading or structural imperfections. He considers the stochastic integral equations which satisfy the boundary conditions. By modifying the Wiener field the Markov property in space is achieved and Itô's integral can be used.

The method of stochastic averaging due to Stratonovich and Khasminskii has aroused much interest in recent years. In this volume the paper by Ariaratnam and Tam provides an example of its application to a multi-degree-of-freedom system problem in parametric excitation in which the excitation is mixed deterministic (periodic) and broad band noise. Under hypotheses on the orders of magnitude of the excitations (the deterministic excitation must be $O(\varepsilon)$ and the stochastic $O(\varepsilon^{\frac{1}{2}})$; $\varepsilon \ll 1$) one can write an equivalent Itô equation with Wiener (white noise) processes. The latter system has for solution a Markov (vector) process $X(t)$ to which the solution $X_\varepsilon(t)$ of the original equations tends weakly as $\varepsilon \to 0$. As Papanicolaou and Kohler have noted, the theorem provides a stochastic analogue of the well known Krylov-Bogolivbov-Mitropolsky deterministic averaging theorem, of which it is in a sense a generalisation.

In some situations the stochastic averaging theorem provides an alternative to the method of equivalent linearisation. In the study of non-linear oscillators with random excitations, the paper by Piszczek presents an application of the latter method, while Ariaratnam's subsequent contribution in the discussion shows that different results are obtained by the use of stochastic averaging. A conclusive settlement to this controversy was not reached.

In a second averaging method, also due to Khasminskii, and represented here in the contribution of Sunahara et al, one starts with an Itô equation with white noise coefficients and when the Fokker-Planck equation corresponding to this has time varying coefficients, averaging is applied to the latter. Thus a version of the usual averaging theorem for ordinary differential equations is applied to a partial differential equation. In a second contribution in which this method was invoked, that of Schmidt, the problem of a pair of coupled, non-linear oscillators was studied. In this case a number of further approximating assumptions appear necessary in order that the resulting (averaged) equations can be solved.

In one of two interesting papers which fell outside the main areas covered in the symposium, Kozin and Sugimoto considered the use of computer methods for statibility analyses of systems not amenable to analytical solutions. Here linear systems with parametric excitations were studied and some surprising results were shown which demonstrated the care with which stochastic stability criteria must be formulated. In the second paper, Bensoussan et al developed a theory of homogenization for replacing the constitutive parameters of a complicated medium by 'effective' (not simply average) values. Although mainly concerned with deterministic problems, examples being presented from acoustics and electromagnetism in which the parametric variation is periodic, the method is also applicable to certain stochastic problems.

The stochastic treatment of dynamical systems usually assumes that the probabilistic structure of the force and response is known. There are certain situations where the parameter of interest is very sensitive to small changes in that structure. Drenik considers this case and suggests a method of treatment which gives bounds on the desired quantity. Earthquake resistance of buildings is quoted as an example.

The first passage problem is common to the contributions by Roberts and Grossmayer. In the paper by Roberts the prediction of first passage probabilities for lightly damped oscillators (with linear and non-linear damping and restoring forces) excited by white noise is approached by considering the energy envelope of the process which is modelled as a one dimensional Markov process and comparisons are made with simulation results. In the Grossmayer paper upper and lower bounds for the first passage density are obtained and applied to the problem of calculating the reliability of structures under earthquake excitation. A feature of the approach is the use of a frequency-time description in the nonstationary envelope process.

Time domain methods are adopted by Sobczyk and McVean for the calculation of the nonstationary random response of vehicles travelling with variable (deterministic) velocity over rough roads whose profiles are assumed spatially statistically homogeneous. When the vehicle velocity is randomly varying the problem formulation is altered to include sinusoidal and random road profiles.

Shinozuk considers the application of time domain system identification methods to the problem of estimating coefficients in the equations of motion for a two dimensional model of a suspension bridge. The methods used include least squares, instrumental variables and maximum likelihood.

The problem of estimation of a spectrum is reduced by Akaike to essentially one of system identification again. An observed time series is expressed as the output of a linear dynamic system driven by white noise and a time domain maximum likelihood parametric model fitting approach is explored. The problem of the determination of the order of the model is solved by use of an information theoretic criterion. The results of this method are contrasted with those obtained by smoothing the Fourier transform of the sample autocovariance function.

In many practical problems structures having a large number of degrees of freedom are excited by random forces. The papers by Bendat and Robson present new methods of processing the response signals. The methods are complementary and rely on conditioning the output by removing coherent components. In Robson's contribution coherency exists through the normal modes of the structure whilst in Bendat's approach coherency is related directly to the input forces.

Crandall shows that when a uniform structure such as a plate with parallel edges is excited by wide band random excitation the response shows some surprisingly regular features. Lines of maximum response appear which form easily recognisable patterns. Similar results have also been obtained for plates having circular or triangular boundaries. The patterns disappear when the boundaries become irregular in shape.

Frýba considers the response of bridges to moving random loads. The cumulative damage hypothesis is used to estimate the fatigue life of bridges of different spans carrying typical trains. The prediction of fatigue load cycles of an aircraft flying through turbulence is considered by Coupry. The major contribution of this paper is the inclusion of the effect of the spanwise distribution of isotropic turbulence. Comparison is made with some full scale flight results.

The interaction of a form of instability such as flutter with random vibration of a plate in supersonic flow is discussed by Elishakoff. The normal mode approach is used in conjunction with Galerkin's method of solving the stochastic boundary problem. The response of a periodically supported beam to supersonic boundary layer pressure fluctuations is considered in the paper by Lin. Here a more realistic model of an aircraft fuselage side is used and the results compared with a large scale wind tunnel experiment.

Narayanan and Nigam demonstrate how the problem of minimum weight design for sheet stringer panels subjected to jet noise with constraints on stresses, fatigue life, natural frequencies and other design variables may be reduced to a non-linear programing problem and to provide an algorithm for automated optimum structural design.

During the closing discussion period significant contributions were made by Van Marke, Schiehlen and Murzewski. These are included at the end of the contributed papers.

Several additional contributions on ship dynamics were made during the symposium. These are also included in this volume. The paper by Yamanouchi discusses the non-linear response of ships on the sea. Kitagawa tabled a paper on the identification of ship's steering dynamics and the paper by Otsu and Kitagawa describes the stochastic control of ship's course keeping motion.

In preparing an edited version of the discussion on each paper we have tried to maintain a narrative style by referring to the questioner or contributor by name in brackets at an appropriate place in the sentence. We hope that the several controversies and unexplained effects which were highlighted in the discussion will form the basis of much further fruitful research.

As editors we wish to express our very sincere thanks to Mrs. Margaret Newton and Mrs. Janet Ward for their great contribution in typing the manuscript. The final pleasing and uniform presentation of the many varied contributions is a result of their diligence and care. We also wish to thank Mr. Peter Ward for redrawing the diagrams and thereby bringing a consistent clarity to the illustrations.

F KOZIN and S SUGIMOTO

Decision criteria for stability of stochastic systems from observed data

INTRODUCTION

In recent years a significant amount of research activity has been directed towards the study of properties of the solutions of stochastic differential equations. Boundedness as well as asymptotic behaviour of moments and sample solutions has been the main concern of many of these studies. Determination of parameter regions which guarantee the desired asymptotic behaviour of moments or samples has comprised the major portion of these studies. This is, of course, the central problem of stability of stochastic systems. The studies have been mainly analytical in nature and have been based upon approximations as well as exact analysis employing the sophisticated tools and techniques of stochastic process theory and in particular Markov diffusion theory.

In some cases exact regions of stability for the sample solutions of linear stochastic differential equations have been obtained [1], [2] and in other cases moment or sample stability regions have been approximated through sufficiency conditions often based upon Lyapunov techniques [3], and circle criterion techniques [4]. Very recently a number of problems of significance for linear stochastic differential equations have been solved by small parameter asymptotic methods that lead to diffusion approximations [5]. The development of the entire field is the result of significant ideas and contributions by various researchers from many countries.

Even with the rapid development of this topic, during the past decade one problem which was recognized rather early, at least as early as 1955, has apparently not been given sufficient attention in the literature.

We refer to the problem of determining by observation of sample solutions of stochastic systems whether or not the system is stable. This is the problem of the engineer, it is a problem of computation and simulation, it is a problem that one meets when attempting to determine stability boundaries by experimentation or simulation for verification of analytical results or when no analytical results are available. This problem was met twenty years ago before we could appreciate the analytical tools available to study stochastic stability problems theoretically. Even though in the ensuing years theoretical studies have been fruitful, it appears that we still have with us this old practical problem.

Those of us who have simulated solutions of stochastic differential equations will almost certainly agree that a simple visual observation of a sample solution is not sufficient to ascertain its asymptotic properties. Indeed, it is possible to observe sample solutions that appear to decay during the period of observation and yet are generated from unstable systems and, on the other hand, one can observe samples that appear to grow during the period of observation but are generated from stable systems. One would like to have a reliable method of ascertaining these properties. It appears, at least to the present authors, that this problem demands attention.

Many questions concerning this problem come to mind, among them are

1 How should one process the sample output in order to make a "reasonable" decision as to its asymptotic behaviour?

2 How long should one observe our sample solution before we make a decision concerning its asymptotic behaviour?

Clearly a related question is,

3 How should one simulate solutions of stochastic differential

equations for stability studies?

In the present preliminary study, we present what we feel to be a first look at this problem from a statistical point of view. We propose certain measures that appear to have promise as useful tools to distinguish between stable and unstable systems from their observed sample solutions.

The ideas presented in this manuscript have come as the result of recent advances in theoretical studies of stochastic differential equations, especially due to Khasminskii [6].

We include a number of examples as well as a discussion of the problems that remain open. It is our hope that we can generate sufficient interest in the reader to take up this problem and, hopefully, provide new insights and answers.

THE FIRST ORDER STOCHASTIC DIFFERENTIAL EQUATIONS

In order to illustrate the problem at hand, let us consider a simple first order differential equation

$$\frac{dx(t)}{dt} = (a + f(t))\, x(t) \;, \tag{1}$$

for some initial value $x(o) = x_o$.

We consider three cases for the coefficient f,

1 f is a sample from a known class of deterministic functions

2 f represents the fictitious, Gaussian white noise

3 f represents a sample from a physical noise process

For each case we are concerned with asymptotic stability for almost every sample.

That is, we wish to determine whether or not we have

$$P\{\lim_{t\uparrow\infty} ||x(t:x_o,t_o)|| = 0\} = 1 \tag{2}$$

For the first case we shall illustrate the problem by an example. We assume that the constant, a, in (1) is zero and that the coefficient f is drawn from the family

$$\{f_n(t) = \frac{-2(t-n)}{n^{-6}+(t-n)^2}, \quad n=1,2,\ldots,t\varepsilon(o,\infty)\} \tag{3}$$

The associated family of solutions to (1) will be

$$\{x_n(t) = x_o \frac{n^{-6}+n^2}{n^{-6}+(t-n)^2}, \quad n=1,2,\ldots,t\varepsilon(o,\infty)\} \tag{4}$$

A typical curve from the family of solutions (4) will appear as illustrated in Figure 1.

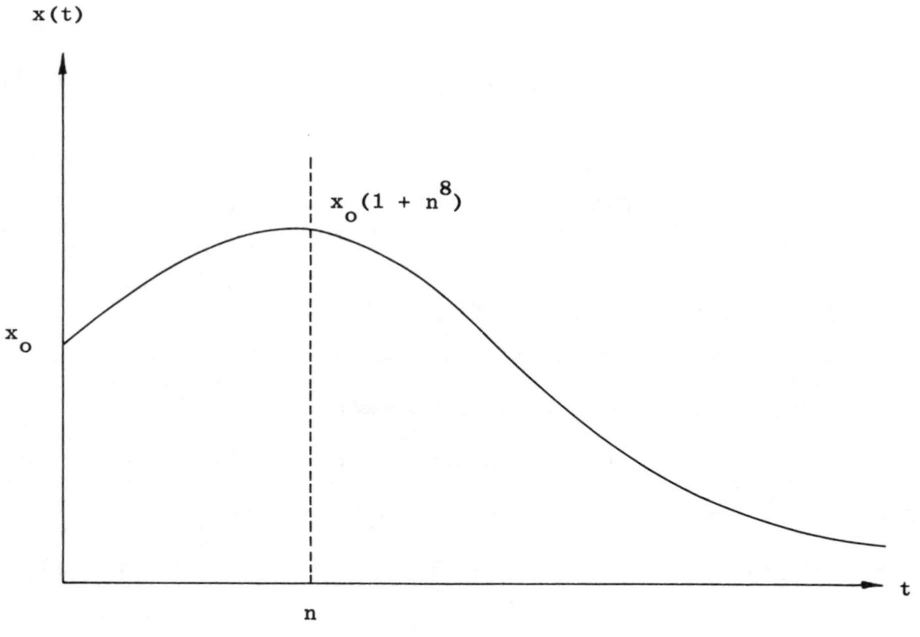

Figure 1 Type 1 Solution

Thus every x_n represents a stable solution, indeed this example is stable for every coefficient in the family (3). On the other hand, if we can only observe the sample solution for some fixed period of time, say T = 50, then the observer would be compelled to classify most solutions $x_n(t)$ for, say, $n > 75$ as unstable since they appear to become unbounded. Naturally, this would also depend upon the probability of obtaining solutions for $n > 75$ as compared to the probability of obtaining solutions for $n < 75$. We will not consider this point at present. In any case, this simple example clearly brings up the question of how long should one observe a record before making a decision. Naturally, for this simple example, the exact solutions are known to us so that any question can ultimately be answered in sufficient detail.

Let us now consider the more important case, 2, of the Gaussian white noise coefficient. In this case we rewrite (1) in the Itô form as

$$dx(t) = (adt + dB(t)x(t) \tag{5}$$

where the B-process is the Brownian motion with zero mean and variance $\sigma^2 t$. The solution to (5) is very well known to be

$$x(t) = x_0 e^{(a - \frac{\sigma^2}{2}) t + B(t)} \tag{6}$$

Hence, because the Brownian motion samples grow like $\sqrt{t \log \log t}$ as t approaches infinity, with probability one, then the stability properties of the solution process are determined by the coefficient $a - \frac{\sigma^2}{2}$.

Indeed the region of almost sure sample stability is $a - \frac{\sigma^2}{2} < 0$.

In Figure 2, we present typical simulated solutions for case 2. They clearly do not exhibit uniform characteristics that would allow a simple stability decision. In view of this fact, how shall we proceed? We have elected to approach our goal as a statistical estimation problem based upon the following observation. Knowing that the sample was generated by a first

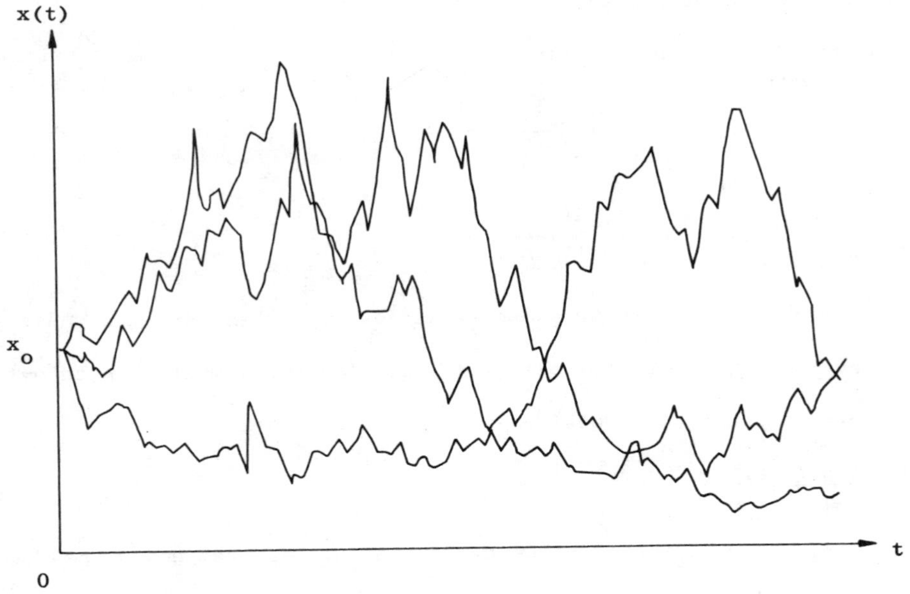

Figure 2 Type 2 Solution

order Itô equation and knowing that $q = a - \frac{\sigma^2}{2}$ is the parameter that determines stability for the solution process, we therefore will direct our attention to estimating the parameter q from the observed sample solution. If our estimate $\hat{q}$ is negative we will declare "stability" and if the estimate is positive, we will declare "instability".

We assume for $t_1 < t_2 < \ldots < t_L$, the observed sample values $x_j \equiv x(t_j)$, $j = 1, .., L$. Without loss of generality, we set $x_o = 1$, and define from (6)

$$r_j = \frac{\log x_j}{t_j} = q + \frac{B_j}{t_j} \tag{7}$$

where

$$B_j = B(t_j), \quad j = 1, \ldots, L.$$

This yields a classical statistical estimation problem; given observations r_1,

$r_2, \ldots, r_L$ of a Gaussian process, estimate the mean $q = E\{r_j\}$.

The covariances of the r_j's are

$$E\{(r_i-q)(r_j-q)\} = E\left\{\frac{B_iB_j}{t_it_j}\right\} = \sigma^2 \frac{\min(t_i,t_j)}{t_it_j}$$

$$= \frac{\sigma^2}{\max(t_i,t_j)} = d_{ij} \tag{8}$$

Since we are dealing with Gaussian variables, it makes sense to apply the maximum likelihood appraoch for estimating q. The maximum likelihood estimate of q is the solution of the equation

$$\frac{\partial \log p(r|q)}{\partial q} = 0 \tag{9}$$

where $r^T = (r_1, \ldots, r_L)$, the transposed L-vector of observations. The Gaussian density function $p(r|q)$ is

$$\left[(2\pi)^L \det D\right]^{-\frac{1}{2}} \exp\left[-\frac{1}{2}(r-q\mu)^T D^{-1}(r-q\mu)\right], \tag{10}$$

where $D = (d_{ij})$ is the covariance matrix (8) and μ^T is the L-vector $[1,1,..1]$.

Upon solution of (9) for the density (10) we obtain the somewhat surprising result

$$\hat{q}(r) = r_L, \tag{11}$$

which is the last observed value of r_j's.

It is in fact, easy to show that this estimation is unbiased as well as efficient. That is, it is the minimum variance estimate. This is due to the fact that

$$\frac{\partial \log p(r|q)}{\partial q} = (\hat{q}(r)-q)\, f(q) = 0 \tag{12}$$

where f is independent of the observed data. Furthermore, the variance of the estimate $\hat{q}$ is σ^2/t_L .

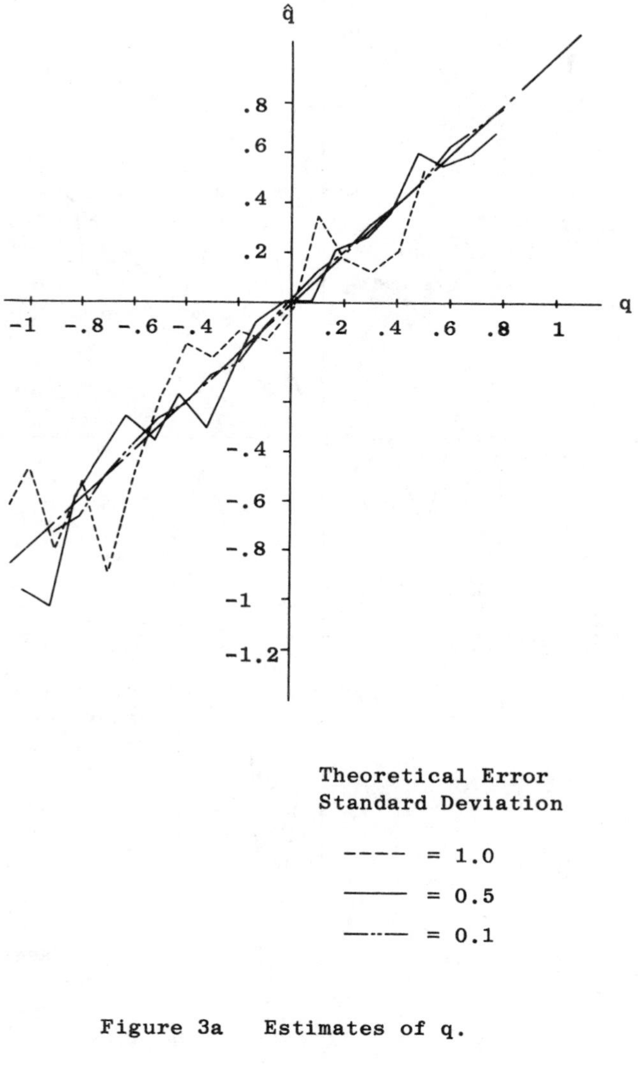

Figure 3a Estimates of q.

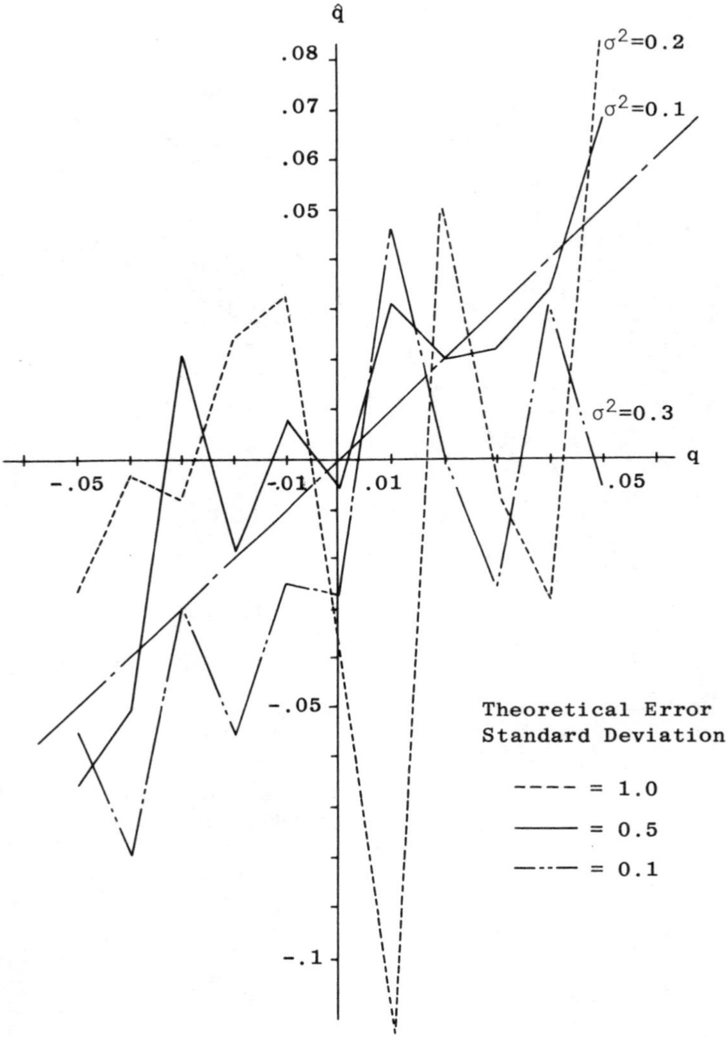

Figure 3b   Estimates of q.

In Figures 3a and 3b we show results of estimations of q for a number of samples generated with different q-values, and different σ values, where t_L is fixed at 25. It is clear that for q values greater than .1, in Figure 3a, the estimates are of the correct sign and are closer to the correct values of q for smaller values of σ. On the other hand Figure 3b, shows for $q < .1$ quite a variation in the estimated values, although again smaller values of yield better estimates.

Let us now look at case 3 for the physical noise coefficient. We shall write the equation as

$$\frac{dx(t)}{dt} = (q + y(t))\, x(t), \quad x(o) = 1 , \tag{13}$$

where the y-process is a zero mean stationary ergodic Gaussian process with covariance function $R_y(\tau)$. We know that the necessary and sufficient condition for almost sure sample stability is $q < 0$.

Therefore, as in case 2 we write for $t_1 < t_2 < \ldots < t_L$,

$$r_j = \frac{\log x_j}{t_j} = q + \frac{\int_o^{t_j} y(t)\,dt}{t_j} \quad j = 1,2,\ldots,L, \tag{14}$$

Again we can determine that $E\ \{r_j\} = q$ and

$$E\ \{(r_i-q)(r_j-q)\} = \frac{\int_o^{t_i} d\tau \int_o^{t_j} ds R_y(\tau-s)}{t_i t_j} = \gamma_{ij} \quad i,j = 1,\ldots,L. \tag{15}$$

If we define the matrix $(\lambda_{ij}) = (\gamma_{ij})^{-1}$ then it can be shown that the maximum likelihood estimation of q from equations (14) and (15) is

$$\hat{q} = W^T r , \tag{16}$$

where the row vector W^T is

$$W^T = \frac{1}{\sum_{i,j} \lambda_{ij}} \left(\sum_i \lambda_{i1}, \sum_i \lambda_{i2}, \ldots, \sum_i \lambda_{iL} \right), \tag{17}$$

and the λ_{ij}'s are defined above.

This estimation is unbiased and efficient and its variance is

$$\left(\sum_{i,j} \lambda_{ij} \right)^{-1} = \sigma_{\hat{q}}^2 .$$

Clearly, the $W_j = \sigma_{\hat{q}}^2 \sum_{i=1}^{L} \lambda_{ij}$ must be evaluated in order to obtain the estimation. This estimation (16) depends upon all of the past information in contrast to case 2. This is reasonable, since in equation (14), the Gaussian noise terms are correlated with the past.

In the next section, we shall discuss the directions we can take in order to extend these ideas to higher order stochastic differential equations.

HIGHER ORDER STOCHASTIC DIFFERENTIAL EQUATIONS

In the previous section our estimates were not in terms of the $x(t_j)$'s directly but, instead, the estimates were linear combinations of the logarithms of the observed sample solution, divided by their appropriate times, t_j. There is a natural reason to consider the logarithm which we explain below. Let us consider the linear system

$$\frac{dx(t)}{dt} = A(t)\, x(t),\ x(o) = x_o, \tag{18}$$

where x is an n-vector, and A(t) is an n x n matrix whose elements are physical stochastic processes (i.e. not Gaussian white noises).

Thus, we may apply ordinary calculus to (18) to obtain an equation for the logarithm of $||x||^2 = x^T x$,

$$\log||x(t)||^2 - \log||x_o||^2 = \int_o^t Q(\tau)d\tau , \tag{19}$$

$$Q(t) = \frac{x^T(t)\left[A(t) + A^T(t)\right]x(t)}{||x(t)||^2} \tag{20}$$

Upon dividing (19) by t, we obtain exactly the function r(t) that we studied in the previous section, for the first order stochastic differential equation. Suppose now, we study the symptotic behaviour of $\frac{1}{t}\int_0^t Q(\tau)d\tau$ as t approaches infinity.

If the time average exists and is negative, then (19) implies that $||x(t)||\downarrow 0$ as $t\uparrow\infty$, and if the time average is positive, then $||x(t)||\uparrow\infty$ as $t\uparrow\infty$.

Thus it is clear that this integral average is basic in determining the stability properties of linear stochastic differential equations.

The first study of the Q-function and its time average as a necessary and sufficient condition for sample stability of linear Itô differential equations was performed by Khasminskii [6]. Applications of these ideas to specific second order Itô stochastic differential equations was made by this author and his students [1], [2]. We have found, furthermore, that the Q-function is basic in determining the exact relationship between stability of moments and sample stability.

Our objective in this section is to show how the Q-function may be used in estimation procedures for distinguishing between stable and unstable linear Itô stochastic differential equations from observed sample solution data.

The general linear Itô stochastic differential equation may be written as

$$dx(t) = Fx(t)dt + \sum_{r=1}^{m} G_r x(t)dB_r(t), \tag{21}$$

where x is an n-vector, F and G_r, $r=1,\ldots, m$, are n x n constant matrices and the B_r - processes are independent identically distributed Brownian motions, with zero means and variances t.

From equation (21) and Itô's differential rule it can be established that

$$d \log||x(t)|| = Q(\lambda(t))dt + \sum_{r=1}^{m} R_r(\lambda(t))dB_r(t), \tag{22}$$

where $\lambda(t) = \frac{x(t)}{||x(t)||}$ is a unit vector, $R_r(\lambda)$ is a polynomial in the components of the unit vector λ and is therefore bounded for each r, and $Q(\lambda)$ is given as

$$Q(\lambda) = (F\lambda,\lambda) + \frac{1}{2}\sum_{i=1}^{n} A_{ii}(\lambda) - \sum_{i,j=1}^{n} A_{ij}(\lambda)\lambda_i\lambda_j \tag{23}$$

In equation (23) the terms A_{ij} are defined as

$$A_{ij}(\lambda) = \sum_{k,s=1}^{n}\sum_{r=1}^{m} G_{r,ik}G_{r,js}\lambda_k\lambda_s .$$

We note here that $Q(\lambda)$ is, in fact, $\mathcal{L}\log||x||$, where $\mathcal{L}$ is the backward diffusion operator generated by the linear equation (21).

Upon integrating equation (22) and dividing by t, we obtain

$$\frac{\log||x(t)|| - \log||x_o||}{t} = \frac{1}{t}\int_o^t Q(\lambda(s))ds + \frac{1}{t}\sum_{r=1}^{m}\int_o^t R_r(\lambda(s)dB_r(s) \tag{24}$$

Upon letting $t\uparrow\infty$, the second integral on the right-hand side of equation (24) becomes zero with probability one, due to the boundedness of R_r and because the integral is a martingale. If the time average of the Q-integral in (24) exists with probability one, then its arithmetic sign determines whether or not the system (21) is stable. We notice that the unit vector λ is defined on the surface of the n-dimensional unit sphere. Therefore, if the λ-process is ergodic on the sphere then

$$\lim_{t\uparrow\infty} \frac{1}{t}\int_o^t Q(\lambda(s))ds = E\{Q(\lambda)\} , \tag{25}$$

with probability one.

Determining the expectation in (25) is very difficult even for second order stochastic equations as can be seen in [1] and [2].

For higher order systems there is not a sufficient theory available at this time to calculate the integral or expectation analytically for those systems most significant to engineers. The reason is that in many important parametically excited systems, the λ-process will not be ergodic on the n-dimensional sphere, but will be ergodic on sub-domains of the surface, depending upon whether or not the process starts in that sub-domain. The following theorem due to Khasminskii clarifies this problem.

Theorem [6],

If there exist n-linearly independent vectors $\lambda^1,\dots,\lambda^n$ in E^n such that with probability one $\overline{\lim_{t\uparrow\infty}} \frac{1}{t} \int_0^t Q(\lambda(s;\lambda^i))ds < 0$ for $i = 1,\dots,n$, where λ^i denotes the initial value, then the system (21) is almost surely asymptotically stable.

If, however, there is a $\lambda_o \varepsilon E^n$ for which $\underline{\lim_{t\uparrow\infty}} \frac{1}{t} \int_0^t Q(\lambda(s;\lambda_o))ds > 0$ then the system (21) is unstable in the sense that $P\{\lim_{t\uparrow\infty} ||x(t:x_o)|| = \infty\} = 1$.

Based upon the theorem above, we can now present an estimation procedure by which we shall decide whether given sample solutions are generated by stable or unstable systems of the form (21).

The procedure is as follows:

1 Observe the sample solution vector of a known linear Itô equation (21) with a known initial condition, x_o', over some fixed time interval $[0,T]$.

2 Form the unit vector $\lambda = x/||x||$, with initial condition $\lambda^1 = x_o^1/||x_o^1||$.

3 Calculate $J_T = \frac{1}{T} \int_0^T Q(\lambda(s))ds$ from the known function Q given in equation (23).

4 If J_T is positive, then declare the system to be unstable.

If J_T is negative, repeat the procedure 1-3 with a new initial condition, x_o^2, linearly independent of x_o^1. If for n-linearly independent initial values, J_T is found to be negative, then declare the system to be stable.

For computation of the integral J_T, it is convenient to note that, as a function of t, J_t satisfies

$$\frac{dJ_t}{dt} = -\frac{1}{t} J_t + \frac{1}{t} Q(\lambda(t)) \quad , \tag{26}$$

where $J_o = Q(\lambda(o))$, $\lambda(o)$ is the initial vector. For digital computation, the discrete version of (26) may be written as,

$$J_{i+1} = \frac{i}{i+1} + \frac{Q(\lambda_{i+1}) + Q(\lambda_i)}{2(i+1)} \tag{27}$$

where $J_{i+1} = J_{t_{i+1}}$, and $t_{i+1} = (i+1)\ \Delta t$.

COMPUTATIONAL RESULTS

In this section we shall present the results of three examples to which the procedure of the previous section has been applied. The three examples are comprised of two second order Itô equations, for which analytical results exist to allow a check on the estimated stability characteristics, and a fourth order system for which no explicit analytical results are available.

Example I

Consider the system

$$d\begin{pmatrix} x_1 \\ x_2 \end{pmatrix} = \begin{pmatrix} 0 & 1 \\ -1 & -2\zeta \end{pmatrix} \begin{pmatrix} x_1 \\ x_2 \end{pmatrix} dt + \begin{pmatrix} \sigma & k \\ -k & \sigma \end{pmatrix} \begin{pmatrix} x_1 \\ x_2 \end{pmatrix} dB, \tag{28}$$

for $\sigma k = 1$.

For this system the desired time average of the Q-function can be evaluated explicitly as

$$\lim_{t\uparrow\infty} \frac{1}{t} \int_o^t Q(\lambda(s))\,ds = E\ \{Q(\lambda)\}$$

$$= -\ \zeta \left[1 - \frac{Y_1(\zeta\sigma^2)}{Y_o(\zeta\sigma^2)}\right] - \frac{1}{2} \left[\sigma^2 - \frac{1}{\sigma^2}\right] \tag{29}$$

with probability one, where Y_1, Y_o are Y-Bessel functions.

In Figure 4 the exact stability boundary $E\{Q(\lambda)\} = 0$ is shown for the system (28).

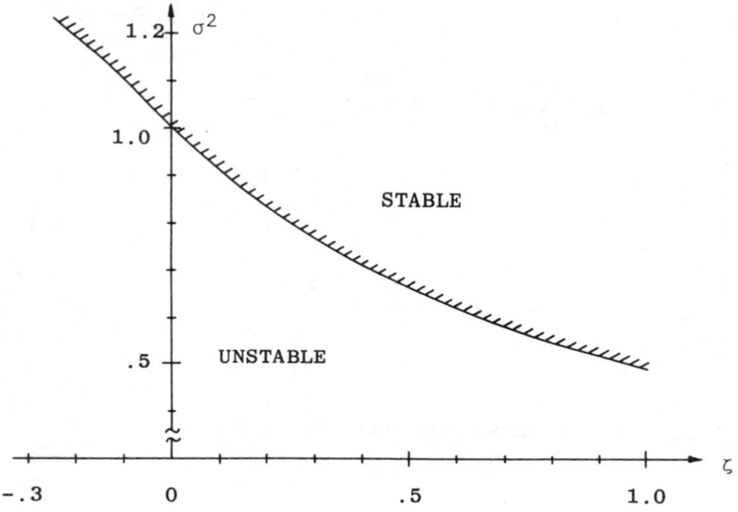

Figure 4 Stability Domain of Example 1.

Figure 5 illustrates a sample solution generated in the stable region $\zeta = 0.5$, $\sigma^2 = 0.8$ for a duration of 10 seconds and Figure 6 illustrates the estimator J_t which converges very rapidly to its true value which is about -0.3. We notice from Figure 5 that, even though the sample behaviour does not immediately convey a "stable" behaviour, the estimator J_t declares "stability" in about five seconds.

In Figure 7 samples generated in the unstable region $\zeta = 0.5$, $\sigma^2 = 0.6$ are illustrated. Clearly, their unstable behaviour is not immediately apparent and yet, again, in Figure 8 the estimator J_t climbs toward the true value .0107 quite rapidly establishing an unstable system.

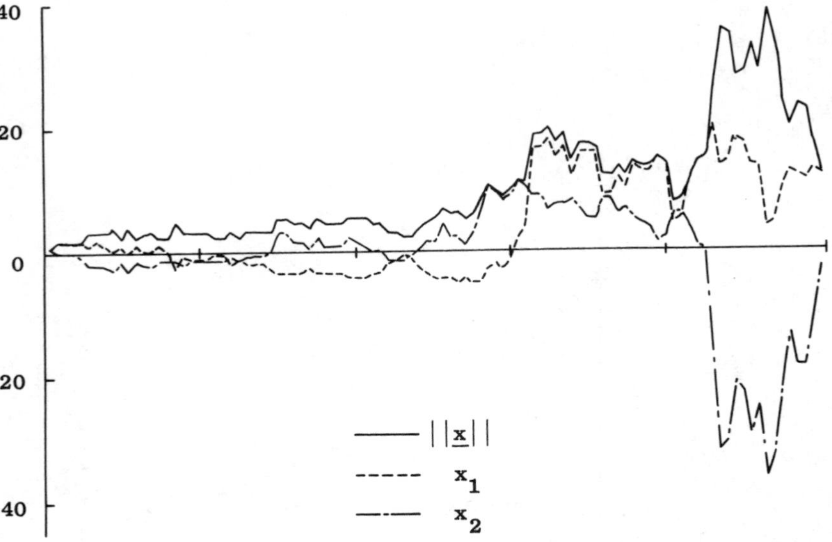

Figure 5 A Sample Process with ζ = 0.5, σ^2 0.8

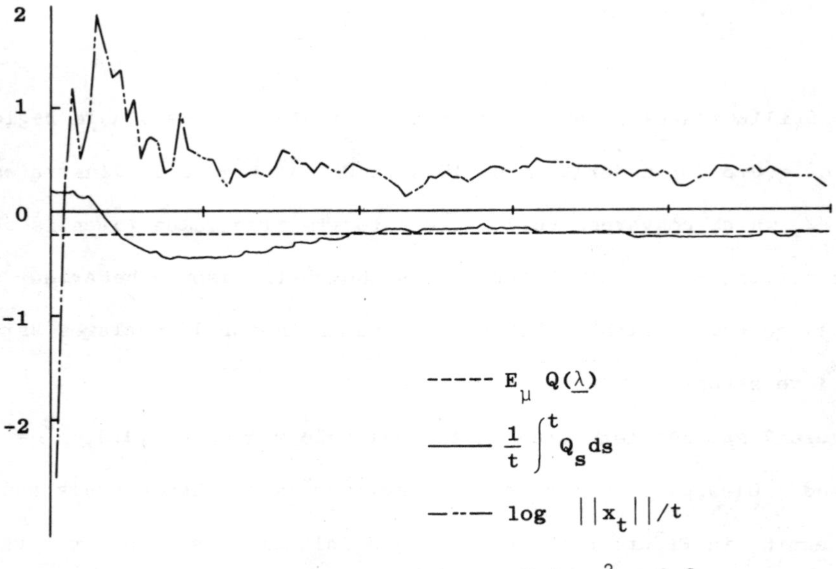

Figure 6 A J(t) process with ζ = 0.5, σ^2 = 0.8

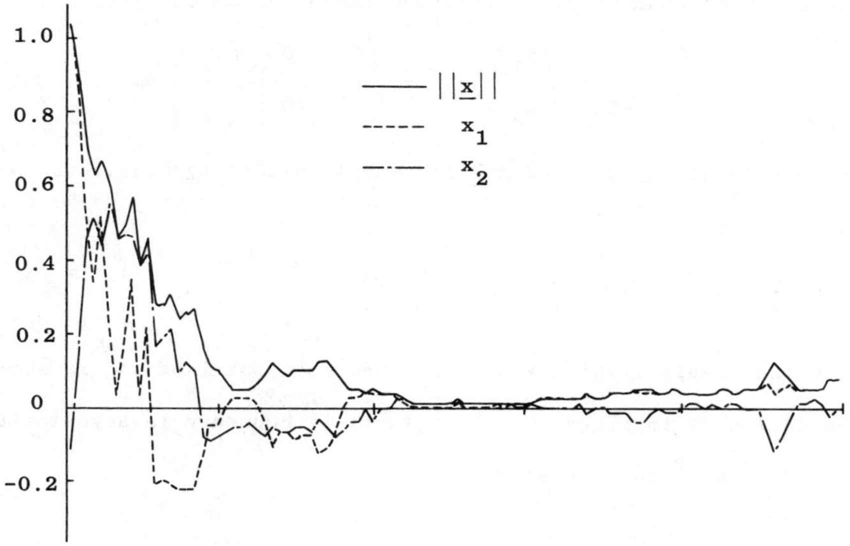

Figure 7 A Sample Process with $\zeta = 0.5$, $\sigma^2 = 0.6$

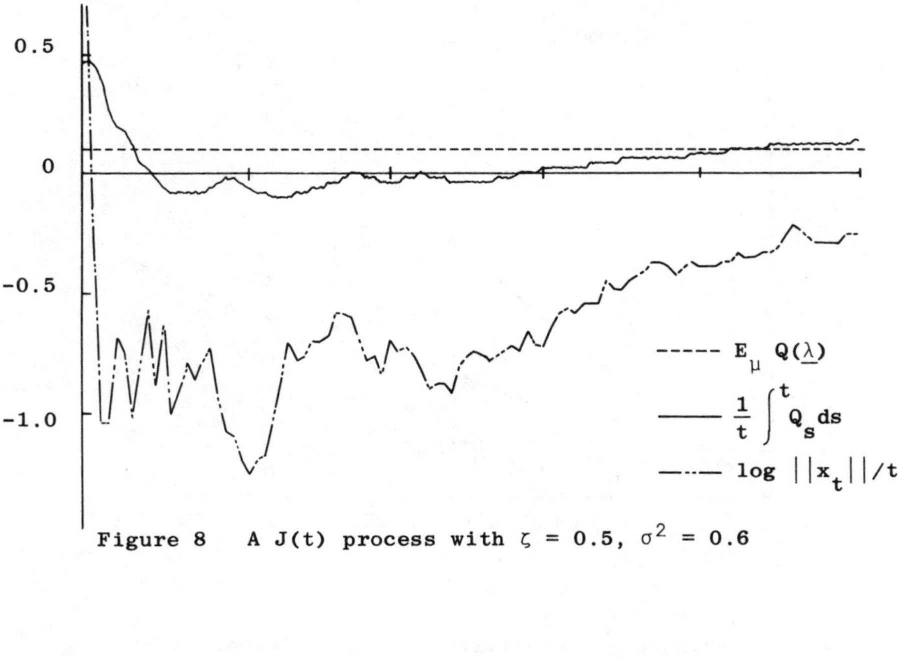

Figure 8 A J(t) process with $\zeta = 0.5$, $\sigma^2 = 0.6$

Example II

This example is the second order oscillator with a random coefficient,

$$d\begin{pmatrix} x_1 \\ x_2 \end{pmatrix} = \begin{pmatrix} 0 & 1 \\ -1 & -2\zeta \end{pmatrix}\begin{pmatrix} x_1 \\ x_2 \end{pmatrix} dt + \begin{pmatrix} 0 & 0 \\ \sigma & 0 \end{pmatrix}\begin{pmatrix} x_1 \\ x_2 \end{pmatrix} dB \qquad (30)$$

which has been extensively studied for its sample stability behaviour in [1], [2]. The Q-function is

$$Q(\lambda) = \frac{\sigma^2}{2} \lambda_1^2 - 2\zeta \lambda_2^2 - \sigma^2 \lambda_1^2 \lambda_2^2 \qquad (31)$$

For this case the sample stability boundary has been obtained and is shown for small values of (ζ,σ^2) in Figure 9. The stability boundary is asymptotic to the line $\sigma^2 = 8\zeta$ as σ^2 approaches zero.

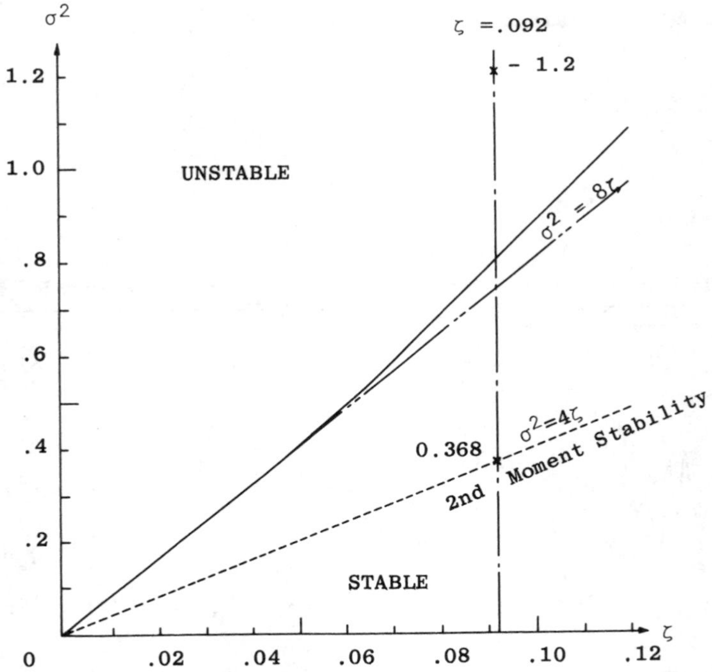

Figure 9 Stability Domain of Example 2

In Figures 10 and 11 we illustrate samples generated in the unstable region $\zeta = 0.092$, $\sigma^2 = 1.2$, for twenty seconds and the associated estimator J_t for these samples. Although the samples appear to be decaying, the estimator is steadily growing to a positive value as it should for unstable systems.

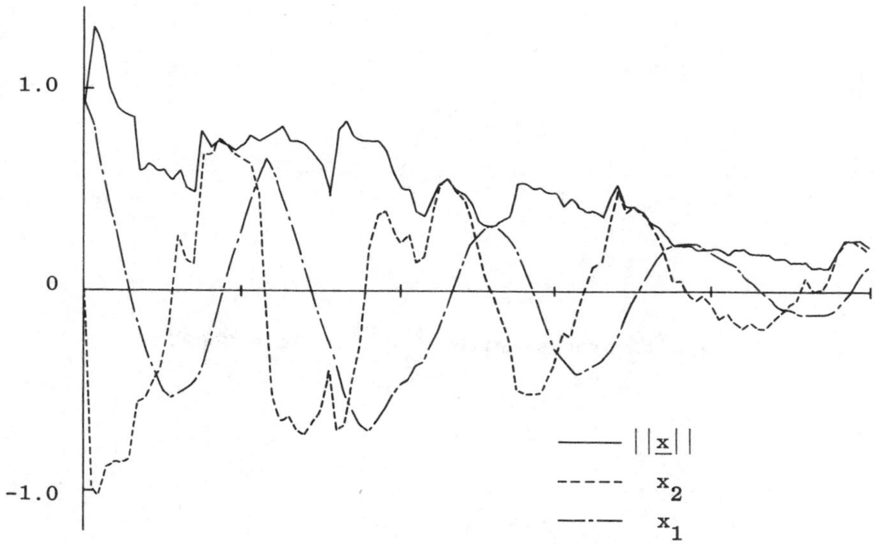

Figure 10 A Sample Process with $\sigma^2 = 1.2$, $\zeta = 0.002$

In Figure 12, samples are presented that were generated in the stable region $\zeta = 0.092$, $\sigma^2 = 0.368$, again for twenty seconds.

The estimator shown in Figure 13, rapidly drops to a negative value declaring "stability".

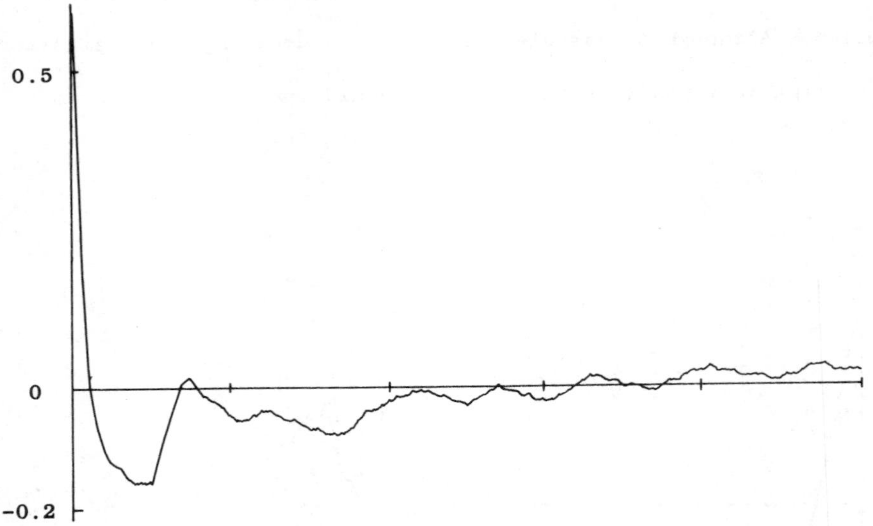

Figure 11 A J(t) process with $\sigma^2 = 1.2$, $\zeta = 0.092$

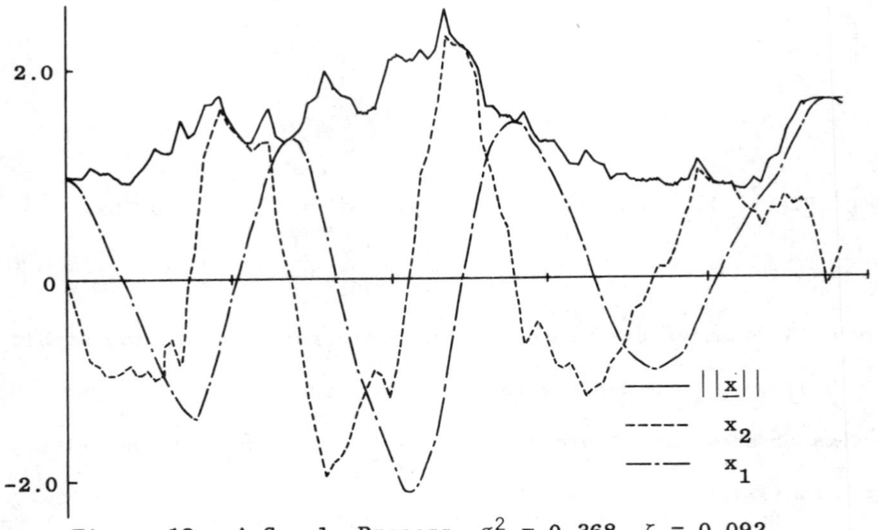

Figure 12 A Sample Process, $\sigma^2 = 0.368$, $\zeta = 0.092$

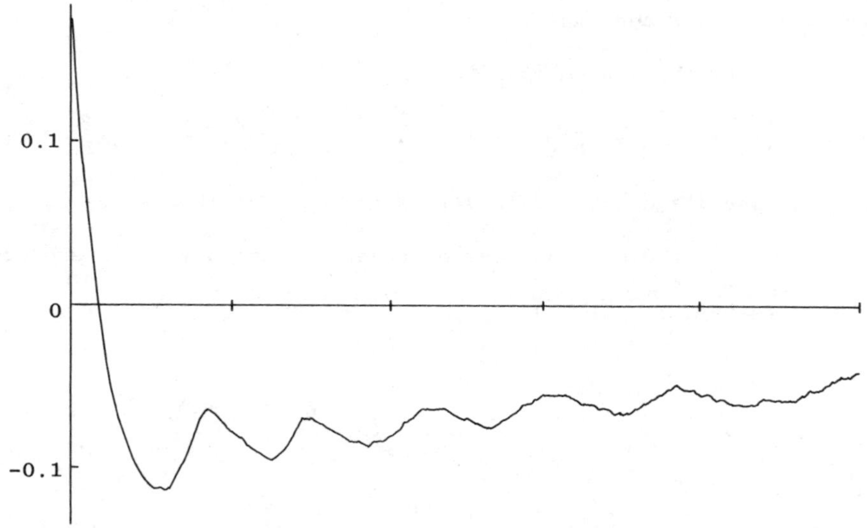

Figure 13 A J(t) process with $\sigma^2 = 0.368$, $\zeta = 0.092$

Example III

We consider the fourth order system

$$d\begin{pmatrix} x_1 \\ x_2 \\ x_3 \\ x_4 \end{pmatrix} = \begin{pmatrix} 0 & 1 & 0 & 0 \\ 0 & 0 & 1 & 0 \\ 0 & 0 & 0 & 1 \\ -1 & 0 & -\alpha & 0 \end{pmatrix} \begin{pmatrix} x_1 \\ x_2 \\ x_3 \\ x_4 \end{pmatrix} dt + \begin{pmatrix} 0 & 0 & 0 & 0 \\ 0 & 0 & 0 & 0 \\ 0 & 0 & 0 & 0 \\ \sigma & 0 & 0 & 0 \end{pmatrix} \begin{pmatrix} x_1 \\ x_2 \\ x_3 \\ x_4 \end{pmatrix} dB \tag{32}$$

which formally corresponds to the fourth order equation.

$$\frac{d^4x(t)}{dt^4} + \alpha \frac{d^2x(t)}{dt^2} + (1 + \sigma B(t))\, x(t) = 0 \tag{33}$$

There is no explicit stability result for this system (although one could obtain sufficient conditions for sample stability from second moment properties which can be obtained explicitly).

The Q-function for the system (31) is

$$Q(\lambda) = \lambda_1\lambda_2 + \lambda_2\lambda_3 + \lambda_3\lambda_4 + \lambda_1\lambda_4 - \alpha\lambda_3\lambda_4 + \frac{1}{2}\sigma^2\lambda_1^2 - \sigma^2\lambda_1^2\lambda_1^4 \tag{34}$$

In Figures 14 and 15 we illustrate J(t) processes for the two cases $\sigma = 2.0$, $\alpha = 3.0$ and $\sigma = 0.1$, $\alpha = 3.0$. The four component solution vectors are not shown for these cases.

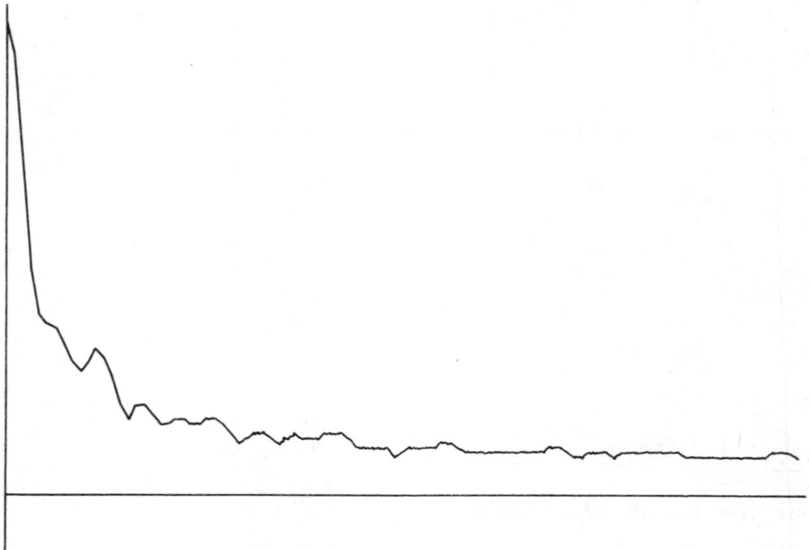

Figure 14 A J(t) process with $\sigma = 2.0$, $\alpha = 3.0$

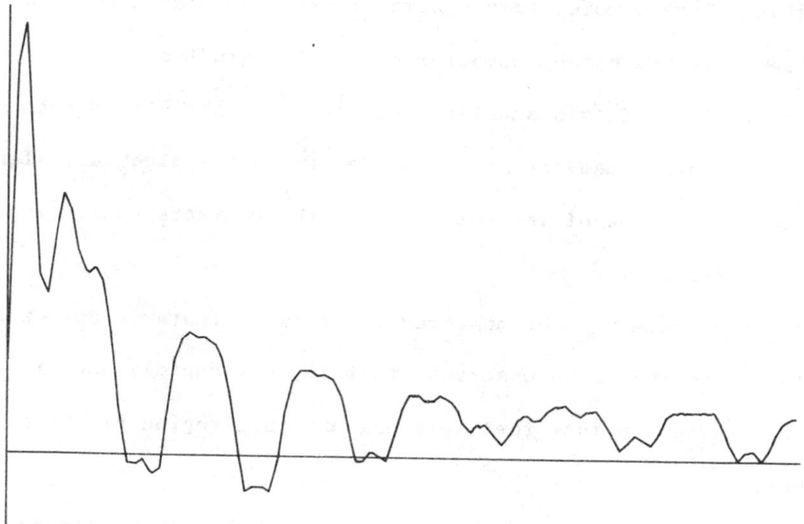

Figure 15   A J(t) process with $\sigma = 0.1$, $\alpha = 3.0$

For the larger variance J_t remains well into the positive region clearly signifying an unstable system. For the smaller variance the samples are more periodic and J_t, although still an unstable system, begins to touch the origin. A further detailed computer study would be required in order to obtain a reliable picture of the stability domain. A first try in this case could be made from sufficient conditions in terms of second moments. This last condition can be obtained analytically.

CONCLUSION

We have presented an estimator that appears to have potential practical use for determining whether a given linear stochastic system is operating in a stable region. Furthermore, this estimator can also play a role in determination of stability regions by computer simulation studies.

The major feature of this stability estimator is that it is much smoother than a sample solution and one need only determine its algebraic sign. The actual theoretical value of the estimator is the characteristic exponent for the given linear system.

Many questions remain to be answered for this estimator. For example, how long should the estimator be observed; that is, how rapidly does it converge to $E\{Q(\lambda)\}$, when we know that λ is ergodic in a region of the surface of the n-sphere.

Although we have a reasonably good understanding of the estimator for the Itô equation, we still have very little knowledge or experience with it in the physical noise coefficient case. That is, what are the ergodic properties of the λ-process in that case ?

We have presented only one, potential, estimator. It is conceivable that others may be equally as good or perhaps, better for stability determination. As we mentioned in the Introduction, we hope that others may find this problem sufficiently interesting to take up its challenge, and generate the needed experience and answers.

REFERENCES

1 F. Kozin and S. Prodromou, Necessary and sufficient conditions for almost sure sample stability of linear Itô equations. SIAM J. Appl. Math., Vol. 21 No.3. (1971).

2 R.R. Mitchell and F. Kozin, Sample stability of second order linear differential equations with wide band noise coefficients. SIAM J. Appl. Math., Vol. 27 No.4 (1974).

3 F. Kozin, Stability of the linear stochastic systems. Lecture notes in Mathematics No. 294 Springer-Verlag, New York (1972).

4 J.L. Willems, Lyapunov functions and global frequency domain stability criteria for a class of stochastic feedback systems. Ibid 131-146.

5 G.C. Papanicolaou, Stochastic equations and their applications. SIAM Review Vol. 15, 526-545. (1973).

6 R.Z. Khasminskii, Stability of systems of differential equations with random parametric excitation (in Russian). Nauka Moskow. (1969).

ACKNOWLEDGEMENT

The support of the National Science Foundation under Grant ENG75-03576 is gratefully acknowledged.

Professor F. Kozin, Polytechnic Institute of New York, Route 110, Farmingdale, New York, 11735.

Professor S. Sugimoto, Osaka Universif, Japan.

DISCUSSION

In response to a question on applications (Mark) Professor Kozin mentioned that the approach should be useful in a wide range of problems involving parametric excitations including structural vibrations, electronic circuitry and plasma dynamics. To a later question on phase locked loops (Nightingale), Professor Kozin indicated that the non parametric nature of the excitation would make the approach inapplicable; its simplicity depended upon the excitations appearing as coefficients in the equation. A discussion then developed (Crandall, Ariaratnam, Holmes) on the meaning of the Q(.) function (equations (20) and (23)) which Professor Kozin identified as the phase process. In a two degree of freedom or planar system this corresponds to the θ process in polar coordinates (r, θ); in a n-degree of freedom system it becomes a process on the surface of the n-1 sphere. However, in the n dimensional case diffusion theory is insufficiently developed to permit the identification of regions of ergodicity and non ergodicity. In the one dimensional (unit circle) case Mitchell and Kozin (the authors reference [2]) had made considerable progress, but simulations such as the present ones seemed most suitable in the n-dimensional case. It was pointed out that in the planar case the θ equation can normally be solved explicitly and that the r solution then depends only on θ. Moreover, one is taking a non-ergodic solution process and, by a suitable transformation, obtaining an ergodic process which can then be averaged to obtain an estimator J_T which behaves smoothly and from which one can estimate sample properties with probability 1. It was mentioned (Ariaratnam) that a limited class of nonlinear systems in which the right hand side of the equation is homogeneous and of degree 1 can also be tackled by this method but in response to a subsequent question (Barr), Professor Kozin said that he had not performed any simulations of nonlinear systems. The distinction between the expressions for $Q(.)$ in the case of non-white "physical" noise (20) and white noise (23) was pointed out (Lin) and it was stressed that Itô calculus was used in the derivation of (23).

Some discussion then followed on what knowledge one should have of a system before attempting simulations. In contrast to some suggestions (Lin, Shinozuka) Professor Kozin was of the opinion that foreknowledge of system stability was unnecessary and that simulations of unstable systems could be useful. It was noted (Ariaratnam) that the present J_T indicator was an improvement on previous methods, for example that of Bolotin, in which instability was declared if the solution norm exceeded some multiple of the variance of some response, but Professor Kozin stressed that bounds may still play an important role. Finally, in reply to a question (Schiehlen) on the determination of observation times, it was noted that the rate of convergence of quantities such as J_T was indeed a difficult question and that there seemed to be no generally useful rules although a knowledge of the (unforced) system dynamics might be helpful.

A KISTNER

On the moments of linear systems excited by a coloured noise process

INTRODUCTION

Linear systems

$$\dot{x} = \{A+B\xi\}x, \quad x = x_t \varepsilon R^n, \quad \xi = \xi_t \varepsilon R^1, \quad t \geq 0 \tag{1}$$

with parameters excited by a coloured noise process ξ_t are considered. The disturbance ξ_t is assumed to be stationary and Gaussian and to have zero mean $E\xi_t \equiv 0$.

In applications, the vector $m_x(t) = Ex_t$ of the first moments and the matrix $m_{xx}(t) = Ex_t x_t^*$ of the second moments of the state variables are currently used for judging the behaviour of system (1). It is well known that the moments $m_x(t)$ and $m_{xx}(t)$ satisfy infinite dimensional systems of ordinary differential equations. For practical use these systems have to be closed at some finite stage. For this purpose several approximative procedures have been proposed, e.g. [1,2,3]. But they seem to give good results only for the case of small disturbances and with appropriate parameters [4]. Therefore the question for strictly closed moment equations and rigorous stability criteria arises.

The results available until now depend on the structure of the Lie algebra L generated by the matrices A and B [5]. Closed moment equations are given in [6] for the case of an Abelian Lie algebra L (with AB = BA). Strict stability criteria for the wider class of systems (1) which lead to a solvable Lie algebra L [5] are derived in [7].

Subsequently, closed moment equations will be derived for the case of a general Lie algebra L. For this purpose a series expansion for the solution

x_t of (1) is given first. This expression will be used to calculate the equations for the first and the second moments of x_t, respectively. The application of these equations is then illustrated by a simple example.

THE SOLUTION PROCESS x_t

Using the matrizant of (1) and its Peano-Baker series [8] we have

$$x_t = \sum_{k=0}^{\infty} P_k(t).x_o, \quad P_k(t) = \int_0^t \{A+B\xi_\tau\}P_{k-1}(\tau)d\tau ,$$
$$k = 1,2,3,\ldots , \quad P_o(t) \equiv I. \tag{2}$$

Considering the assumptions made for the noise process ξ_t, the series will converge almost surely. Thus, the following equations will hold with probability 1.

Integrating by parts and using complete induction, one can show that

$$P_k(t) = \sum_{q=0}^{k} \frac{1}{q!} (B\eta_t)^k \left[\frac{1}{(k-q)!}(At)^{k-q} + f_{k-q}(t)\right] , \tag{3}$$
$$k = 1,2,3,\ldots .$$

Here η_t denotes the stochastic process

$$\eta_t = \int_0^t \xi_\tau d\tau \tag{4}$$

and the matrix valued functions $f_\ell(t)$ may be obtained from the recursion system

$$f_{\ell+1}(t) = \sum_{p=1}^{\ell} \left[\sum_{r=0}^{p-1} B^r [A,B] B^{p-1-r}\right] A^{\ell-p} \int_0^t \frac{1}{p!} \eta_\tau^p \frac{1}{(\ell-p)!} \tau^{\ell-p} d\tau$$
$$+ \sum_{p=0}^{\ell} AB^p \int_0^t \frac{1}{p!} \eta_\tau^p f_{\ell-p}(\tau) d\tau$$

$$- \sum_{p=0}^{\ell} B^{p+1} \int_0^t \frac{1}{(p+1)!} \eta_\tau^{p+1} \frac{\partial}{\partial \tau} f_{\ell-p} (\tau)\, d\tau \; ,$$

$$\ell=1,2,3,\dots ,$$

$$f_0(t) \equiv 0, \quad f_1(t) \equiv 0, \tag{5}$$

where $[A,B] = AB-BA$ is the commutator product of the matrices A and B.

Together with equations (2) and (3) we obtain a first but rather formal expression for the solution of (1)

$$x_t = e^{B\eta_t} \left[e^{At} + \sum_{k=0}^{\infty} f_k(t) \right] .x_0 \; . \tag{6}$$

Keeping in mind that the influence of the noise process ξ_t in (1) is governed by matrix B, one can obtain a more physical representation of the solution if terms of the functions $f_\ell(t)$ are collected which have the same intensity of noise, i.e. which are of the same order in B and η, respectively. Let $F_r(t)$ denote the sum of terms of order r in B included in $\Sigma f_k(t)$. The final series expansion of the solution of (1) then writes as

$$x_t = e^{B\eta_t} \left[e^{At} + \sum_{r=1}^{\infty} F_r(t) \right] .x_0 \tag{7}$$

A somewhat lengthy analysis with the recursion formula (5) finally leads to a system of first order differential equations

$$\dot{F}_r = AF_r + \sum_{q=0}^{r-1} \frac{1}{(r-q)!} \eta_t^{r-q} \left[AB^{r-q}F_q - B^{r-q}\dot{F}_q \right] ,$$

$$F_r(0) = 0, \qquad r = 1,2,3,\dots , \tag{8}$$

$$F_0(t) = e^{At}$$

from which the functions $F_r(t)$ may be computed successively. For example we have

$$F_1(t) = e^{At} \cdot \int_0^t \eta_u e^{-Au} [A,B] e^{Au} du ,$$

$$F_2(t) = e^{At} \left[\int_0^t \eta_u e^{-Au} [A,B] e^{Au} \int_0^u \eta_v e^{-Av} [A,B] e^{Av} dv du \right. \tag{9}$$

$$\left. + \int_0^t \frac{1}{2} \eta_u^2 e^{-Au} [[A,B],B] e^{Au} du \right] .$$

Continuing the procedure, commutator products with the matrix B of increasing order appear in the functions $F_r(t)$.

It is readily seen that the functions $F_r(t)$, $r = 1,2,\ldots$, vanish identically if the matrices A and B commute, i.e. $[A,B] = 0$. In this case the solution (7) reduces to the one given in [6].

THE FIRST MOMENTS OF x_t

The series expansion (7) of the solution of (1) may be used to get an expression for the first moments of the state variables applying the expectation operator. If we assume the initial condition x_0 to be independent of the process η_t , $t \geq 0$, we obtain

$$m_x(t) = \sum_{j=0}^{\infty} \sum_{q=0}^{2j} \frac{1}{(2j-q)!} B^{2j-q} E \left[\eta_t^{2j-q} F_q(t) \right] . m_x(0) \tag{10}$$

considering that ξ_t is a Gaussian process with $E\xi_t \equiv 0$.

In more detail we have

$$m_x(t) = e^{\frac{1}{2}\sigma^2(t)B^2} \left[e^{At} + Be^{At} \int_0^t R_{\eta\eta}(t,\tau) e^{-A\tau} [A,B] e^{A\tau} d\tau \right. \tag{11}$$

$$\left. + \ldots\ldots \right] . m_x(0)$$

by taking into account the influence of $F_0(t)$ and $F_1(t)$, respectively. Here $\sigma^2(t)$ denotes the variance of the process η_t, and $R_{\eta\eta}(t,\tau) = E\eta_t\eta_\tau$ denotes its autocorrelation function.

Equations (10) and (11) are useful in computing the evolution of the first moments of x_t with increasing time t. But the asymptotic behaviour of $m_x(t)$ is difficult to detect from (10) and (11). For this purpose a first order differential equation can be derived.

Taking the expectation of system equation (1), we have

$$\dot{m}_x = Am_x + Bm_{x\xi} \quad . \tag{12}$$

Here unknown correlation moments $m_{x\xi}(t) = Ex_t\xi_t$ between x_t and the noise process ξ_t appear. Using equation (7), $m_{x\xi}(t)$ may be computed from an equation similar to (10). In comparison with $m_x(t)$ it can be shown that

$$m_{x\xi}(t) = \frac{1}{2}\frac{d\sigma^2(t)}{dt} B.m_x(t) + \Phi(t)$$

where

$$\Phi(t) = \sum_{j=0}^{\infty}\sum_{q=0}^{2j} \frac{1}{(2j-q)!} B^{2j-q}\phi(\xi_t\eta_t^{2j-q}F_{q+1}(t)).m_x(0)$$

$$\phi(\xi_t\eta_t^pF_r(t)) = E\left[\xi_t\eta_t^pF_r(t)\right] - pE\left[\xi_t\eta_t\right]E\left[\eta_t^{p-1}F_r(t)\right] \; .$$

Taking $F_1(t)$ into account explicitly, we have

$$\Phi(t) = e^{\frac{1}{2}\sigma^2(t)B^2}\left[e^{At}\int_0^t R_{\xi\eta}(t,\tau)e^{-A\tau}\left[A,B\right]e^{A\tau}d\tau + ..\right] . \; m_x(0) \tag{14}$$

where $R_{\xi\eta}(t,\tau) = E\xi_t\eta_\tau$ denotes the crosscorrelation function of the processes ξ and η.

Inserting this into (12), we finally obtain an inhomogeneous differential equation with time varying coefficients for the first moments of x_t.

$$\dot{m}_x = \left[A + \frac{1}{2}\frac{d\sigma^2(t)}{dt} B^2\right] m_x + B\Phi(t) \ . \tag{15}$$

For wide band noise ξ_t which is generated by the first order shaping filter

$$\dot{\xi} + \gamma\xi = \dot{w} \ , \quad \gamma > 0 \tag{16}$$

from standard white noise $\dot{w}_t$ with $E\dot{w}_t \equiv 0$ and $E\dot{w}_t\dot{w}_\tau = \delta(t-\tau)$, we may obtain standard white noise itself going to infinity with γ, the bandwidth of the process ξ_t. Then it can be readily shown that the function $\Phi(t)$ vanishes identically. In addition we get $d\sigma^2(t)/dt \equiv 1$. Thus, we obtain in the limit the well known equation

$$\dot{m}_x = \left[A+\tfrac{1}{2}B^2\right] m_x \tag{17}$$

corresponding to the Stratonovich type system (1) with white noise excitation [9]. Consequently, a new derivation of the Stratonovich equation is given without using Itô calculus and correction terms.

Furthermore, as the functions $F_r(t)$, $r = 1,2,\ldots$, vanish identically if the matrices A and B commute, $\Phi(t)$ vanishes also in this case. Then, equation (15) reduces to the corresponding homogeneous equation already given in [6].

THE SECOND MOMENTS OF x_t

A similar procedure as above leads to an expression for the second moments of the state variables which corresponds to (10). But in general this equation is rather difficult to deal with.

A more convenient way to get a closed equation for the second moments of x_t is to use the first order differential equation

$$\dot{m}_{xx} = Am_{xx} + m_{xx}A^* + Bm_{xx\xi} + m_{xx\xi}B^* \tag{18}$$

which is obtained from (1) considering $(xx^*)^\cdot = \dot{x}x^* + x\dot{x}^*$ and taking the expectation. Here $m_{xx\xi}(t) = Ex_t x_t^* \xi_t$ denotes unknown correlation moments with the noise process ξ_t. These may be computed from an equation similar to (10). In comparison with $m_{xx}(t)$ it can be shown that

$$m_{xx\xi}(t) = \frac{1}{2}\frac{d\sigma^2(t)}{dt} Bm_{xx}(t) + m_{xx}(t)\frac{1}{2}\frac{d\sigma^2(t)}{dt} B^* + \psi(t) + \psi^*(t)$$

where

$$\psi(t) = \sum_{i=0}^{\infty}\sum_{p=0}^{2i+1}\sum_{j=0}^{\infty}\sum_{q=0}^{2j} \frac{1}{p!} B^p\Psi\,(\ldots)\,\frac{1}{q!} B^{*q} \ ,$$

$$\Psi(\ldots) = \Psi(\xi_t\eta_t^{p+q}\ F_{2i+1-p}(t)x_o x_o^*\ F^*_{2j-q}(t)) \ , \tag{19}$$

$$\Psi(\xi_t\eta_t^u F_v(t)x_o x_o^*\ F_w^*(t)) = E\left[\xi_t\eta_t^u F_v(t)x_o x_o^* F_w^*(t)\right]$$

$$- uE\left[\xi_t\eta_t\right].E\left[\eta_t^{u-1} F_v(t)x_o x_o^*\ F_w^*(t)\right] \ .$$

Taking $F_1(t)$ into account explicitly, we obtain in more detail

$$\psi(t)+\psi^*(t) = \sum_{j=0}^{\infty}\frac{1}{j!}\,(\tfrac{1}{2}\sigma^2(t))^j \sum_{q=0}^{2j}\binom{2j}{q} B^q G_1(t) B^{*2j-q}$$

$$+ \sum_{j=0}^{\infty}(\tfrac{1}{2}\sigma^2(t))^j \sum_{q=0}^{2j+1}\binom{2j+1}{q} B^q G_2(t) B^{*2j+1-q} + \ldots$$

$$G_1(t) = e^{At}\,\left[H_1(t) + H_1^*(t)\right]\,e^{A^*t} \ , \tag{20}$$

$$G_2(t) = e^{At}\,\left[H_1(t) + H_2(t)\ H_1^*(t)\right]\,e^{A^*t} \ ,$$

$$H_1(t) = \int_o^t R_{\xi\eta}(t,\tau)e^{-A\tau}\,\left[A,B\right]e^{A\tau}\,d\tau.m_{xx}(0) \ ,$$

$$H_2(t) = \int_o^t R_{\eta\eta}(t,\tau)e^{-A\tau}\,\left[A,B\right]\,e^{A\tau}\,d\tau \ .$$

Inserting this into (18), we finally get an inhomogeneous differential equation with time varying coefficients for the second moments of x_t.

$$\dot{m}_{xx} = \left[A+\tfrac{1}{2}\frac{d\sigma^2(t)}{dt}B^2\right]m_{xx} + m_{xx}\left[A+\tfrac{1}{2}\frac{d\sigma^2(t)}{dt}B^2\right]^* + \frac{d\sigma^2(t)}{dt}Bm_{xx}B^*$$

$$+ B(\psi(t) + \psi^*(t)) + (\psi(t) + \psi^*(t))B^* \,. \tag{21}$$

Going from the wide band noise ξ_t of shaping filter (16) to standard white noise $\dot{w}_t$ in the limit, it can be shown that the function $\psi(t)$ vanishes identically. Thus we obtain the well known equation

$$\dot{m}_{xx} = \left[A+\tfrac{1}{2}B^2\right]m_{xx} + m_{xx}\left[A+\tfrac{1}{2}B^2\right]^* + Bm_{xx}B^* \tag{22}$$

corresponding to the Stratonovich type system (1) with white noise excitation, just as for the first moments $m_x(t)$.

Furthermore, $\psi(t)$ vanishes also if the matrices A and B commute, because the functions $F_r(t)$, r=1,2,.., do so. Then, equation (21) reduces to the corresponding homogeneous equation already given in [6].

An Example

As an application for the moment equations given above, a second order system is considered consisting of two distinct randomly disturbed first order systems coupled by a noisy term

$$\dot{x} = \{A+B\xi\}x \quad , \quad x=x_t = \begin{bmatrix} y_t \\ z_t \end{bmatrix} , \quad A = \begin{bmatrix} a & 0 \\ 0 & b \end{bmatrix} , \; a \neq b \; ,$$

$$B = \begin{bmatrix} c & 0 \\ d & c \end{bmatrix} \quad d \neq 0 \; . \tag{23}$$

Two cases are considered concerning the noise ξ_t:

(i) ξ_t is assumed to be the wide band process generated by the shaping filter (16) from standard white noise $\dot{w}_t$.

(ii) ξ_t is assumed to be the stationary output of the second order oscillating shaping filter

$$\ddot{\xi} + 2\gamma\delta\dot{\xi} + \gamma^2\xi = \gamma^2\dot{w} \quad , \quad \gamma>0 \quad , \quad 1>\delta>0 \tag{24}$$

excited by standard white noise $\dot{w}_t$. In the following this process ξ_t will be called 'oscillating noise'.

A simple test shows that the Lie algebra L generated by the matrices A and B of system (23) is solvable. Thus, the stability criteria given in [7] apply to system (23). After some short calculations they prove for both cases of noise ξ_t that the first moments of y_t and z_t are stable if

$$\max\{a,b\} < -\tfrac{1}{2}c^2 \quad , \tag{25}$$

and that their second moments are stable if

$$\max\{a,b\} < -c^2 \ . \tag{26}$$

It is worth noting that neither the coupling parameter d nor the coefficients of the noise processes are involved in these conditions. Moreover, both conditions require the corresponding undisturbed system to be stable.

Turning to the moment equations of system (23), one finds that only $F_1(t)$ remains in the solution (7) whereas $F_r(t)\equiv 0$, $r = 2,3,\ldots$. Thus, the differential equation for the first moments is given by (14) and (15) without any additional terms.

Calculations for both cases of noise ξ_t show that the inhomogeneous part of the moment equation tends to 0 as time t goes to infinity if the system matrix of the homogeneous part tends to a stability matrix. Thus the stability of the first moments of system (23) is governed by the limit

$$A+\frac{1}{2}\,\frac{d\sigma^2(t)}{dt}\,B^2 \xrightarrow{t\to\infty} A+\tfrac{1}{2}S_{\xi\xi}(0)B^2 \tag{27}$$

where $S_{\xi\xi}(0)$ is the value of the spectral density of the noise process ξ_t at point 0. This again leads to stability condition (25).

After some analysis which may be carried out at least for the noise processes ξ_t considered here, we obtain homogeneous equations for the first moments

$$\dot{m}_y = \{a+\tfrac{1}{2}\frac{d\sigma^2(t)}{dt}c^2\}m_y \quad ,$$

$$\dot{m}_z = \{b+\tfrac{1}{2}\frac{d\sigma^2(t)}{dt}c^2\}m_z + h_1(t)m_y \quad ,$$

$$h_1(t) = cd\{\tfrac{1}{2}\frac{d\sigma^2(t)}{dt} + h_2(t)\} \quad , \tag{28}$$

$$h_2(t) = e^{(b-a)t}\int_0^t R_{\xi\xi}(t,\tau)e^{(a-b)\tau}d\tau \; .$$

The corresponding equation for the second moments is given by (20) and (21) without additional terms. An argumentation as above again leads to stability condition (26). Once more homogeneous equations for the second moments may be obtained.

$$\dot{m}_{yy} = 2\{a+\frac{d\sigma^2(t)}{dt}c^2\}m_{yy} \quad ,$$

$$\dot{m}_{yz} = \{a+b+2\frac{d\sigma^2(t)}{dt}c^2\}m_{yz} + 2h_1(t)m_{yy} \quad , \tag{29}$$

$$\dot{m}_{zz} = 2\{b+\frac{d\sigma^2(t)}{dt}c^2\}m_{zz} + 4h_1(t)m_{yz} + 2d^2h_2(t)m_{yy}$$

where the functions $h_1(t)$ and $h_2(t)$ are defined by (28).

In the case of wide band noise ξ_t one gets

$$\frac{d\sigma^2(t)}{dt} = 1-e^{-\gamma t} \quad ,$$

$$h_2(t) = \frac{\gamma}{2(b-a-\gamma)}\left[e^{(b-a-\gamma)t}-1\right] \; . \tag{30}$$

Similarly for oscillating noise ξ_t one obtains

$$\frac{d\sigma^2(t)}{dt} = 1-g_1(t) \quad ,$$

$$g_1(t) = e^{-\gamma\delta t}(\cos\gamma\omega t + \frac{2\delta^2-1}{2\delta\omega}\sin\gamma\omega t) \;, \quad \omega = \sqrt{1-\delta^2} \quad ,$$

$$h_2(t) = \frac{\gamma}{2\{(b-a)^2-2\gamma\delta(b-a)+\gamma^2\}}\left[\frac{b-a}{2\delta}\left[e^{(b-a)t}g_2(t)-1\right]\right.$$

$$\left. -\gamma\left[e^{(b-a)t}g_1(t)-1\right]\right] \quad , \tag{31}$$

$$g_2(t) = e^{-\gamma\delta t}(\cos\gamma\omega t + \frac{\delta}{\omega}\sin\gamma\omega t).$$

As expected from system equation (23), the first and second moments of y_t are not influencedby z_t, but those of z_t are driven by y_t. Therefore, it is worth looking at $m_z(t)$ and $m_{zz}(t)$ in more detail. From numerical computations some graphs are given. The parameters a, b and c, the initial conditions, and the coefficient δ of shaping filter (24) were chosen as

$$\begin{aligned} &a = -10, \quad b = -2, \quad c = 1, \quad \delta = 0.05, \\ &m_y(0) = 0.1, \quad m_{yy}(0) = 0.01, \\ &m_z(0) = 0.01, \quad m_{zz}(0) = 0.0001, \quad m_{yz}(0) = 0.001. \end{aligned} \tag{32}$$

Obviously, the values (32) guarantee the first and second moments of system (23) to be stable.

In general, it may be seen from the graphs that $m_z(t)$ as well as $m_{zz}(t)$ have peaks but nevertheless tend to zero very rapidly with increasing time t. Figure 1 gives $m_z(t)$ for the wide band noise ξ_t corresponding to $\gamma = 1000$. As indicated by (28), the curves plotted for d and -d are not quite symmetric with respect to the time axis. In Figure 2 we have the same configuration, but now the noise coefficient γ varies whereas d=30. Here the difference between curves obtained for further increasing γ-values is below the accuracy of drawing.

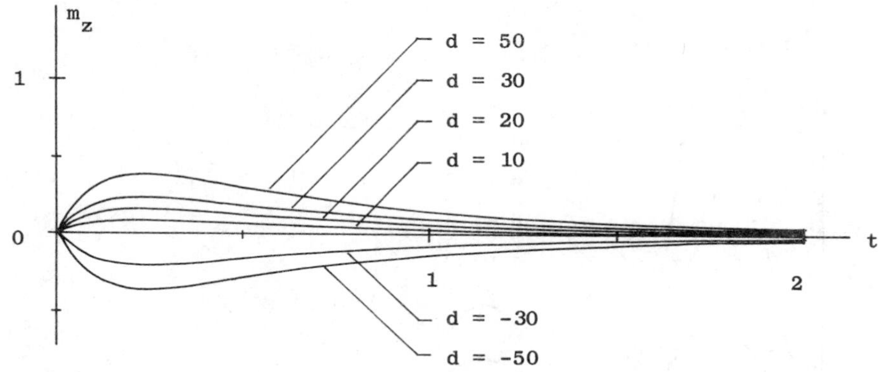

FIG. 1 $m_z(t)$ for the case of wide band noise with $\gamma=1000$.

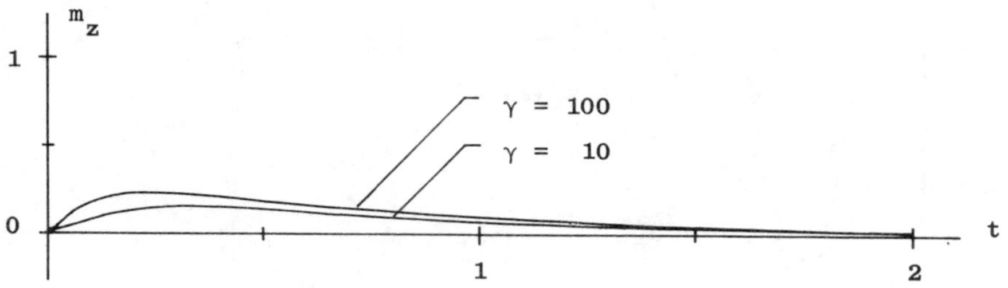

FIG. 2 $m_z(t)$ for the case of wide band noise; d=30.

Figures 3 and 4 show how the situation is for the case of oscillating noise ξ_t. The frequency of the oscillations of $m_z(t)$ is equal to the frequency of the noise.

Figures 5 and 6 give plots of $m_{zz}(t)$ for wide band noise. According with (29), the curves obtained for d and -d are not quite the same: the lower ones correspond to the negative values of d.

Finally, in Figures 7 and 8 we have $m_{zz}(t)$ for oscillating noise ξ_t. Here the difference between curves obtained for d and -d is below the accuracy of drawing.

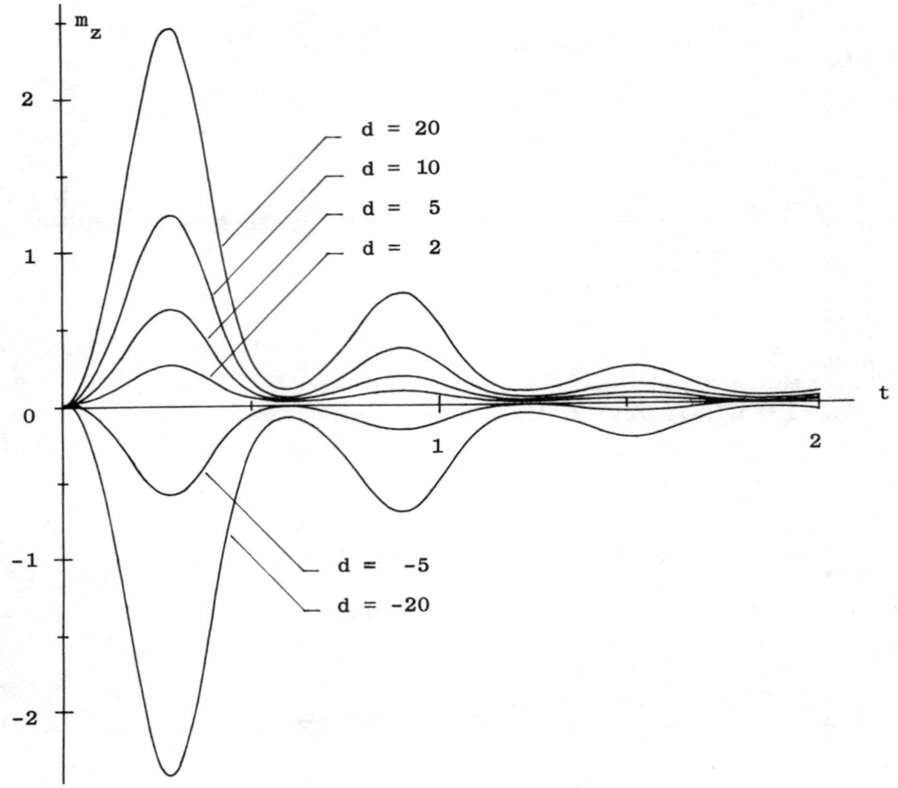

FIG. 3 $m_z(t)$ for the case of oscillating noise with $\gamma=10$.

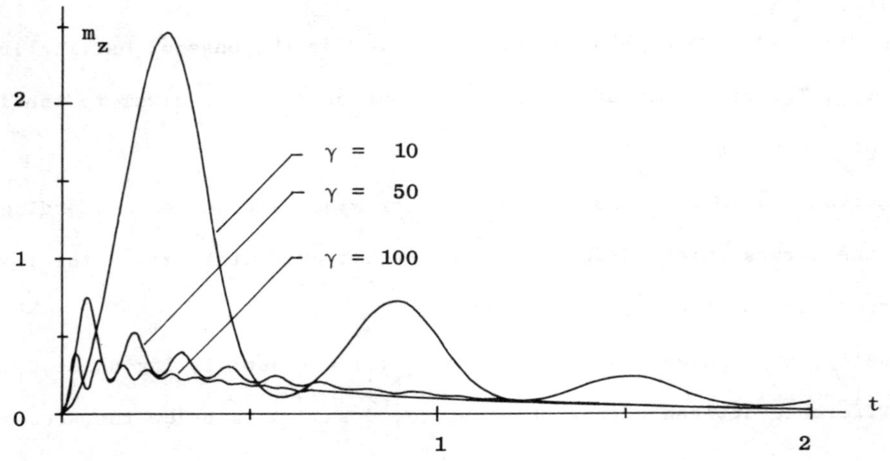

FIG. 4 $m_z(t)$ for the case of oscillating noise; d=20.

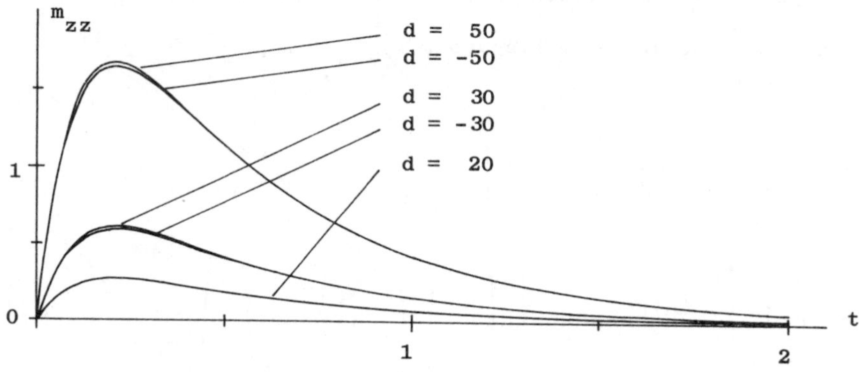

FIG. 5 $m_{zz}(t)$ for the case of wide band noise with γ =1000.

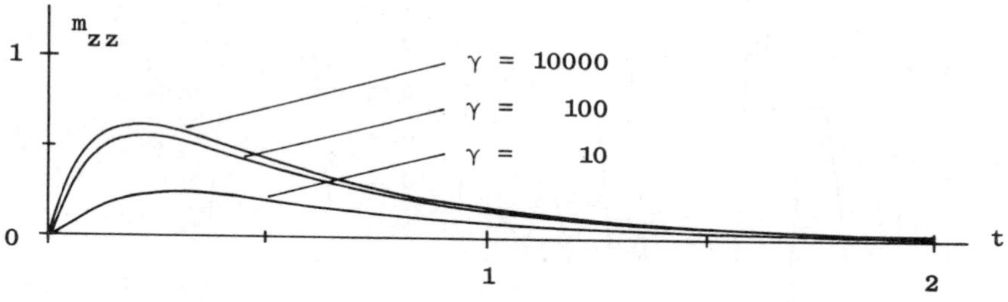

FIG. 6 $m_{zz}(t)$ for the case of wide band noise; d=30.

Summarizing, it may be noted that the size of the peaks increases with growing absolute value of the coupling parameter d for some fixed noise ξ_t. On the other hand, one finds that the curves of $m_z(t)$ and $m_{zz}(t)$ approach the corresponding ones for the Stratonovich type system (23) if the noise coefficient γ increases, i.e. if both wide band noise and oscillating noise get more and more white.

Comparing both types of noise processes considered here, oscillating noise obviously gives worse peaks than wide band noise. For example, in Figures 7 and 8 the maximum peak of $m_{zz}(t)$ reaches 216 for γ=10 and d=20; this is 2 160 000 times the value of the initial condition $m_{zz}(0)$. Thus, from a practical point of view system (23) is not well behaved in this case although it is stable in the sense of Liapunov.

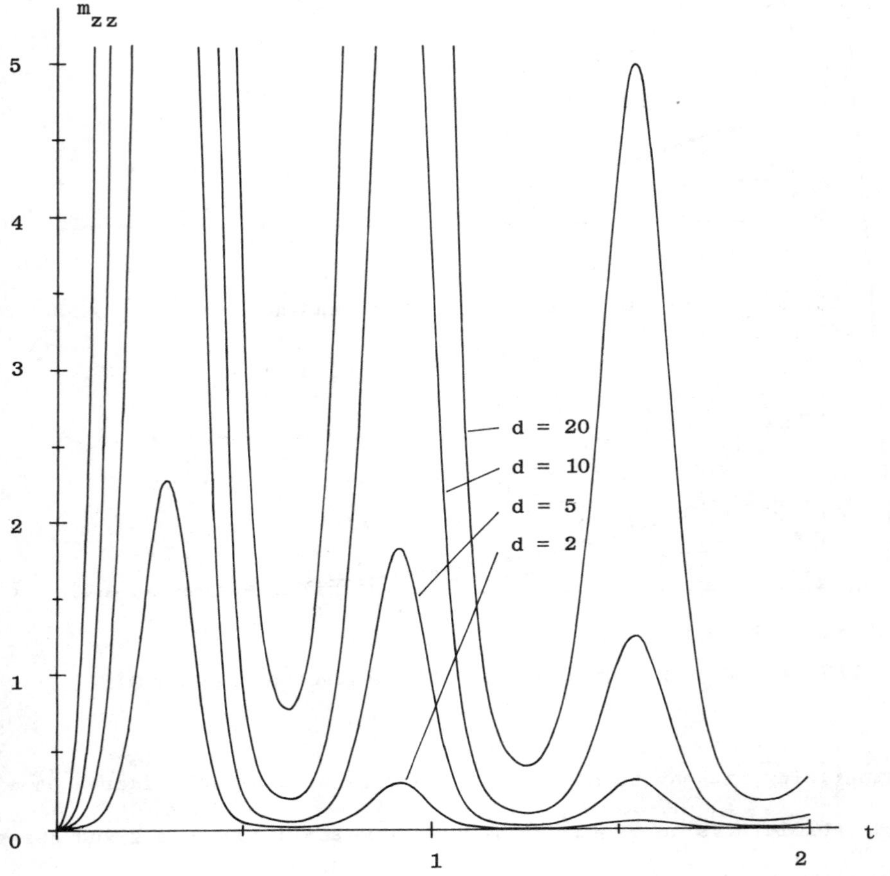

FIG. 7 $m_{zz}(t)$ for the case of oscillating noise with $\gamma=10$.

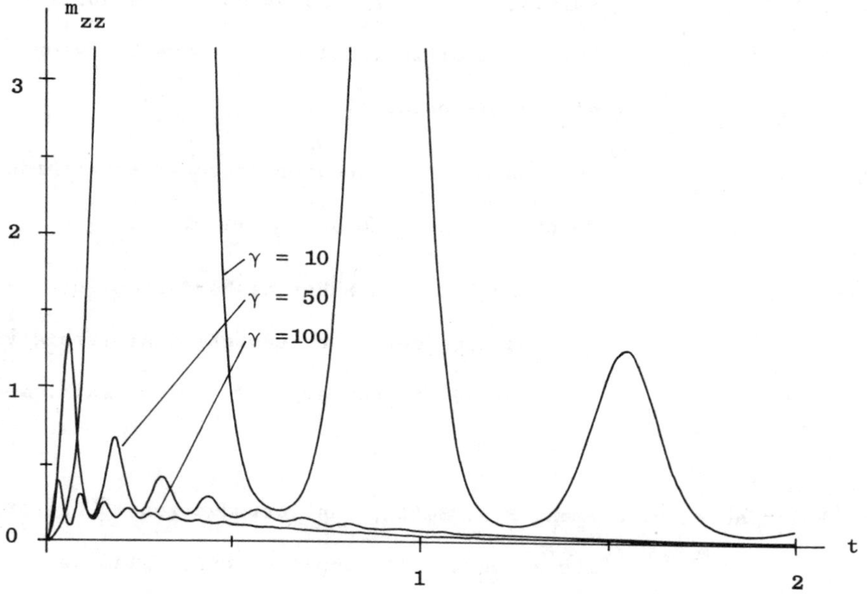

FIG. 8 $m_{zz}(t)$ for the case of oscillating noise; d=10.

CONCLUSION

Strictly closed equations for the first and second moments of system (1) have been derived for the case of a general Lie algebra L generated by the matrices A and B. Applying these equations to a simple example has shown that not only the asymptotic behaviour of the moments is of great importance but also the behaviour for finite time intervals. Here the closed equations obtained above may give a tool for detailed judging of systems (1).

REFERENCES

1 J.B. Keller, Stochastic equations and wave propagation in random media. Proc. Symposia Appl. Math. Vol. XVI, 1964.

2 V.V. Bolotin, Reliability theory and stochastic stability. Study No. 6, Stability, Solid Mechanics Division, University of Waterloo, 1971.

3 J.L. Zeman, Zur Lösung nichtlinearer stochastischer Probleme der Mechanik. Acta Mechanica 14, 1972.

4 A. Kistner, Über die Güte von Näherungsverfahren zur Untersuchung der Momentenstabilität farbig verrauschter Systeme (to appear). Z. Angew. Math. Mech. 57, 1977.

5 J.G. Belinfante, B. Kolman, H.A. Smith, An Introduction to Lie groups and Lie algebras with applications. SIAM Review 8, 1966.

6 P. Sagirow, Zur AbschlieBung der Momentengleichungen linearer Systemme mit stochastischer Parametererregung. Z. Angew. Math. Mech. 56, 1976.

7 J.L. Willems, Stability criteria for stochastic systems with colored multiplicative noise. Acta Mechanica 23, 1975.

8 R.W. Brockett, Finite dimensional linear systems; Wiley, New York, 1970.

9 L. Arnold Stochastische Differentialgleichungen; Oldenbourg, München, 1973.

Dipl.-Math. A. Kistner, Institut A für Mechanik, Universitat, Stuttgart 1, Keplerstr. 17.

DISCUSSION

In response to questions on the functions $\Phi(t)$ and $\psi(t)$ in the moment equations (13) and (21) (Grossmayer) Mr. Kistner explained that in general $\Phi(t)$ and $\psi(t)$ may be computed from the infinite series (13) and (19), respectively. General statements about the convergence of these series could not be given. The possibility of obtaining closed form expressions from the series depended on the solvability of the Lie algebra generated by the matrices A and B of system (1) and, moreover, on A and B themselves. For example with the system (23) the series of $\Phi(t)$ and $\psi(t)$ consist of a single term, but they consist of infinitely many terms if the diagonal elements of B are modified to be distinct. Nevertheless in both cases the Lie algebra is solvable, and closed form expressions of $\Phi(t)$ and $\psi(t)$ may be obtained.

Asked about physical examples of systems which directly lead to homogeneous moment equations (Wedig), Mr. Kistner mentioned the important class of dynamical systems transformed to normal coordinates (matrix A) and with appropriate matrix B, e.g. systems in normal coordinates and with linear disturbances of the real and imaginary parts of the eigenvalues. Here the matrices A and B commute (AB=BA), thus the Lie algebra generated by A and B is Abelian and solvable, and the corresponding moment equations are homogeneous.

W WEDIG

Stochastic boundary and eigenvalue problems

INTRODUCTION

When replacing the time variable of the Wiener process by a space coordinate, we obtain a Wiener field which may be regarded as a base model for randomly distributed loadings or imperfections of continuous systems. Analogous to initial value problems in the dynamical case, the solutions of such boundary problems may be described by integral equations defined on the Wiener field. But, as the consequence of the boundary conditions, they are not Markovian and therefore dependent on the increment of the Wiener field. It is shown, how the Wiener field has to be modified in order to get the Markov property in space, as well. Then, the application of Itô's integral is possible and leads to associated moments' equations which will allow to analyse inhomogeneous characteristics of such problems or their stability in mean square.

BOUNDARY AND INITIAL VALUE PROBLEMS

Let us regard two simple examples, in order to show essential differences between stochastic boundary and initial value problems. The first one is given as an initial value problem by

$$\dot{X}_{1t} = X_{2t}, \qquad \text{I.C.:} \quad X_{1,t_o=0} = X_o,$$
$$\dot{X}_{2t} = \sigma \dot{W}_t, \qquad \qquad X_{2,t_o=0} = V_o \tag{1}$$

and may describe e.g. the motion of a free mass under the action of a stochastic $\sigma\dot{W}_t$ force. Herein, σ is an intensity parameter and W_t is the normed Wiener process with the moments

$$E(W_t) = 0\ , \qquad E(W_t W_s) = \min(t,s) \tag{2}$$

Since the time derivation $\dot{W}_t$ of the Wiener process is not declared in the classical sense of the probability theory, we write the differential equation (1) in the form of integral equations

$$X_{1t} = X_o + \int_o^t X_{2\tau}\, d\tau\ ,$$

$$X_{2t} = V_o + \sigma \int_o^t dW\tau \tag{3}$$

using the above given initial conditions of the position coordinate X_{1t} and the velocity X_{2t}.

Due to Itô's integral definition and the fact, that the state vector of (1) is independent on the increment dW_t of the Wiener process, it is possible to go over to stochastic differential equations as a short hand notation of (3) and to apply Itô's calculus [1] in order to set up the associated equations of the second order moments $P_{ij}(t) = E(X_{it}X_{jt})$ $(i,j = 1,2)$.

$$\dot{P}_{11} = 2P_{12}\ , \qquad P_{22}(t) = \sigma^2 t + V_o^2\ ,$$

$$\dot{P}_{12} = P_{22}\ , \qquad P_{12}(t) = \sigma^2 t^2/2 + V_o^2 t + X_o V_o\ ,$$

$$\dot{P}_{22} = \sigma^2\ , \qquad P_{11}(t) = \sigma^2 t^3/3 + V_o^2 t^2 + 2X_o V_o t + X_o^2\ .$$

The solutions of them are simple polynomials in the time t.

Now, let us consider the corresponding boundary value problem, given by

$$Y_x' = \frac{1}{H_o} V_x\ , \qquad \text{B.C.:} \quad Y_o = Y_\ell = 0\ ,$$

$$V_x' = -\sigma\, W_x'\ , \qquad 0 \leq X \leq \ell\ , \tag{4}$$

which may describe e.g. the deflection Y_x and the transverse force V_x of a string with the length ℓ and preloaded by the deterministic axial support force H_o. As the external transverse loading of the string we assume a white field W'_x , which is the derivation of a normed Wiener field W_x normally distributed in space with the initial value $W_{x=0}=0$ and the expectation values

$$E(W_x) = 0 , \qquad E(W_x W_z) = \min (x,z). \tag{5}$$

The parameter σ characterizes the intensity of the external load.

Evidently, the differential equations (4) have a similar form as in (1) and may be transformed in integral equations, as well.

$$\begin{aligned} Y_x &= Y_o + \frac{1}{H_o} \int_0^x V_\xi \, d\xi , \\ V_x &= V_o - \sigma \int_0^x dW_\xi . \end{aligned} \tag{6}$$

But, there is an essential difference consisting in the fact, that the boundary value V_o is the support force at the left end of the string calculable by integrating its external load

$$V_o = \int_0^\ell (1 - \xi/\ell) \, dW_\xi \tag{7}$$

and therefore not independent on the increment dW_x of the Wiener field. Accordingly, it is not allowed to apply Itô's calculus to this boundary problem in order to set up the associated moments' equations.

Naturally, this fact is only important for more general eigenvalue problems. In the given case the integrals defined on dW_x don't contain any state coordinate and can therefore be evaluated, by use of classical methods [2]. We insert the boundary value (7) into the equation (6) and obtain by partial integration

$$V_x = \frac{1}{\ell} \int_0^\ell W_\xi d\xi - \sigma W_x \tag{8}$$

having used the property $W_{x=0}=0$. Thus, the stochastic transverse force V_x

at $x = 0$ is an ordinary Riemann's integral, calculable for each realization of W_x. The equation (8) leads to the correlation function

$$E(V_x V_z) = \sigma^2 \left[\frac{\ell}{3} - x(1 - \frac{x}{\ell}) - z(1 - \frac{z}{\ell}) + \min(x,z) \right]$$

and finally to

$$E(Y_x^2) = \frac{1}{H_o^2} \int_0^x \int_0^x E(V_\xi V_\zeta)\, d\xi\, d\zeta = \frac{\sigma^2 \ell^3}{3H_o^2} (\frac{x}{\ell})^2 (1 - \frac{x}{\ell})^2 \tag{9}$$

as the variance function of the inhomogeneous deflection of the spring [3]. It is positive definite and in opposite to (3) a polynomial of fourth degree in the space coordinate x.

MARKOV PROPERTY IN BOUNDARY VALUE PROBLEMS

The above performed calculation of the variance function is simplified drastically by use of the Green's function [4]. It is determined by

$$G(x,\zeta) = \begin{cases} \zeta(1 - \frac{x}{\ell}) , & 0 \leq \zeta \leq x , \\ x(1 - \frac{\zeta}{\ell}) , & x \leq \zeta \leq \ell . \end{cases} \tag{10}$$

It is symmetric and solves the given boundary problem in the form

$$Y_x = \frac{\sigma}{H_o} \int_0^\ell G(x,\zeta)\, d\zeta , \tag{11}$$

if we replace $W_x'\, dx$ by dW_x as usual. Inserting the Green's function (10) into the equation (11) we obtain the form

$$Y_z = \frac{\sigma}{H_o} \left[\int_0^x \zeta(1 - \frac{x}{\ell})\, dW_\zeta + \int_x^\ell x(1 - \frac{\zeta}{\ell})\, dW_\zeta \right] , \tag{12}$$

which obviously satisfies the boundary conditions, that there exists no deflection at both ends of the string. The first derivation of (12)

$$Y'_x = \frac{\sigma}{H_o} \left[-\int_0^x \frac{\zeta}{\ell}\, dW_\zeta + \int_x^\ell (1 - \frac{\zeta}{\ell})\, dW_\zeta \right] \tag{13}$$

is not vanishing at $x = 0, \ell$ and its second derivation leads to

$$Y''_x = - \frac{\sigma}{H_o} W'_x \tag{14}$$

that is equivalent to the two first order differential equations in (4).

As already mentioned, the use of $G(x,\zeta)$ Green's function has the advantage that the statistical characteristics of Y_x are easier to be calculated. For the variance function e.g. we simply take the expectation of the squared equation (12)

$$E(Y_x^2) = \frac{\sigma^2}{H_o^2} \left[\int_0^x \zeta^2 (1 - \frac{x}{\ell})^2 d\zeta + \int_x^\ell x^2 (1 - \frac{\zeta}{\ell})^2 d\zeta \right], \tag{15}$$

where we have applied the property of the Wiener field, that its increments at two different positions are independent. Therefore, the expectation of squared integrals in (15) reduces to single ones and the expectations of the mixed integral are vanishing. Naturally, the evaluation of (15) leads to the same result as in (9).

But there is a second simplification, which has been achieved by the representation (12). It is clear, that the first integral in (12)

$$A_x = \int_0^x \zeta(1 - \frac{x}{\ell})\, dW_\zeta, \qquad E(A_x dW_x) = 0 \tag{16}$$

has the Markov property, i.e. it is independent on the increment of the Wiener field. Only, the second integral is not Marcovian

$$B_x = \int_x^\ell x(1 - \frac{\zeta}{\ell})\, dW_\zeta, \qquad E(B_x dW_x) = (\ell - x)dx, \tag{17}$$

since the expectation of B_x and dW_x results in the non-vanishing term (17). To remove this dependence and to get the Markov property in B_x as well, we introduce the new coordinate

$$\xi = \ell - \zeta \quad \text{for} \quad \ell - x \geq \xi \geq 0 \tag{18}$$

into the second integral B_x, which may then be written in the form

$$B_x = \int_{\ell-x}^{0} x \frac{\xi}{\ell} \, dW_{\ell-\xi} = \int_{0}^{\ell-x} x \frac{\xi}{\ell} \, dW_{\xi} \, . \tag{19}$$

Herein we have replaced the Wiener field at the position $\ell - \xi$ by the difference of two Wiener fields at the positions ℓ and ξ

$$W_{\ell-\xi} = W_{\ell} - W_{\xi} \, , \qquad dW_{\ell-\xi} = -dW_{\xi} \tag{20}$$

and accordingly the corresponding increments.

It is clear, that the relations (20) don't hold with probability one and are therefore not applicable to study sample properties of the boundary problem (4). These relations are much more valid in the n-th mean, that is easily proved by the coincidence of the correlation functions

$$E(W_{\ell-\xi} W_{\ell-\eta}) = E((W_{\ell} - W_{\xi})(W_{\ell} - W_{\eta})) \tag{21}$$

on both sides of the equation (20). Thus, if we restrict our interest to statistical characteristics of boundary problems such as mean square values or density functions, we surely can make use of the transformed second integral (19), in order to get e.g. the same result as in (9). Furthermore, the transformed integral has obviously the Markov property, and this fact will be important in the more general case of eigenvalue problems.

STOCHASTIC EIGENVALUE PROBLEMS

According to the above mentioned relations let us now consider a second order eigenvalue problem given by the integral equation

$$Y_x = \sigma \left[\int_0^x \zeta \, (1 - \frac{x}{\ell}) Y_{\zeta} \, dW_{\zeta} + \int_0^{\ell-x} x \frac{\xi}{\ell} \, Y_{\ell-\xi} \, dW_{\xi} \right] , \tag{22}$$

which is valid in the range $0 \leq x \leq \ell$ and obviously satisfies the boundary conditions $Y_0 = Y_\ell = 0$ with probability one. The equation (22) is homogeneous with two different multiplicative fields. The first one may be given by the Wiener field W_ζ acting in the range $0 \leq \zeta \leq x$ with the initial value $W_{\zeta=0} = 0$. The second one may start at the right hand side with $W_{\xi=0} = 0$ and is valid in the range $0 \leq \xi \leq \ell - x$. Both Wiener fields are normed having the expectation values (5). Obviously, each integral of this eigenvalue problem has the form, to which we are able to apply Itô's integral definition, in order to study sample properties or statistical characteristics of its solution.

Restricting our interest to the variance function of Y_x, we simply have to take the expectation of its squared form and obtain the moment's equation

$$E(Y_x^2) = \sigma^2 \left[\int_0^x \zeta^2 \left(1 - \frac{x}{\ell}\right)^2 E(Y_\zeta^2)\, d\zeta + \int_0^{\ell - x} x^2 \left(\frac{\xi}{\ell}\right)^2 E(Y_{\ell-\xi}^2)\, d\xi \right] \tag{23}$$

using the Markov property, that Y_ζ is independent on dW_ζ and $Y_{\ell-\xi}$ is independent on the increment dW_ξ. The first derivation of (23) is

$$E(Y_x^2)' = \sigma^2 \left[\int_0^x \zeta^2\, 2\left(1 - \frac{x}{\ell}\right)^2 \left(-\frac{1}{\ell}\right) E(Y_\zeta^2)\, d\zeta + \int_0^{\ell - x} 2x \left(\frac{\xi}{\ell}\right)^2 E(Y_{\ell-\xi}^2)\, d\xi \right]. \tag{24}$$

The equations (23) and (24) lead to the four boundary conditions

$$E(Y_0^2) = E(Y_\ell^2) = E(Y_0^2)' = E(Y_\ell^2)' = 0 , \tag{25}$$

meanwhile higher derivatives of $E(Y_x^2)$ are not vanishing at $x = 0, \ell$. This coincides with the sample properties $Y_0 = Y_\ell = 0$ and Y_0' , $Y_\ell' \neq 0$.

Consequently, the differential equation associated to the boundary conditions (25) is of the fourth order and then of the form

$$E(Y_x^2)'''' + 2\sigma^2 \Big[x(1-\frac{x}{\ell}) E(Y_x^2)'' + (1 - 2\frac{x}{\ell} - 2\frac{x^2}{\ell^2}) E(Y_x^2)' - \frac{4x}{\ell^2} E(Y_x^2) \Big] = 0 . \tag{26}$$

It is linear and homogeneous. It has to be solved, in order to determine such values σ^2, for which nonvanishing solutions $E(Y_x^2)$ exist and satisfy the given boundary conditions (25).

Before doing it, we are looking for a physical interpretation of the associated stochastic eigenvalue problem. For this purpose the integral equation (22) has to be transformed into a single second order differential equation. This is only possible, if we replace the Wiener field W_ξ in the second integral of (22) by the difference $W_\ell - W_{\ell-\xi}$ in the sense, that according to (21) the covariance functions of both different fields coincide. Then the equation (22) goes over into

$$Y_x = \sigma \Big[\int_0^x \zeta (1-\frac{x}{\ell}) Y_\zeta \, dW_\zeta - \int_0^{\ell-x} x \frac{\xi}{\ell} Y_{\ell-\xi} \, dW_{\ell-\xi} \Big] \tag{27}$$

and leads by a twice differentiation to the single second order differential equation

$$Y_x'' + \sigma W_x' Y_x = 0 , \qquad \text{B.C.:} \quad Y_0 = Y_\ell = 0 \tag{28}$$

Finally, in order to find a technical application, we slightly modify this eigenvalue problem to

$$Y_x'' + (\alpha^2 + \sigma W_x') Y_x = 0 , \qquad \text{B.C.:} \quad Y_0 = Y_\ell = 0 \tag{29}$$

assuming, that the multiplicative white field in (28) has a nonvanishing mean value α^2. In case of small intensities σ this set up is physically meaningful e.g. for the stability investigation of an elastic, non perfectly formed column subjected to a deterministic axial force. Its imperfections result in a stochastically distributed flexural rigidity, which may be described by the mean value α^2 and the random variation $\sigma W'_x$.

In the above mentioned sense the modified eigenvalue problem (29) can be identified by the integral equation

$$Y_x = \int_0^x \zeta(1 - \frac{x}{\ell})\, Y_\zeta\, (\alpha^2\, d\zeta + \sigma\, dW_\zeta)$$

$$+ \int_0^{\ell-x} x\, \frac{\xi}{\ell}\, Y_{\ell-\xi}\, (\alpha^2\, d\xi + \sigma\, dW_\xi) \tag{30}$$

and accordingly be investigated. Taking the expectation in (30) we obtain

$$E(Y_x) = \alpha^2 \int_0^x \zeta(1 - \frac{x}{\ell})\, E(Y_\zeta)\, d\zeta$$

$$+ \alpha^2 \int_0^{\ell-x} x\, \frac{\xi}{\ell}\, E(Y_{\ell-\xi})\, d\xi \tag{31}$$

for the mean value of Y_x, respectively for its variance function

$$E(Y_x^2) = \alpha^4 \Big[\int_0^x \int_0^x \zeta\eta\, (1 - \frac{x}{\ell})^2\, E(Y_\zeta Y_\eta)\, d\eta\, d\zeta + \int_0^{\ell-x}\int_0^{\ell-x} x^2\, \frac{\xi\eta}{\ell^2}$$

$$\times E(Y_{\ell-\xi} Y_{\ell-\eta}) d\eta d\xi + 2 \int_0^x \int_0^{\ell-x} \zeta\, \frac{\xi}{\ell}\, x(1 - \frac{x}{\ell}) E(Y_\zeta Y_{\ell-\xi}) d\xi d\zeta \Big]$$

$$+ \sigma^2 \Big[\int_0^x \zeta^2 (1 - \frac{x}{\ell})^2\, E(Y_\zeta^2) d\zeta + \int_0^{\ell-x} x^2 (\frac{\xi}{\ell})^2 E(Y_{\ell-\xi}^2)\, d\xi \Big] \tag{32}$$

Evidently, the mean value equation (31) coincides with the corresponding deterministic eigenvalue problem and results therefore in the well-known eigenvalues and eigenfunctions

$$\alpha_k = \frac{\pi}{\ell}\, k, \qquad \sin \frac{\pi}{\ell}\, kx, \qquad k = 1,2,... \tag{33}$$

The eigenvalue problem (32) of the mean square value is more complicated and may approximately be solved using the moment's method. We restrict our attention to a first approximation evaluated by means of the one-termed series

$$E(Y_x Y_z) = C \sin \frac{\pi}{\ell} x \sin \frac{\pi}{\ell} z , \qquad 0 \leq x, z \leq \ell , \tag{34}$$

which fulfils the boundary conditions (25). According to the moment's method, we multiply the equation (32) by the set up (34) and integrate it over the domain $0 \leq x \leq \ell$. This procedure results in

$$1 = (\frac{\alpha \ell}{\pi})^4 + \frac{1}{3} \sigma^2 \ell^3 \left(\frac{1}{45} + \frac{1}{12\pi^2} + \frac{29}{16\pi^4}\right) \tag{35}$$

as the condition for a non-vanishing coefficient C and finally to

$$\alpha^2 = (\frac{\pi}{\ell})^2 \sqrt{1 - 0.0164 \, \sigma^2 \ell^3} \tag{36}$$

as the lowest eigenvalue of the variance equation (32).

According to the corresponding deterministic stability problem we can interprete it as the critical value, which may not be exceeded, in order to secure the mean square stability of the equilibrium position of the system (29). It is physically reasonable, that it will be as higher, the smaller the intensity σ^2 of the multiplicative white field is. For vanishing imperfections $\sigma = 0$, it finally coincides with the critical value (33) of the mean value function. Naturally, such investigations can be extended to higher order eigenvalue problems such as beams under stochastically distributed axial forces as well as to kinetic stability problems [5] e.g. in case of stochastic follower forces in order to study divergence or flutter properties of non-conservative stochastic systems.

CONCLUSION

In the present paper we consider stochastic integral equations satisfying boundary conditions with probability one. If their integrals are defined on two different independent Wiener fields they possess the Markov property and allow to apply Itô's integral definition in order to calculate mean value and variance functions. It is shown, that such integral equations may be interpreted in the mean square sense by boundary and eigenvalue problems of elastic structures with stochastically distributed imperfections or loadings.

REFERENCES

1 P. Sagirow, Stochastic methods in the dynamics of satellites, courses and lectures No. 57, CISM Udine, Springer-Verlag, 1970.

2 L. Arnold, Stochastische Differentialgleichungen, Oldenbourg-Verlag, München, 1973.

3 W. Wedig, Moments and probability densities of parametrically excited systems and continuous systems, to appear in Schriftenreihe des Zentralinstituts für Mathematik und Mechanik, Akademie-Verlag, Berlin.

4 L. Collatz, Eigenwertaufgaben mit technischen Anwendungen, Greest u. Portig-Verlag, Leipzig, 1963.

5 H. Leipholz, Stability Theory, Academic Press, New York, 1970.

Professor Dr. Ing. Walter Wedig, Institut für Tech. Mechanik, Universität Karlsruhe, 75 KARLSRUHE 1, BRD.

DISCUSSION

During the discussion it was stated clearly that the quantity α^2 (equation (36)) obtained by means of Galerkin's method using the first two terms (question Elishakoff) is a mean value which may be interpreted as a critical boundary for the static load concerning mean stability of the system (29) (question Holmes). A comparison with values calculated for periodic deterministic imperfections (question Schiehlen) had not yet been carried out.

The main discussion arose from the Wiener fields W_ζ, W_ξ and the derived white noise fields W'_ζ, W'_ξ used with the stochastic eigenvalue problems (22) and (28), (29) (Ariaratnam, Kozin, Lin). Professor Wedig stated clearly that W_ζ, W_ξ are two distinct Wiener fields, the first running forward along the beam or string with initial value $W_{\zeta=0}=0$ and the second running backward with $W_{\xi=0}=0$ ($\xi=0$ corresponds to $\zeta=\ell$). Both fields are assumed to be uncorrelated. Answering a question on discontinuity problems with this construction (Kozin), Professor Wedig explained that these problems had prevented from studying sample properties but had not affected his mean value considerations.

In a comment Professor Ariaratnam supposed that it had not been allowed to make use of stationarity, ergodicity, the Markov property etc. when applying the Wiener fields W_ζ, W_ξ and the derived white noise fields to beams or strings of finite length ℓ. These properties were affected by the necessity of putting the values of the fields to zero outside the physical object, but the properties would be applicable to infinite beams and strings. Thus the calculations carried out could be understood as some approximations only. Professor Wedig did not agree to this because one had not had to restrict the white noise fields but only the Wiener fields at point 0 and point ℓ, respectively.

Asked about a generalization to other boundary values (Schmidt), Professor Wedig supposed that there were no difficulties in principle. But problems would arise generalizing the methods to nonlinear problems, e.g. with non-linear elastic foundations (question Elishakoff). Concerning generalizations to higher dimensional problems, Dr. Sobczyk pointed out in a comment that there will be difficulties in understanding the Markov property of higher dimensional Wiener fields depending on spatial coordinates.

J L WILLEMS

Moment stability of linear white noise and coloured noise systems

INTRODUCTION

The mathematical model of a dynamic system often requires the consideration of stochastic elements. Additive stochastic elements represent e.g. measurement errors, external perturbing inputs, or round-off errors in fixed point arithmetic; the analysis of systems with additive noise is well developed, and is not essentially different from the analysis of corresponding deterministic dynamic systems. Multiplicative noise may account e.g. for parameter uncertainties or inaccuracies, for variations of the actuator and amplifier gains, or round-off errors in floating point arithmetic. Such models occur in the analysis of systems involving human operators, in mechanical systems subject to random vibrations, or in economic system theory where often one introduces random time delays. The presence of the multiplicative noise has an essential effect on the system properties. This is in particular true for the stability behaviour of such stochastic systems. An important question is how to define the convergence of the solutions; indeed, any boundedness or convergence property used in deterministic system theory can be translated for stochastic systems in different ways, depending on the type of stochastic convergence one wishes to consider. Stability definitions very frequently used in stochastic system theory are almost sure stability (almost all sample solutions are stable) and stability of the moments of a given order. In the present contribution some aspects of the relationships between these stability properties are dealt with. In particular some classes of stochastic systems with coloured or white multiplicative noise are considered, for which explicit criteria for stability are obtained.

MATHEMATICAL PRELIMINARIES

Linear dynamic systems with stochastic parameters are described by the differential equation

$$dx(t)/dt = \left[A + \sum_{i=1}^{m} B_i f_i(t) \right] x(t) \tag{1}$$

for coloured noise parameters, or by the Itô differential equation

$$dx(t) = Ax(t)dt + \sum_{i=1}^{m} \sigma_i B_i x(t) dW_i(t) \tag{2}$$

for white noise parameters. In these equations x(t) is the state vector, of which the dimension will be denoted by n and the components by $x_1, x_2, \ldots,$ x_n. $A, B_1, \ldots, B_m$, are constant square matrices of order n. The stochastic processes are assumed Gaussian and stationary, and the mean is zero. The correlation matrix is

$$R_{ff}(T) = E\left[f(t)f(t+T)^T\right]$$

where the superscript T denotes matrix transposition, where E denotes the expected value, and where f is the vector

$$f(t) = \left[f_1(t)\ f_2(t)\ \ldots\ f_m(t)\ \right]^T .$$

The spectral density matrix is

$$S_{ff}(s) = \mathcal{L}\ \left[R_{ff}(T)\right]$$

where $\mathcal{L}$ denotes the Laplace transform. The processes $W_i(t)$ in (2) are normalized Wiener processes with

$$E\ \left[W_i(t)\right] = 0$$

$$E\ \left[W_i(t)\ W_i(s)\right] = \mathrm{Min}(t,s)$$

The Wiener processes are assumed independent; this entails no loss of generality. The non-negative constants σ_i represent the noise intensities. The Itô differential equation (2) is a rigorous description of the stochastic system (1) for white noise processes (infinite bandwidth). If the stochastic processes $f_i(t)$ in (1) have very large bandwidth, then the system should be described by an Itô equation of the type (2), where however the system matrix A is replaced by

$$A + \frac{1}{2} \sum_{i=1}^{m} \sigma_i^2 B_i^2 \quad .$$

This is commonly called the Stratonovitch correction. Hence for the analysis of wide band stochastic parameters the Itô equation (3) should be considered

$$dx(t) = \left[A + \frac{1}{2} \sum_{i=1}^{m} \sigma_i^2 B_i^2\right] x(t)dt + \sum_{i=1}^{m} \sigma_i B_i x(t) dW_i(t) \tag{3}$$

The convergence properties of the sample solutions of the stochastic dynamic system are frequently characterized by the following stability properties [2]:

<u>Definition 1</u> The null solution of a stochastic dynamic system is called <u>almost surely exponentially stable</u>, if the sample solutions converge exponentially to zero as t tends to infinity, with probability one.

<u>Definition 2</u> The null solution of a stochastic dynamic system is called <u>p-th moment exponentially stable</u>, where p is a positive integer, if for all initial states the p-th moments $E[x_1(t)^{p_1} x_2(t)^{p_2} \ldots x_n(t)^{p_n}]$ of the state variables converge exponentially to zero as t tends to infinity, for any set of non-negative integers $p_1, \ldots p_n$, such that $p_1 + p_2 + \ldots + p_n = p$.

Definition 3 The null solution of a stochastic dynamic system is called p-th mean exponentially stable, with p positive, if for all initial states $E[\ ||x(t)||^p\ ]$ where $||x||$ denotes the norm of x, converges exponentially to zero as t tends to infinity.

It is clear from the definitions that p-th moment stability actually only expresses a convergence property to zero of the state vector if p is an even integer. Moreover, for even p the properties of definitions 2 and 3 are equivalent. The p-th mean exponential stability is for any p at least as strong as almost sure exponential stability. In fact the condition for p-th mean exponential stability tends to the condition for almost sure exponential stability as $p \to 0$. It is also easy to show that p^*-th mean exponential stability implies p-th mean exponential stability for all p less than p^*.

To simplify this introductory discussion consider (2) with a single stochastic element $dx(t) = Ax(t)dt + \sigma Bx(t)dW(t)$. If this system is p-th mean exponentially stable for some noise intensity σ, then it has this property also for any smaller noise intensity. Hence there exists a critical noise intensity $\sigma^*(p)$, depending on p, such that the system is p-th mean exponentially stable iff σ is less than $\sigma^*(p)$. From the discussion above it follows that $\sigma^*(p)$ is a non-increasing function of p; moreover using the Lyapunov function [3] $V(x) = (x^T Px)^{p/2}$, where P satisfies $A^T P + PA + (2p-1)\sigma^2 B^T PB = -Q$, where Q is an arbitrary positive definite matrix, it is shown that for $p \geq 1$, $\sigma^*(p) \geq \sigma^*(2)/\sqrt{p-1}$, where $\sigma^*(2)$ follows from a mean square or second mean stability analysis. Thus an upper and a lower bound for $\sigma^*(p)$ are obtained. In general these bounds cannot be improved upon. Indeed, the lower bound yields the exact $\sigma^*(p)$ for the first order stochastic system, $dx(t) = ax(t)dt + \sigma x(t)dW(t)$, with x(t) scalar. For this system the criterion for p-th mean exponential stability is

$2a + (p-1)\,\sigma^{*}(p)^{2} < 0$. The upper bound yields the exact stability boundary for the second order system

$$dx(t) = \begin{bmatrix} a & c \\ -c & a \end{bmatrix} x(t)dt + \sigma \begin{bmatrix} 0 & 1 \\ -1 & 0 \end{bmatrix} x(t)dW$$

with a negative. Itô calculus yields $d(x_1^2 + x_2^2)/dt = (2a + \sigma^2)(x_1^2 + x_2^2)$. Hence $\sigma^{*}(p) = \sqrt{-2a}$ is true for all positive p. In this paper some classes of systems are discussed for which explicit expressions of $\sigma^{*}(p)$ can be derived.

COLOURED NOISE STOCHASTIC SYSTEMS

First we consider the coloured noise system described by the differential equation (1) [4]. In general the transition matrix cannot analytically be computed. However for the first order case

$$dx(t)/dt = ax(t) + \sum_{i=1}^{m} b_i f_i(t)x(t) \tag{4}$$

with x(t) scalar, the explicit solution is

$$x(t) = \exp\{at + \sum_{i=1}^{m} b_i \int_0^t f_i(s)ds\}\, x(0).$$

The noises and the initial state are assumed independent. The stochastic processes are stationary , zero-mean, and Gaussian, with correlation matrix $R_{ff}(T)$ and spectral density matrix $S_{ff}(s)$. Then

$$E\left[x(t)^p\right] = \exp\{pat + \frac{p^2}{2} b^T \left[\int_0^t \int_0^t R_{ff}(u-v)\, du\, dv\right] b\}$$

where b is the column vector with components $b_1, b_2, \ldots, b_m$. From this expression we conclude that a necessary and sufficient condition for p-th moment and p-th mean exponential stability is

$$a < -\frac{p}{2} b^T S_{ff}(0) b.$$

If the processes $f_i(t)$ are ergodic, then

$$\lim_{T\to\infty} \frac{1}{2T} \int_{-T}^{T} f_i(s)\,ds = 0 \qquad (i = 1,\ldots,m)$$

holds with probability one; a necessary and sufficient condition for almost sure exponential stability is hence that a be negative.

If a, b_1, ..., b_m are complex numbers, then the same analysis yields the following criterion, where Re denotes the real part of a complex quantity:

<u>Criterion 1</u> The null solution of (4) is p-th moment exponentially stable if

$$\mathrm{Re}(a) < -\frac{p}{2} \mathrm{Re}(b^T S_{ff}(0) b)$$

It is p-th mean exponentially stable iff

$$\mathrm{Re}(a) < -\frac{p}{2} \mathrm{Re}(b^T)\ S_{ff}(0)\ \mathrm{Re}(b)$$

It is almost surely exponentially stable iff $\mathrm{Re}(a) < 0$.

The question now is to what extent this result can be generalized. A straightforward analysis shows that first moment and even order moment exponential stability, p-th moment exponential stability, and almost sure exponential stability of the n-th order system (1), with <u>diagonal</u> matrices A, B_1, ..., B_m, are equivalent to the same properties for the n first order systems,

$$dx_j(t)/dt = a_j x_j(t) + \sum_{i=1}^{m} b_{ij} f_i(t) x_j(t)$$

where a_j, b_{1j},..., b_{mj}, are <u>corresponding</u> eigenvalues of the matrices A, B_1, ..., B_m, which means that they are the diagonal entries on the same row. This result can be applied if the matrices A, B_1,..., B_m, can be diagonalized by means of the same similarity transformation. A necessary condition for this property is that the matrices be pairwise commutative [5]; a sufficient condition is that moreover at least one of the matrices has n distinct eigenvalues.

This result can be extended to the cases where the matrices A, $B_1, \ldots, B_m$, are upper triangular, or where they can be transformed into such a form by means of a similarity transformation. Consider indeed the stochastic system equation

$$\frac{dx}{dt} = \begin{bmatrix} a_1 & & * \\ & a_2 & \\ & & \ddots & \\ 0 & & & a_n \end{bmatrix} x + \sum_{i=1}^{m} f_i \begin{bmatrix} b_{i1} & & * \\ & b_{i2} & \\ & & \ddots & \\ 0 & & & b_{in} \end{bmatrix} x \tag{5}$$

where the elements (*) above the main diagonal are not essential for the further analysis, and where all elements below the main diagonal vanish. Then it can be proved that first moment exponential stability, even moment exponential stability, p-th mean exponential stability, and almost sure exponential stability are equivalent to the same properties for the n first order systems

$$dx_j(t)/dt = a_j x_j(t) + \sum_{i=1}^{m} f_i b_{ij} x_j(t)$$

for which necessary and sufficient conditions have been derived above.

This result can be expressed in concise form by means of Lie algebra theory [6]. A subspace L of square matrices of order n is called a Lie algebra if for all A and B in L, the commutator product $[A,B] \triangleq AB - BA$ also belongs to L. Let $L(A,B_1,\ldots,B_m)$ denote the smallest Lie algebra containing the matrices A, $B_1,\ldots, B_m$. This is often called the Lia algebra generated by the matrices A, $B_1,\ldots, B_m$. The derived series of Lie algebra L is defined as follows:

$$L^{(0)} = L$$

$$L^{(1)} = [L,L]$$

$$L^{n+1} = [L^{(n)}, L^{(n)}]$$

$$\vdots$$

The Lie algebra is called <u>solvable</u> if $L^{(n)} = \{0\}$ for some n. Note that an abelian Lie algebra, where all matrices of the algebra are pairwise commutative, is a special case of a solvable Lie algebra since $[L,L] = \{0\}$. A very interesting feature of solvable Lie algebras is expressed in the following lemma [6]:

<u>Lemma</u> A matrix Lie algebra is solvable if and only if there exists a nonsingular matrix P such that PMP^{-1} is upper triangular for all M in the Lie algebra.

This yields the following result:

<u>Criterion 2</u> [4] Suppose $L(A, B_1 ..., B_m)$ is a solvable Lie algebra. Then the system equation can be transformed into the form given by (5).

(a) The null solution is first moment exponentially stable, iff for j=1,..., m $\text{Re}(a_j) < -\frac{1}{2}\text{Re}(b_j^T S_{ff}(0) b_j)$ where $b_j = [b_{1j}\ b_{2j}\ \cdots\ b_{mj}]^T$.

(b) The null solution is p-th mean exponentially stable iff for j=1,...,m, $\text{Re}(a_j) < -\frac{p}{2}\ \text{Re}(b_j^T)\ S_{ff}(0)\ \text{Re}(b_j)$.

(c) The null solution is almost surely exponentially stable, iff for j=1,.. .,m, $\text{Re}(a_j) < 0$.

A rather inconvenient feature of this result is that the application of the stability criterion requires the triangularization of the matrices A, B_1, ..., B_m. It would be much more convenient if the conditions could be explicitly expressed in terms of the system characteristics and the noise data. It is shown below that this can be achieved. There is obviously no problem for almost sure exponential stability, since condition (c) of criterion 2 is equivalent to the Hurwitz character of the matrix A. The condition (a) of criterion 2 requires that the eigenvalues of the matrix

$$\begin{bmatrix} a_1 & & & \\ & a_2 & & * \\ & & \ddots & \\ 0 & & & a_n \end{bmatrix} + \frac{1}{2} \sum_{r,s=1}^{m} S_{ff}(0)_{rs} \begin{bmatrix} b_{r1} & & & \\ & b_{r2} & & * \\ & & \ddots & \\ 0 & & & b_{rn} \end{bmatrix} \begin{bmatrix} b_{s1} & & & \\ & b_{s2} & & * \\ & & \ddots & \\ 0 & & & b_{sn} \end{bmatrix}$$

lie in the left half plane, where $S_{ff}(0)_{rs}$ denotes the element of $S_{ff}(0)$ on the r-th row and the s-th column. Hence this condition is equivalent to the criterion that the matrix

$$A_1 = A + \frac{1}{2} \sum_{r,s=1}^{m} S_{ff}(0)_{rs} B_r B_s \tag{6}$$

be Hurwitz. The condition for p-th mean exponential stability can be dealt with in the same way. Therefore we define the matrix

$$A_p = A + \frac{p}{2} \sum_{r,s=1}^{m} S_{ff}(0)_{rs} B_r B_s \tag{7}$$

and the mapping $Q \to L_1(Q)$ from the space of symmetric matrices of order n into itself

$$\begin{aligned} L_1(Q) = {} & (A + \frac{p}{4} \sum_{r,s=1}^{m} S_{ff}(0)_{rs} B_r B_s) Q \\ & + Q(A + \frac{p}{4} \sum_{r,s=1}^{m} S_{ff}(0)_{rs} B_r B_s)^T \\ & + \frac{p}{2} \sum_{r,s=1}^{m} S_{ff}(0)_{rs} B_r Q B_s^T \end{aligned} \tag{8}$$

Criterion 3 If the Lie algebra $L(A,B_1,...,B_m)$ is solvable then

(a) the null solution of (1) is first moment exponentially stable, iff the matrix A_1 defined in (6) is Hurwitz;

(b) the null solution of (1) is p-th mean exponentially stable, iff the mapping L_1, defined in (8), is Hurwitz; if all eigenvalues of the matrices $B_1,...,B_m$ are real, then this is equivalent to the Hurwitz character of the matrix A_p, defined in (7);

(c) the null solution of (1) is almost surely exponentially stable iff the matrix A is Hurwitz.

The above conditions can be checked by means of straightforward techniques developed for the stability analysis of deterministic dynamic systems and for the mean square stability analysis of stochastic systems [7]. Note that the eigenvalue of L_1 which has largest real part is a real eigenvalue; this can be exploited to simplify the analysis of the Hurwitz character of the mapping.

An additional result can be obtained by means of Lyapunov's direct method for stability analysis. Consider the stochastic system (1), with all matrices B_i (i=1,...,m) skew-symmetric:

$$B_i = - B_i^T$$

The Lyapunov function [3]

$$V(x) = (x^T x)^{p/2}$$

with

$$dV(x)/dt = (x^T x)^{p/2-1} x^T (A+A^T) x$$

proves

Criterion 4 The null solution of the dynamic system (1) with stochastic parameters is almost surely exponentially stable, and p-th mean exponentially stable, for any p, if the matrices B_1, B_2, ..., B_m are skew-symmetric, and if

$$\frac{1}{2}(A+A^T)$$

the symmetric part of A, is Hurwitz (or, equivalently, negative definite).

This criterion yields a sufficient condition; it is valid for any spectral density matrix. In general the condition is not necessary for almost sure exponential stability, and also not for p-th mean exponential stability for a given p. It remains an open question whether the condition is necessary if one wants all moments to be exponentially stable for arbitrary spectral densities. Note that criterion 4 remains valid if the processes $f_i(t)$ do not have zero means, and even if the processes are non-stationary.

Remark 1 Criteria similar to criteria 1, 2, and 3 can also be derived if the stochastic processes are not Gaussian. The condition for almost sure exponential stability is clearly valid for any zero-mean ergodic processes. The condition for p-th mean exponential stability of the null solution of (4) is,in the case of non-Gaussian processes

$$p\ \mathrm{Re}(a) < -\lim_{T\to\infty}\frac{1}{T}\,E\left\{\exp\left[p\sum_{i=1}^{m}\mathrm{Re}(b_i)\int_0^T f_i(s)ds\right]\right\}$$

This readily yields a generalisation of criterion 2. It is obvious that criterion 4 also holds for non-Gaussian stochastic processes.

Remark 2 A very important and remarkable property of criteria 2 and 3 is that the stability conditions only depend on the spectral density matrix at zero frequency. An interesting question is whether the solvability of the Lie algebra, generated by the system matrices $A, B_1, B_2, \ldots, B_m$, is a necessary condition for this property to hold.

Remark 3 The ideas of criteria 3 and 4 can be combined to derive further results, such as the following. Let us assume that the system matrices A,

$B_1, \ldots, B_m$ have the following form

$$A = \begin{bmatrix} A_1 & A_2 \\ 0 & A_3 \end{bmatrix} \qquad B_i = \begin{bmatrix} B_{i1} & B_{i2} \\ 0 & B_{i3} \end{bmatrix}$$

where A_1 and B_{i1} $(i=1,\ldots,m)$ are square matrices of order k, and where A_3 and B_{i3} $(i=1,\ldots,m)$ are square matrices of order (n-k); it is also sufficient that the matrices can be transformed into such a form by means of a similarity transformation. Suppose the Lie algebra $L(A_1, B_{11}, \ldots, B_{m1})$ is solvable, and that the matrices $B_{13}, \ldots, B_{m3}$ are skew-symmetric. Then p-th mean exponential stability conditions can be obtained by combining criterion 3, applied to

$$dx(t)/dt = A_1 x(t) + \sum_{i=1}^{m} f_i(t) B_{i1} x(t)$$

with criterion 4, applied to

$$dx(t)/dt = A_3 x(t) + \sum_{i=1}^{m} f_i(t) B_{i3} x(t)$$

<u>Remark 4</u> Consider the stochastic differential equation (4). Then from the solution it can readily be derived that the p-th moment and the p-th mean satisfy the following time-varying differential equations:

$$dE(x^p)/dt = g(t)\ E(x^p)$$

where

$$g(t) = pa + p^2 b^T \left(\int_0^t R_{ff}(u)\ du \right) b$$

and

$$dE\ (|x|^p)/dt = h(t)\ E(|x|^p)$$

where

$$h(t) = pRe(a) + p^2\ Re(b^T) \left(\int_0^t R_{ff}(u)\ du \right) Re\ (b)$$

These expressions also readily lead to the stability conditions for mean and moment stability of the null solution of system (4). This approach can be considered as the proof of the stability criteria by means of Lyapunov's direct method. From the above expression it is seen that the rate of convergence of the p-th mean decreases with time. For small time, the rate of convergence is not affected by the presence of the stochastic elements. Then the time constants are the same as for the deterministic dynamic system which would be obtained from (4) by deleting the stochastic elements. For increasing time the time constants decrease; this transition is fast if the correlation times of the random processes are small, that is, if the correlation matrix $R_{ff}(T)$ strongly decays as a function of T.

WIDE BAND NOISE STOCHASTIC SYSTEMS

If the spectral density matrix $S_{ff}(j\omega)$, evaluated on the imaginary axis, is constant from zero frequency up to a cut-off frequency ω_c, and if one lets ω_c tend to infinity, then the system description should be the Itô differential equation (3). Since the conditions of the criteria derived in the previous section only depend on the spectral density at zero frequency, stability criteria for system (3) can immediately be derived from the results of the previous section. The results could also directly be obtained from an analysis of the Itô equation itself. For simplicity the Wiener processes in (3) are assumed to be independent. We obtain:

<u>Criterion 5</u> If the Lie algebra $L(A,B_1,\dots,B_m)$ is solvable, then the stability analysis of (3) can be reduced to the stability analysis of n first order systems. A necessary and sufficient condition for almost sure exponential stability is the Hurwitz character of the system matrix A. The null solution of (1) is first moment exponentially stable iff the matrix

$$A + \frac{p}{2} \sum_{i=1}^{m} \sigma_i^2 B_i^2$$

is Hurwitz. It is p-th mean exponentially stable if the mapping $Q \to L_2(Q)$ in the space of symmetric matrices defined as

$$L_2(Q) = (A + \frac{p}{4} \sum_{i=1}^{m} \sigma_i^2 B_i^2)Q + Q(A + \frac{p}{4} \sum_{i=1}^{m} \sigma_i^2 B_i^2)^T + \frac{p}{2} \sum_{i=1}^{m} \sigma_i^2 B_i Q B_i^T$$

has only eigenvalues with negative real parts. If all eigenvalues of the matrices $B_1, B_2, \ldots, B_m$, are real, then this is equivalent to the Hurwitz character of the matrix $A + \frac{p}{2} \sum_{i=1}^{m} \sigma_i^2 B_i^2$.

<u>Criterion 6</u> The null solution is almost surely exponentially stable and p-th mean exponentially stable, for any p, for all noise intensities σ_i, if the matrices B_i are skew symmetric, and if the matrix $\frac{1}{2}(A+A^T)$ is Hurwitz.

WHITE NOISE STOCHASTIC SYSTEMS

In this section white noise systems governed by the Itô equation (2) are considered. The following results are obtained, which are similar to the criteria of the previous sections.

<u>Criterion 7</u> If the Lie algebra $L(A, B_1, \ldots, B_m)$ is solvable, then first moment exponential stability, even order moment exponential stability, p-th mean exponential stability, and almost sure exponential stability of the null solution of (2) are equivalent to the same properties for the n first order Itô systems $dx_j(t) = a_j x_j(t)dt + \sum_{i=1}^{m} \sigma_i b_{ij} x_j(t) dW_i(t)$. The null solution is first moment exponentially stable iff the matrix A is Hurwitz. The null solution is p-th mean exponentially stable, if the mapping $Q \to L_3(Q)$ in the space of symmetric matrices of order n, defined by

$$L_3(Q) = (A + \frac{p-2}{4} \sum_{i=1}^{m} \sigma_i^2 B_i^2)Q + Q(A + \frac{p-2}{4} \sum_{i=1}^{m} \sigma_i^2 B_i^2)^T + \frac{p}{2} \sum_{i=1}^{m} \sigma_i^2 B_i Q B_i^T$$

has only eigenvalues with negative real parts. If all eigenvalues of the matrices B_i are real, then this is equivalent to the Hurwitz character of the matrix $A - \frac{1}{2} \sum_{i=1}^{m} \sigma_i^2 B_i^2$. The null solution of (1) is then almost surely exponentially stable if the matrix $A - \frac{1}{2} \sum_{i=1}^{m} \sigma_i^2 B_i^2$ is Hurwitz.

This criterion can be derived from the results of the third section, or they can be obtained from a direct analysis of the Itô equations. The results on p-th mean stability could be obtained from a Lyapunov approach, using the Lyapunov function

$$V(x) = \sum_{j=1}^{m} r_j (x_j^* x_j)^{p/2}$$

for suitable positive constants r_j. The moments are governed by a set of linear differential equations which can directly be analysed [9]. The matrix of the system of equations for the p-th moments is

$$K_p = \left[A - \frac{1}{2} \sum_{i=1}^{m} \sigma_i^2 B_i^2 \right]_{[p]} + \frac{1}{2} \sum_{i=1}^{m} \sigma_i^2 (B_{i\,[p]})^2$$

where the symbols are defined in the appendix.

<u>Remark 5</u> If the eigenvalues of the matrix B_i are not all imaginary, then there is no non-zero value of the noise intensity σ_i for which the system is p-th mean exponentially stable for all p. This result is also true in more general cases. It can indeed be proved from the consideration of the matrix K_p, that this matrix has at least one positive real eigenvalue for sufficiently high values of p, if B_i has an eigenvalue with non-zero real part and if the noise intensity σ_i is non-zero.

<u>Remark 6</u> If the Lie algebra $L(A, B_1, \ldots, B_m)$ is solvable, and if the eigenvalues of all matrices B_i are on the imaginary axis, then the criterion for p-th mean exponential stability is independent of p. Hence in this case

there exist noise intensities for which all moments are exponentially stable. This is also possible in other cases, as follows from the generalization of criterion 4 to white noise stochastic systems.

Criterion 8 The null solution of (2) is almost surely exponentially stable and p-th mean exponentially stable for any p, if all matrices B_i are skew-symmetric and if the matrix

$$\frac{1}{2}(A+A^T) - \frac{1}{2}\sum_{i=1}^{m} \sigma_i^2 B_i^2 \qquad \text{is Hurwitz.}$$

It would be interesting to know whether the above criterion also yields a necessary condition, i.e. does criterion 8 yield all noise intensities for a given matrix A and given skew symmetric matrices B_i such that all moments are exponentially stable ? We have not been able to solve this problem completely. However, the following particular case has been resolved. If A is symmetric and B_i skew symmetric, then for any set of noise intensities such that

$$A - \frac{1}{2}\sum_{i=1}^{m} \sigma_i^2 B_i^2$$

is not Hurwitz, there exists a p^*, such that all moments of even order higher than p^* are unstable.

Remark 7 If the Lie algebra $L(A, B_1, \ldots, B_m)$ is solvable, and if the matrices B_i are nilpotent (all eigenvalues vanish), then criterion 7 shows that the system is p-th mean exponentially stable, for any p, however large the noise intensities be. This condition is also necessary, because the following results were proved in [10] for mean square exponential stability, and it is therefore clearly also true for higher order moments.

Criterion 9   If the null solution of (2) is p-th moment exponentially stable for all noise intensities, where p is an arbitrary even integer, then the Lie algebra $L(A, B_1,\dots,B_m)$ is solvable and the matrices $B_1, B_2,\dots,B_m$ are nil-potent.

Remark 8   The condition for almost sure exponential stability of criterion 7 is

$$\mathrm{Re}(a_j) < -\frac{1}{2}\sum_{i=1}^{m} \sigma_i^{\,2}\,\{\mathrm{Im}(b_{ij})^2 - \mathrm{Re}(b_{ij})^2\}$$

where a_j and b_{ij} are corresponding eigenvalues as defined in the third section, and where Im denotes the imaginary part of a complex number. This condition is less demanding than the exponential stability of the deterministic system obtained by deleting the noise element, iff the real parts of the eigenvalues b_{ij} are larger than the imaginary parts in absolute value.

Remark 9   First moment and first mean exponential stability are only equiv-alent if all eigenvalues of the matrices B_i are real. Otherwise first mean exponential stability is strictly stronger than first moment exponential stability.

EXAMPLES

Example 1   Consider the second order stochastic system with coloured noise parameters.

$$\dot{x}(t) = (A+B_1f_1(t) + B_2f_2(t))x(t)$$

with

$$A = \begin{bmatrix} 2a & -a \\ a & 0 \end{bmatrix}, \qquad B_1 = \sqrt{\frac{1}{2}}\begin{bmatrix} 1 & 0 \\ 0 & 1 \end{bmatrix},$$

$$B_2 = \sqrt{\frac{1}{2}} \begin{bmatrix} -1 & 2 \\ 0 & 1 \end{bmatrix}$$

It is readily checked that $L(A, B_1, B_2)$ is solvable, and that the eigenvalues of B_1 and B_2 are real. Let the correlation matrix of the processes be given as

$$R_{ff}(T) = \frac{1}{2}\sigma_1^2 \begin{bmatrix} 1 & 1 \\ 1 & 1 \end{bmatrix} \exp(-k_1|T|) + \frac{1}{2}\sigma_2^2 \begin{bmatrix} 1 & -1 \\ -1 & 1 \end{bmatrix} \exp(-k_2|T|)$$

where k_1 and k_2 are positive constants. The spectral density matrix is at zero frequency

$$S_{ff}(0) = \begin{bmatrix} \sigma_1^2/k_1 + \sigma_2^2/k_2 & \sigma_1^2/k_1 - \sigma_2^2/k_2 \\ \sigma_1^2/k_1 - \sigma_2^2/k_2 & \sigma_1^2/k_1 + \sigma_2^2/k_2 \end{bmatrix}$$

To apply criterion 3 we compute

$$A + \frac{p}{2}\sum_{k,1}^{2} S_{ff}(0)_{k1} B_k B_1 = \begin{bmatrix} 2a + p\sigma_2^2/k_2 & -a + p(\sigma_1^2/k_1 - \sigma_2^2/k_2) \\ a & p\,\sigma_1^2/k_1 \end{bmatrix}$$

This matrix is Hurwitz if and only if

$$a < -p\sigma_1^2/k_1$$

$$a < -p\sigma_2^2/k_2$$

This agrees with the result obtained by Willsky et al. [11] for the first moment stability condition by explicit computation of the first moment.

<u>Example 2</u> Consider the white noise stochastic system

$$dx(t) = Ax(t)dt + \sigma x(t)dW$$

The Lie algebra involved here is solvable, since the identity matrix commutes with any matrix A. The condition for p-th mean exponential stability is that the eigenvalues of A have real parts less than

$$-(p-1)\ \sigma^2/2$$

For almost sure exponential stability the eigenvalues of A should have a real part less than $\sigma^2/2$.

If the Stratonovitch correction term is introduced, the Itô equation

$$dx(t) = (A + \frac{1}{2}\ \sigma^2 I)\ x(t)dt + \sigma x(t)dW(t)$$

should be considered, where I denotes the identity matrix, in this case the condition for p-th mean exponential stability is that all eigenvalues of A have a real part smaller than $-\ p\sigma^2/2$; for almost sure exponential stability the real parts should be negative.

APPENDIX

Let $x^{[p]}$ denote the list of all p-th forms of the components of the vector x

$$x^{[p]} = \begin{bmatrix} a\ x_1^{\ p} \\ b\ x_1^{\ p-1} x_2 \\ c\ x_1^{\ p-2} x_2^{\ 2} \\ \cdot \\ \cdot \\ \cdot \\ z\ x_n^{\ p} \end{bmatrix}$$

with coefficients such that

$$||x^{[p]}|| = ||x||^p$$

where the Euclidean norms are used

$$||x||^2 = x^T x$$

Hence the coefficient of $x_1^{\ p_1} x_2^{\ p_2} \ldots x_n^{\ p_n}$ should be

$$\sqrt{\binom{p}{p_1} \binom{p-p_1}{p_2} \quad \cdots \quad \binom{p-p_1 - \ldots -p_{n-1}}{p_n}}$$

If the vector x satisfies the linear time-invariant differential equation

$$dx(t)/dt = Ax(t)$$

then the vector $x^{[p]}$ also satisfies a linear time-invariant differential equation [12]

$$dx^{[p]}/dt = A_{[p]}x^{[p]}$$

Let this procedure define the operator $A_{[p]}$. If x(t) satisfies the white noise stochastic Itô equation (2), then it can be shown that the list of all p-th moments of the state variables satisfies

$$dE(x^{[p]})/dt = K_p \; E(x^{[p]})$$

where

$$K_p = \{ A - \frac{1}{2} \sum_{i=1}^{m} \sigma_i^2 \; B_i^2 \}_{[p]} + \frac{1}{2} \sum_{i=1}^{m} \sigma_i^2 \; (B_{i\,[p]})^2$$

This expression is very useful for analysing moment stability properties.

ACKNOWLEDGEMENTS

The author is grateful to Mr D. Aeyels for discussions on the research described in this paper and for his contribution to the results, and to Mr P. Van Houcke for computer simulations in connection with the criteria; the support from the Belgian Science Research Council (F.K.F.O.).

REFERENCES

1 L. Arnold — Stochastische Differentialgleichungen. Oldenbourg Verlag. (1973).

2 F. Kozin — A survey of stability of stochastic systems. Automatica, 5, 95. (1969).

3 H. Kushner — Stochastic Stability and Control, Academic Press. (1967).

4 J.L. Willems and D. Aeyels An equivalence property for moment stability criteria for parametric stochastic systems and Itô equations. Int. J. Systems Science, Vol. 7, pp 577-590. (1976).

5 F.R. Gantmacher Theory of Matrices, Chelsea Publ. Co. (1959).

6 A.A. Sagle and R.E. Walde Introduction to Lie Groups and Lie Algebras. Academic Press. (1973).

7 J.L. Willems Mean square stability criteria for stochastic feedback systems. Int. J. Systems Science, 4, 545 (1973).

8 M. Loève Probability Theory, 155, Van Nostrand Co. (1963).

9 J.L. Willems Stability of higher moments for linear stochastic systems. Ingenieur-Archiv. 44, 123. (1975).

10 J.L. Willems and J.C. Willems Feedback stabilizability for stochastic systems with state and control dependent noise. Automatica 12. (In press). (1976).

11 A.S. Willsky et al., On the stochastic stability of linear systems containing coloured multiplicative noise. IEEE Trans. Automatic Control, AC-20, 711. (1975).

12 R.W. Brockett Lie theory and control systems defined on spheres. SIAM J. Applied Math., 25, 213. (1973).

Professor J.L. Willems, Laboratorium voor Theoretische Elektriciteit en Toepassingen van de Sterkstroom, 9000 Gent, Sint-Pietersnieuwstraat 41, Belgium.

DISCUSSION

A first question (Lin) referred to the fact that the stability conditions in the criteria given in the paper only depend on the values of the spectral densities of the noise processes at zero frequency whereas conditions obtained by Professor Ariaratnam for other examples include the values of the spectra at twice the natural frequency of the system as well. Professor Willems explained that the property that only $S_{ff}(0)$ is important for the stability analysis of the systems considered in the paper essentially depends on the assumption that the Lie algebra $L(A,B_i)$ is solvable; the systems considered by Professor Ariaratnam do not satisfy this condition. It would be interesting to know whether the solvability assumption on the Lie algebra is also a necessary condition for the property that only the spectral density at zero frequency affects the stability behaviour of the system. Professor Willems said that he had done some research concerning this question but had not yet been able to solve it. He added that the class of systems considered in the paper also included some systems with lightly damped poles close to the imaginary axis; nevertheless the spectral density at twice the corresponding natural frequency does not affect the stability properties, provided the noise enters the system in an appropriate way (i.e. such that the Lie algebra involved is solvable).

In a further discussion (with Kozin) Professor Willems pointed out that the main contribution of Lie algebra theory was the possibility of selecting the class of systems with solvable Lie algebra L where the solution may be obtained as some simple exponential solving a set of first order differential equations one by the other. For those it is well known that stability conditions only depend on the zero values of the spectra of the noise processes.

Asked about a physical interpretation of the fact that the conditions of p-th moment stability get more and more restrictive with increasing p (Wedig), Professor Willems explained that this fact is mathematically well known, but a physical interpretation would be difficult because one could not give a physical meaning to the p-th moment of a stochastic process. Evidently, the more restrictions one puts to higher and higher moments of a process the more moderate the behaviour of the samples of the process becomes, and at the same time a system having this process as its output is squeezed more and more.

In response to a question (Holmes) on the possibility of extending the results in the paper to nonlinear stochastic problems by means of some linearization techniques Professor Willems referred to the main problem when a noise process may be considered to be small enough for linearization. But several averaging methods are available for dealing with nonlinear stochastic problems.

When asked (Wedig) for examples of practical applications of the results, Professor Willems replied that he did not have any application in mind when he derived the results described in the paper. He hoped that somebody analysing a practical stability problem would some day encounter examples where the assumption on the solvability of the Lie algebra is true. On the other hand he thought that results such as the ones described in the paper are also useful for a different purpose. Indeed for the class of systems considered, exact stability conditions can be derived from the criteria of the paper; on the other hand one may apply available approximate stability analysis techniques (such as proposed by Bolotin, Wedig, Zeman) to these systems. In this way it is possible to obtain some idea concerning the validity and the accuracy of the approximation techniques.

S T ARIARATNAM and D S F TAM

Moment stability of coupled linear systems under combined harmonic and stochastic excitation

INTRODUCTION

Problems of parametric excitation of linear and nonlinear dynamical systems are of importance in several branches of engineering. For example, such problems arise naturally in studies of the dynamic stability of elastic structures which exhibit a bifurcational form of instability under applied loads. They also occur in the investigation of the stability of steady-state motions of non-linear dynamical systems.

When the parametric excitation is a deterministic harmonic function, several investigations of the stability behaviour of both linear and non-linear mechanical systems have been carried out, see, for example, the surveys by Mettler [1], [2]. In particular, it is well known that, for linear systems, instabilities of the sub-harmonic and combination types of resonance occur when the excitation frequency is in the neighbourhood of twice the system natural frequencies and the sums and differences of pairs of these frequencies. In the case of stochastic parametric excitation, somewhat parallel results were obtained and reported in [3], [4], [5]. The present paper deals with the case when the parametric excitation consists of a combination of a harmonic term and a stationary stochastic process. Such forms of excitation may be the realistic ones to take in many practical situations where the disturbances arise from both deterministic and non-deterministic sources. Conditions for stability in the first and second moments of the response are derived for small intensity excitations, and the results are applied to the problem of the flexural-torsional stability of a thin, simply-supported, elastic beam subjected to a dynamic transverse load at mid-span.

FORMULATION

Consider a parametrically excited, linear discrete, non-gyroscopic system described by equations of motion of the form:

$$\ddot{q}_r + \omega_r^2 q_r + 2\varepsilon\omega_r \sum_s \beta_{rs}\dot{q}_s + 2\omega_r^2 \left[\varepsilon h \sin \nu t + \varepsilon^{\frac{1}{2}} f(t)\right]\sum_s k_{rs}q_s = 0, \qquad (1)$$

$(r,s=1,2,\ldots n)$

where q_r denote the generalised normal coordinates of free vibration of the system, ω_r the natural frequencies, β_{rs} the viscous damping coefficients, k_{rs}, h, ν are constants, and f(t) is a stationary stochastic process with zero mean value; ε is a small parameter $(0<\varepsilon<<1)$. These equations describe the motion of a linear, elastic, discrete system about its equilibrium configuration; they may also be regarded as describing approximately the motion of certain continuous elastic systems which have been discretised by some suitable technique, e.g. Galerkin's method, finite differences, finite element etc.

Introducing non-dimensional quantities τ, Ω_r, λ by the relations:

$$\tau = \nu t, \quad \Omega_r = \omega_r/\omega_o, \quad \nu = \omega_o(1-\varepsilon\lambda), \qquad (2)$$

equation (1) becomes

$$q_r'' + \Omega_r^2 q_r + 2\varepsilon\lambda\Omega_r^2 q_r + 2\varepsilon\Omega_r \sum_s \beta_{rs}q_s' + 2\Omega_r^2\left[\varepsilon h \sin \tau + \varepsilon^{\frac{1}{2}} f(\tau/\nu)\right]\sum k_{rs}q_s = 0 \qquad (3)$$

$(r,s=1,2,\ldots,n)$,

where the primes denote differentiation with respect to τ.

Using the transformation

$$q_r = z_r e^{i\Omega_r\tau} + \bar{z}_r e^{-i\Omega_r\tau}, \quad q_r' = i\Omega_r(z_r e^{i\Omega_r\tau} - \bar{z}_r e^{-i\Omega_r\tau}) \qquad (4)$$

$(r=1,2,\ldots,n)$,

where $i = \sqrt{(-1)}$, and z_r are complex variables with conjugates $\bar{z}_r$, equations (3) may be replaced by the following first-order equations:

$$z'_r = \varepsilon\sum_s \left[b_{rs} z_s e^{-i(\Omega_r - \Omega_s)\tau} - \bar{b}_{rs}\bar{z}_s e^{-i(\Omega_r+\Omega_s)\tau}\right]$$

$$+ \left[\varepsilon h \sin \tau + \varepsilon^{\frac{1}{2}} f(\tau/\nu)\right] \sum_s p_{rs} \left[z_s e^{-i(\Omega_r-\Omega_s)\tau} + \bar{z}_s e^{-i(\Omega_r+\Omega_s)\tau}\right] \qquad (5)$$

$(r,s=1,2,\ldots,n)$,

where $b_{rr} = -\Omega_r(\beta_{rr} - i\lambda)$, $b_{rs} = -\Omega_s\beta_{rs}$, $p_{rs} = i\Omega_r k_{rs}$.

For small values of ε, the quantities $z_r(t)$ are slowly varying functions of ε. Hence, to a first approximation, they may be replaced by the solutions of 'averaged' equations using the method of 'stochastic averaging' due to Stratonovich [6], Khasminskii [7], and Papanicolaou and Kohler [8]. According to this procedure, the deterministic terms on the right hand sides are averaged in the usual manner (see, e.g. [9]), while the stochastic terms are replaced by their averaged mean values plus equivalent fluctuational parts; the details may be found in the references quoted above. The resulting equations constitute a set of Itô stochastic differential equations for the averaged variables, which therefore form a diffusive Markov process. Applying this procedure to equations (5), it is found that the harmonic excitation contributes to the averaged equations only when the excitating frequency ν is close to any of the values $|\omega_\ell \pm \omega_m|$, $(\ell,m=1,2,\ldots,n)$. Thus, for $\omega_o = |\omega_\ell \pm \omega_m|$, the averaged equations take the form:

$$\left.\begin{aligned}
dz_r &= \varepsilon\Omega_r \left[-\beta^*_{rr} + i(\lambda-\gamma_r)\right] z_r dt + \varepsilon^{\frac{1}{2}} \sum_{j=1}^{2n} \sigma_{rj} dw_j, \\
&\quad r=1,2,\ldots,n; r\neq\ell,m \\
dz_\ell &= \varepsilon\Omega_\ell \left[-\beta^*_{\ell\ell} + i(\lambda-\gamma_\ell) z_\ell + \frac{1}{2} k_{\ell m} h^2 \begin{Bmatrix}\bar{z}_m \\ z_m\end{Bmatrix}\right] dt + \varepsilon^{\frac{1}{2}} \sum_{j=1}^{2n} \sigma_{\ell j} dw_j, \\
dz_m &= \varepsilon\Omega_m \left[-\beta^*_{mm} + i(\lambda-\gamma_m) z_m + \frac{1}{2} k_{m\ell} h^2 \begin{Bmatrix}\bar{z} \\ -z_\ell\end{Bmatrix}\right] dt + \varepsilon^{\frac{1}{2}} \sum_{j=1}^{2n} \sigma_{mj} dw_j
\end{aligned}\right\} \qquad (6)$$

where the upper and lower terms within the double brackets correspond, respectively, to the cases $\omega_o = \omega_\ell + \omega_m$ and $\omega_o = |\omega_\ell - \omega_m|$, and

$$\beta^*_{jj} = \beta_{jj} - \frac{1}{2}\sum_{s=1}^{n} \omega_s k_{js} k_{sj} S^-(\omega_j,\omega_s),$$

$$\gamma_j = \frac{1}{2}\sum_{s=1}^{n} \omega_s k_{js} k_{sj} \Psi^-(\omega_j,\omega_s),$$

$$S(\omega) = 2\int_0^\infty \langle f(t)f(t+u)\rangle \cos \omega u \, du,$$

$$\Psi(\omega) = 2\int_0^\infty \langle f(t)f(t+u)\rangle \sin \omega u \, du,$$

$$S^{\pm}(\omega_r,\omega_s) = S(\omega_r+\omega_s) \pm S(\omega_r-\omega_s),$$

$$\Psi^{\pm}(\omega_r,\omega_s) = \Psi(\omega_r+\omega_s) \pm \Psi(\omega_r-\omega_s),$$

$$[\sigma\sigma']_{rs} = -\left[k_{rr}k_{ss}S(o) + k_{rs}k_{sr}S(\omega_r-\omega_s)\right] \omega_o\Omega_r\Omega_s z_r z_s,$$

$$[\sigma\sigma']_{r,s+n} = \left[k_{rr}k_{ss}S(o) + k_{rs}k_{sr}S(\omega_r+\omega_s)\right] \omega_o\Omega_r\Omega_s z_r \bar{z}_s$$

$$+ \delta_{rs}\,\omega_o\Omega_r^2 \sum_{\substack{j=1 \\ j\neq r}}^{n} k_{rj}^2 S^-(\omega_r,\omega_j) z_j \bar{z}_j, \qquad (r,s=1,2,\ldots,n).$$

The angular brackets denote the expectation or the ensemble average, and δ_{rs} is the Kronecker delta; w_j, $(j=1,2,\ldots,2n)$, are independent Wiener processes of unit intensity. In obtaining equations (6), it is assumed that the natural frequencies do not satisfy a relation of the form $\omega_i \pm \omega_j \pm \omega_k \pm \omega_\ell = 0$, for distinct positive integers $i,j,k,\ell \leq n$.

FIRST MOMENT STABILITY

The differential equations satisfied by the first moments of z_r, $(r=1,2,\ldots,n)$, can be obtained by taking the expectation of both sides of equations (6). The resulting equations will be the same as equations (6) with the stochastic terms absent and the variables z_r, $\bar{z}_r$ replaced by their expectations. Since these equations will be linear, the conditions for stability in the first moments can be found readily by the Routh-Hurwitz criterion.

Sub-harmonic resonance $\omega_o = 2\omega_m$

Setting ℓ=m, the following stability conditions are obtained:

$$\beta^*_{rr} > 0, \quad (r=1,2,\ldots,n) \tag{7a}$$

$$(\lambda-\gamma_m)^2 > \frac{1}{4} k_{mm}^2 h^2 - \beta^{*2}_{mm} \tag{7b}$$

where $\varepsilon\lambda\omega_o = \omega_o - \nu$ denotes the amount of detuning. The results for a single-degree of freedom system found earlier [10] may be deduced immediately from the above. The conditions (7a) depend only on the stochastic part of the excitation, while (7b) is the modification due to the stochastic term of the well-known stability condition for the case of harmonic excitation, Figure 1.

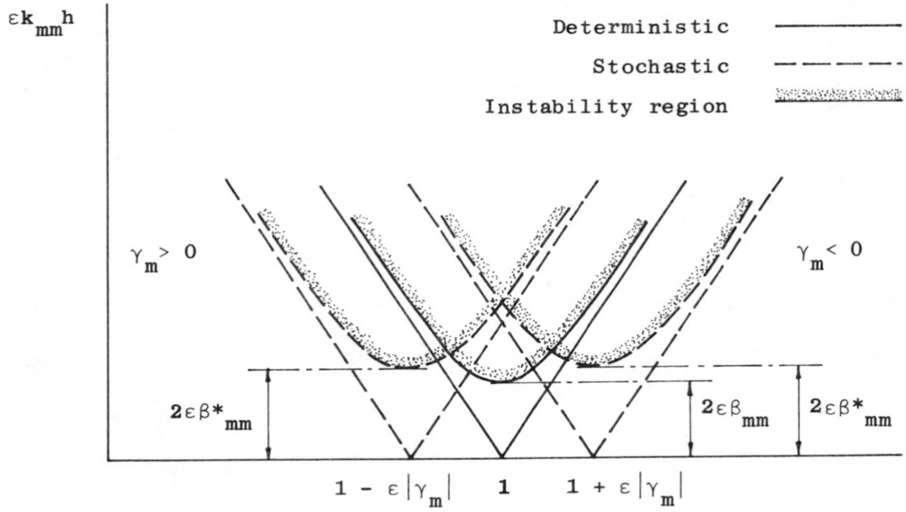

Figure 1

When the stochastic excitation is either absent or a white-noise process, (7a) and (7b) reduce to:

$$\beta^*_{rr} > 0, \quad (r=1,2,\ldots,n), \tag{8a}$$

$$\lambda^2 > \frac{1}{4} k_{mm}^2 h^2 - \beta^2_{mm} \tag{8b}$$

which are identical to the stability conditions for harmonic excitation.

Combination resonance $\omega_o = \omega_\ell \pm \omega_m, \omega_\ell > \omega_m$

The stability conditions for this case are found to be

$$\beta^*_{rr} > 0, \quad (r=1,2,\ldots n) \tag{9a}$$

$$\left[\lambda\omega_o - (\gamma_\ell\omega_\ell \pm \gamma_m\omega_m)\right]^2 > \frac{(\beta^*_{\ell\ell}\omega_\ell \; \beta^*_{mm}\omega_m)^2}{\beta^*_{\ell\ell} \; \beta^*_{mm}} \left(\pm \frac{k_{\ell m}k_{m\ell}h^2}{4} - \beta^*_{\ell\ell}\beta^*_{mm}\right) \tag{9b}$$

where the (+) sign is taken for $\omega_o = \omega_\ell + \omega_n$, and the (-) sign for $\omega_o = \omega_\ell - \omega_m$. Again, when the stochastic part of the excitation is either absent or a white noise process, the conditions (9) reduce to the known results for the case of harmonic excitation.

SECOND MOMENT STABILITY

The differential equations governing the second moments are obtained by applying the Itô differential rule to the quantities $z_r\bar{z}_r$, $(r=1,2,\ldots,n)$, and taking the expectation. This leads to the following equations for the two cases $\omega_o = \omega_\ell + \omega_m$ and $\omega_o = \omega_\ell - \omega_m$.

(i) $\omega_o = \omega_\ell + \omega_m$

$$\frac{d}{dt} \langle z_r\bar{z}_r\rangle = \varepsilon \sum_{s=1}^{n} c_{rs} \langle z_s\bar{z}_s\rangle$$

$$+ \varepsilon(\delta_{r\ell}H_{\ell m}+\delta_{rm}\varepsilon H_{m\ell}) \left[\langle z_\ell z_m\rangle + \langle\bar{z}_\ell\bar{z}_m\rangle\right] \quad , \quad (r=1,2,\ldots,n) \tag{10a}$$

where $\langle z_\ell z_m\rangle$ is governed by

$$\frac{d}{dt} \langle z_\ell z_m\rangle = \varepsilon(a^+_{\ell m}+ib^+_{\ell m}) \langle z_\ell z_m\rangle + \varepsilon H_{m\ell}\langle z_\ell\bar{z}_\ell\rangle + \varepsilon H_{\ell m} \langle z_m\bar{z}_m\rangle \; . \tag{10b}$$

(ii) $\omega_o = \omega_\ell - \omega_m, \quad \omega_\ell > \omega_m$

$$\frac{d}{dt} \langle z_r\bar{z}_r\rangle = \varepsilon \sum_{s=1}^{n} c_{rs} \langle z_s\bar{z}_s\rangle + \varepsilon(\delta_{r\ell}H_{\ell m}-\delta_{rm}H_{m\ell}) \left[\langle z_\ell\bar{z}_m\rangle + \langle z_m\bar{z}_\ell\rangle\right] \tag{11a}$$

$(r=1,2,\ldots,n)$

where $\langle z_\ell \bar{z}_m \rangle$ is governed by

$$\frac{d}{dt} \langle z \bar{z}_m \rangle = \varepsilon(\bar{a}_{\ell m} + i\bar{b}_{\ell m}) \langle z \bar{z}_m \rangle - \varepsilon H_{m\ell} \langle z_\ell \bar{z}_\ell \rangle + \varepsilon H_{m\ell} \langle z_m \bar{z}_m \rangle \qquad (11b)$$
$$+ \varepsilon H_{\ell m} \langle z_m \bar{z}_m \rangle$$

The constants $a^{\pm}_{\ell m}$, $b^{\pm}_{\ell m}$, c_{rs}, $H_{\ell m}$ are defined by:

$$a^{\pm}_{\ell m} = -(\Omega_\ell \beta^*_{\ell\ell} + \Omega_m \beta^*_{mm}) \pm \omega_o \Omega_\ell \Omega_m k_{\ell\ell} k_{mm} S(o)$$
$$\mp (1-\delta_{\ell m}) \omega_o \Omega_\ell \Omega_m k_{\ell m} k_{m\ell} S(\omega_\ell \pm \omega),$$

$$b^{\pm}_{\ell m} = \lambda - (\Omega_\ell \gamma_\ell \pm \Omega_m \gamma_m),$$

$$c_{rs} = -2\delta_{rs} \Omega_r \beta^*_{rr} + \Omega_r \omega_r k^2_{rs} S^+ (\omega_r, \omega_s),$$

$$H_{\ell m} = \frac{1}{2} h\Omega_\ell k_{\ell m} .$$

Equations (10a) together with equation (10b) and its conjugate form a closed system of (n+2) linear equations in the quantities $\langle z_\ell z_m \rangle$, $\langle \bar{z}_\ell \bar{z}_m \rangle$, $\langle z_r \bar{z}_r \rangle$, (r=1,2,...,n). Similarly, equations (11a) with equation (11b) and its conjugate form a closed set in $\langle z_\ell \bar{z}_m \rangle$, $\langle z_m \bar{z}_\ell \rangle$, $\langle z_r z_r \rangle$, (r=1,2,...,n). The conditions of stability in the second moments may therefore be obtained by the Routh-Hurwitz criterion. For a two-degree of freedom system, they are found to be as follows:

<u>Sub-harmonic resonance</u>: $\omega_o = 2\omega_1$

$$\left.\begin{aligned} &\beta^{**}_{11} > 0, \quad \beta^{**}_{22} > 0 \\ &\beta^{**}_{11}\, \beta^{**}_{22} > \frac{1}{4}\, \omega_1 \omega_2 \left[k_{12} k_{21} S^+(\omega_1, \omega_2)\right]^2 \\ &(\lambda-\gamma_1)^2 > \frac{(\beta^{**}_{11} + \Delta_1)\beta^{**}_{22}}{\beta^{**}_{11}\, \beta^{**}_{22} - \Delta} \left(\frac{1}{4}\, k^2_{11} h^2\right) - (\beta^{**}_{11} + \Delta_1)^2 \end{aligned}\right\} \qquad (12)$$

where

$$\beta^{**}_{11} = \beta_{11} - \omega_1 k^2_{11} S(2\omega_1) - \frac{1}{2}\, \omega_2 k_{12} k_{21} S^-(\omega_1, \omega_2),$$

$$\beta_{22}^{**} = \beta_{22} - \omega_2 k_{22}^2 S(2\omega_2) - \frac{1}{2}\,\omega_1 k_{12} k_{21}\, S^-(\omega_1,\omega_2),$$

$$\Delta = \frac{1}{4}\,\omega_1\omega_2\,\left[k_{12}k_{21}\,S^+(\omega_1,\omega_2)\right]^2,$$

$$\Delta_1 = \frac{1}{2}\,\omega_1 k_{11}^2\,\left[2S(o)+S(2\omega_1)\right] .$$

Conditions (12) reduce to the results found earlier [10] for a single-degree of freedom system when k_{12} and k_{21} are set equal to zero.

When the stochastic excitation is a white noise process, $S(\omega) = S_o$ = constant, $\Psi(\omega) = 0$, and the conditions reduce to:

$$\left.\begin{aligned} &\beta_{11} > \alpha_{11}, \quad \beta_{22} > \alpha_{22}, \\ &(\beta_{11} - \alpha_{11})\,(\beta_{22} - \alpha_{22}) > \alpha_{12}^2, \\ &\lambda^2 > \frac{(\beta_{11} + \frac{1}{2}\alpha_{11})\,(\beta_{22} - \alpha_{22})}{(\beta_{11}-\alpha_{11})(\beta_{22}-\alpha_{22}) - \alpha_{12}^2}\,\left(\frac{1}{4}\,k_{11}^2 h^2\right) - \left(\beta_{11} + \frac{1}{2}\,\alpha_{11}\right)^2, \end{aligned}\right\} \qquad (13)$$

where

$$\alpha_{rs} = (\omega_r\omega_s)^{\frac{1}{2}}\, k_{rs}k_{sr}S_o, \quad (r,s=1,2).$$

It can be shown that the region of instability is widened by the presence of the white noise excitation, Figure 2.

$$h_o^2 = \frac{\beta_{11} + \frac{1}{2}\alpha_{11}}{\beta_{22} - \alpha_{22}} \qquad \left|(\beta_{11} - \alpha_{11})(\beta_{22} - \alpha_{22}) - \alpha_{12}^2\right| < \beta_{11}^2$$

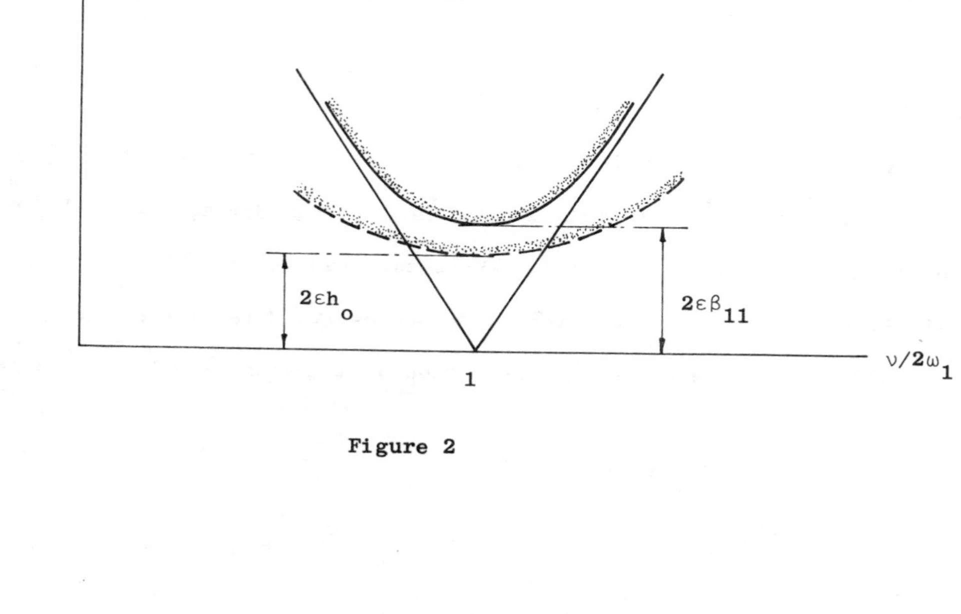

Figure 2

Combination resonance: $\omega_o = \omega_1 \pm \omega_2,\ \omega_1 > \omega_2$

The first three stability conditions are the same as in (12), while the fourth takes the form:

$$\left[\lambda\omega_o - (\omega_1\gamma_1 \pm \omega_2\gamma_2)\right]^2 > \frac{(\omega_1\beta^{**}_{11} + \omega_2\beta^{**}_{22}) \pm 2(\omega_1\omega_2\Delta)^{\frac{1}{2}}}{\beta^{**}_{11}\ \beta^{**}_{22} - \Delta}$$

$$\times\ A^{\pm}\ (\pm \frac{1}{4} k_{12}k_{21}h^2) - (A^{\pm})^2 \qquad (14)$$

where Δ is as defined earlier, and

$$A^{\pm} = \omega_1\beta^{**}_{11} + \omega_2\beta^{**}_{22} + \frac{1}{2}\left[\omega_1^2 k_{11}^2 S(2\omega_1) + \omega_2^2 k_{22}^2 S(2\omega_2)\right]$$

$$+ \frac{1}{2}\left[\omega_1 k_{11} \pm \omega_2 k_{22}\right]^2 S(o) \pm \omega_1\omega_2 k_{12}k_{21} S(\omega_1 \mp \omega_2).$$

In the case of white noise excitation, the condition (14) reduces to:

$$(\lambda\omega_o)^2 > \frac{\omega_1(\beta_{11}-\alpha_{11}) + \omega_2(\beta_{22}-\alpha_{22}) \pm 2(\omega_1\omega_2)^{\frac{1}{2}}\alpha_{12}}{(\beta_{11}-\alpha_{11})(\beta_{22}-\alpha_{22}) - \alpha_{12}^2}$$

$$\times\ A^{\pm}\ (\pm \frac{1}{4} k_{12}k_{21}h^2) - (A^{\pm})^2 \qquad (15)$$

where

$$A^{\pm} = \omega_1\beta_{11} + \omega_2\beta_{22} \pm \omega_1\omega_2(k_{11}k_{22} + k_{12}k_{21})S_o$$

$$= \omega_1\beta_{11} + \omega_2\beta_{22} \pm (\omega_1\omega_2)^{\frac{1}{2}}\ \left[(\alpha_{11}\alpha_{22})^{\frac{1}{2}} + \alpha_{12}\right]$$

Here again the (+) sign is taken for $\omega_o = \omega_1 + \omega_2$ and the (-) sign for $\omega_o = \omega_1 - \omega_2$.

APPLICATION

As an application, the flexural-torsional stability of a thin simply-supported beam subjected to a transverse load at mid-span is considered, Figure 3. Two forms of loading are investigated: (a) where the load maintains a constant vertical direction, and (b) where the load changes its direction so as to always remain in the plan of the cross-section at mid-span, i.e. a 'follower load.

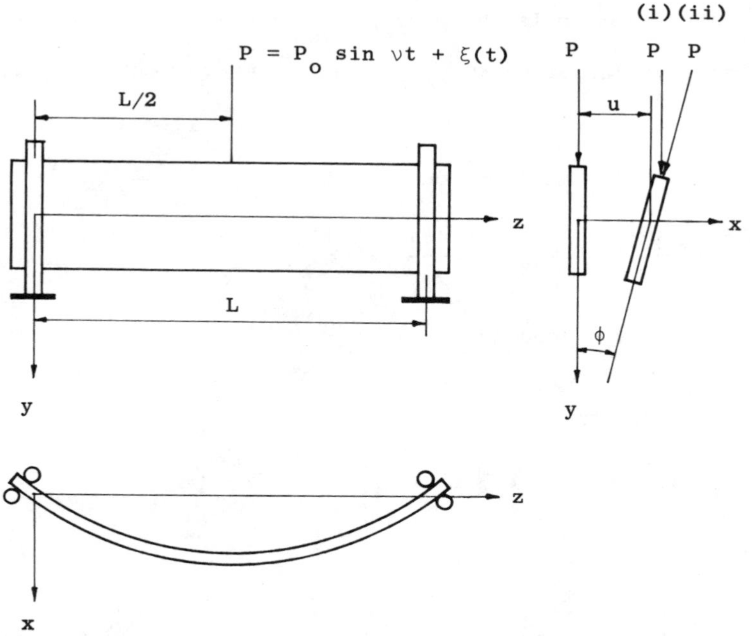

Figure 3

Assuming deflection functions of the form: $u(z,t) = q_1(t)\sin(\pi z/L)$ $\phi(z,t) = q_2(t)\sin(\pi z/L)$, which satisfy the boundary conditions of simple support, and using a Galerkin technique, it can be shown (11) that the equations of motion reduce to:

$$\ddot{q}_1 + 2\varepsilon\omega_1\beta_1\,\dot{q}_1 + \omega_1^2\,q_1 + 2\omega_1^2k_{12}\left[\varepsilon h\sin\nu t + \varepsilon^{\frac{1}{2}}f(t)\right]\,q_2 = 0$$

$$\ddot{q}_1 + 2\varepsilon\omega_2\beta_2\,\dot{q}_2 + \omega_2^2\,q_2 + 2\omega_2^2k_{21}\left[\varepsilon h\sin\nu t + \varepsilon^{\frac{1}{2}}f(t)\right]\,q_1 = 0$$

where

$$\omega_1^2 = \pi^4 EI_y/(mL^4)$$

$$\omega_2^2 = \pi^2(\pi^2EI_w + GJL^2)/(m_{\rho}{}^2L^4)$$

$$\varepsilon h = p_o/p_{cr}, \qquad \varepsilon^{\frac{1}{2}}f(t) = \xi(t)/p_{cr},$$

and EI_y, GJ, EI_w denote, respectively, the flexural, torsional and warping rigidities of the beam; m is the mass per unit length and ρ is the polar radius of gyration. The quantity p_{cr} is the critical load for time-independent loading.

For constant directional (vertical) loading:

$$p_{cr} = \frac{8mL\rho\omega_1\omega_2}{4+\pi^2}, \quad k_{12} = -\frac{\rho\omega_2}{2\omega_1}, \quad k_{21} = -\frac{\omega_1}{2\rho\omega_2}$$

while, for follower-type loading:

$$p_{cr} = \frac{4mL\rho|\omega_1^2-\omega_2^2|}{[(12-\pi^2)(4+\pi^2)]^{\frac{1}{2}}}$$

$$k_{12} = \left[\frac{12-\pi^2}{4+\pi^2}\right]^{\frac{1}{2}} \frac{\rho|\omega_1^2-\omega_2^2|}{4\omega_1^2}$$

$$k_{21} = -\left[\frac{4+\pi^2}{12-\pi^2}\right]^{\frac{1}{2}} \frac{|\omega_1^2-\omega_2^2|}{4\rho\omega_2^2}$$

Considering only the case where $f(t)$ is a white noise process with spectral density S_o, the conditions for stability in the second moments, deduced from (13) and (15), are as follows:

<u>Constant directional load</u>: $\omega_o = \omega_1 + \omega_2$

$$\beta_1 > 0, \quad \beta_2 > 0$$

$$16\beta_1\beta_2 > \omega_1\omega_2 S_o^2$$

$$(\lambda\omega_o)^2 > Ch^2 - D^2,$$

where

$$C = \frac{2(\omega_1\beta_1 + \omega_2\beta_2) + \omega_1\omega_2 S_o}{2(16\beta_1\omega_2 - \omega_1\omega_2 S_o^2)} D,$$

$$D = \omega_1\beta_1 + \omega_2\beta_2 + \frac{\omega_1\omega_2}{4} S_o$$

<u>Follower loading</u>: $\omega_o = |\omega_1 - \omega_2|$

$$\beta_1 > 0, \qquad \beta_2 > 0$$

$$256\beta_1\beta_2 > \frac{(\omega_1^2 - \omega_2^2)^4}{\omega_1^3\omega_2^3} S_o^2 , \qquad (\lambda\omega_o)^2 > Ch^2 - D^2,$$

where

$$C = \frac{(\omega_1^2-\omega_2^2)^2 \left[8\omega_1\omega_2(\omega_1\beta_1+\omega_2\beta_2)+(\omega_1^2-\omega_2^2)^2 S_o\right]}{2\left[256\omega_1^3\omega_2^3\beta_1\beta_2-(\omega_1^2-\omega_2^2)^4 S_o^2\right]} D,$$

$$D = (\omega_1\beta_1 + \omega_2\beta_2) + \frac{(\omega_1^2-\omega_2^2)^2}{16\omega_1\omega_2} S_o .$$

CONCLUSIONS

The effect of an additional stochastic excitation on the parametric instability of harmonically excited linear mechanical systems has been investigated. Explicit stability conditions have been derived for the stability in the first and the second moments of two-degree of freedom systems, and applied to the example of the torsional-flexural stability of a thin elastic beam.

ACKNOWLEDGEMENT

The research for this paper was supported (in part) by the National Research Council of Canada through Grant No. A-1815.

REFERENCES

1 E. Mettler — Dynamic stability of structures. Ed. G. Herrmann, Pergamon, p. 169. Stability and Vibration Problems of Mechanical Systems under Harmonic Excitation (1966)

2 E. Mettler — Combination Resonances in Mechanical Systems under Harmonic Excitation. Proc. Fourth Conf. on Non-Linear Oscillations, Prague, Academia, p. 51. (1968).

3 S.T. Ariaratnam — Dynamic Stability under Random Excitation. Proc. Canadian Cong. Appl. Mech. Vol. 3, 3-163. (1967).

4 S.T. Ariaratnam — Proc. IUTAM Conf. on Instability of Continuous Systems, Herrenalb, Springer Verlag, p. 78. Stability of Structures under Stochastic Disturbances (1969).

5 S.T. Ariaratnam — Proc. IUTAM Symposium on Stability of Stochastic Dynamical Systems, Coventry, Lecture Notes in Mathematics No. 294, Springer-Verlag, p. 291. Stability of Mechanical Systems under Stochastic Parametric Excitation. (1972).

6 R.L. Stratonovich — Topics in the Theory of Random Noise, Vol. 1, Gordon and Breach, New York. (1963).

7 R.Z. Khasminskii — A Limit Theorem for Solutions of Differential Equations with a Random Right-hand Side. Theory Prob. Applications, Vol. 11, p. 390. (1966).

8 G.C. Papanicolaou and W. Kohler — Asymptotic Theory of Mixing Stochastic Ordinary Differential Equations. Communications on Pure and Appl. Maths. XXVII, p 641. (1974).

9 N.N. Bogolyubov and Yu A. Mitropolskii — Asymptotic Methods in the Theory of Non-linear Oscillations, Gordon and Breach, New York. (1961).

10 S.T. Ariaratnam and D.S.F. Tam, Parametric Random Excitation of a Damped Mathieu Oscillator. Zeit Angew Math. u Mech. Vol. 56, No. 11, p. 499 (1976).

11 V.V. Bolotin, Dynamic Stability of Elastic Systems, Holden-Day, San Francisco, p. 337 (1964).

Professor S.T. Ariaratnam, Department of Civil Engineering, University of Waterloo, Waterloo, Ontario, Canada N2L 3G1.

Dr. D.S.F. Tam, Department of Civil Engineering, University of Waterloo, Waterloo, Ontario, Canada N2L 3G1.

DISCUSSION

In the discussion it was stressed that the transformation of equation (4) is exact, each pair of coordinates q_r, q'_r being transformed into a further pair z_r, $\bar{z}_r$, the complex form being essentially similar to the familiar real cartesian or polar forms of the van der Pol transformation:

$$q_r = X_1 \sin \Omega t + X_2 \cos \Omega t \quad \text{and} \quad q_r = X \sin (\Omega t + \emptyset)$$

In the n degree of freedom case the complex form is simpler to use. In response to questions (Barr, Kozin) about the order of coefficients (ε, $\varepsilon^{\frac{1}{2}}$) Professor Ariaratnam confirmed that the method was restricted to the case treated and that the stochastic term must be $O(\varepsilon^{\frac{1}{2}})$ if the damping and deterministic force are $O(\varepsilon)$ because stability depends on the spectral height of the former, which is $O(\varepsilon^{\frac{1}{2}})^2 = O(\varepsilon)$. There might be difficulties in obtaining equations with the correct order coefficients in practical problems but it was pointed out (Kozin) that in many cases "natural" transformations can yield the requisite orders.

Replying to a further question (Barr) it was explained that the other instability zone at ω_o and at higher frequencies would only be detected by going to a second order approximation and that although this was possible in the deterministic case, in stochastic problems the Khasminskii Limit Theorem only caters for first approximations. However the proofs of Khasminskii (author's ref. [7]) and subsequent workers [8] do contain expressions for the error between the original and the limit process and this error is Gaussian (Kozin). A discussion of the physical interpretation of the mathematical assumptions necessary for use of the Theorem then ensued (Sobczyk) and it was noted that the "regularity" condition corresponds to assumption of a finite memory process. In the original treatment of Stratonovich [6] the relationship between the system relaxation time τ_{rel} and the excitation correlation time

τ_{cor} is used. If $\tau_{rel} >> \tau_{cor}$ then the system sees the excitation as (almost) white noise. τ_{rel} and τ_{cor} can of course be related to system and excitation bandwidths and in terms of the small parameter the necessary condition may be stated $\tau_{cor} << 0(1/\varepsilon)$.

In subsequent treatments and applications (Papanicolaou et al. see [8] and the references therein), further physical interpretations have been given. The restrictions of the original proof by Khasminskii include that of bounds on derivatives of the function f(x) in the differential equation $\dot{x} = f(x)$ to be averaged. Some of these seem difficult to satisfy but in practice suitable transformations can be found. In particular Professor Ariaratnam noted that the bound on f(x) can be satisfied by using a transformation of the type $x \to \lambda = x/||x||$, $\rho = \log ||x||$ the theorem can thus be applied to linear systems per se.

In reply to further question (Lin) it was noted that the σ_{ij} coefficients in the Itô equation (6) are complex functions of z_i, $\overline{z}_i$ etc. and that the cross terms σ_{jk} ($j \neq k$) are non zero (cf. the real case of cartesian coordinates treated in the authors' reference [10]: in polar coordinates the cross terms are often identically equal to zero).

Replying to a question on assumptions (Elishakoff) Professor Ariaratnam confirmed that the case of coincident natural frequencies ($\omega_i = \omega_j$, $i \neq j$) had been excluded but mentioned that Professor Schmidt (q.v.) had treated such degenerate problems in the deterministic case.

In a response to a final question (Wedig) it was noted that the presence of parametric (internal) stochastic excitation leads to primary instability at $2\omega_n$ (and 0) while that of external excitation leads to instability at ω_n, where ω_n is the natural frequency of the system.

A BENSOUSSAN, J L LIONS and G C PAPANICOLAOU

Homogenization in deterministic and stochastic problems

INTRODUCTION

This is a brief account of some results in homogenization. In a forthcoming book [1] we give details and many other applications and connections with probability theory, control theory, etc. For references and historical remarks concerning the development of homogenization (it is an old problem) see the papers of Babuska [2]. Some of our own contributions have been announced in [3].

We begin with the homogenization of a diffusion equation and then in the next section we explain how to obtain the results by the usual methods of asymptotic expansions. The formal constructions yield also an elementary proof under enough regularity conditions.

Following this we consider the homogenization of the problem of sound propagation in a periodically inhomogeneous viscous fluid. The next section contains homogenization formulas for Maxwell's equations. Finally we comment on the stochastic problems and refer to [4] for references and the detailed (formal) treatment of electromagnetic waves in a random medium.

HOMOGENIZATION OF DIFFUSIONS

Let D be a bounded open set in R^n with smooth boundary and consider the diffusion equation

$$\frac{\partial u^{\varepsilon}(t,x)}{\partial t} = \sum_{i,j=1}^{n} \frac{\partial}{\partial x_i} \left(a_{ij}\left(\frac{x}{\varepsilon}\right) \frac{\partial u^{\varepsilon}(t,x)}{\partial x_j} \right), \quad x \in D \text{ , } t > 0 \text{ ,} \qquad (1)$$

$$u^{\varepsilon}\ (0,x) = f(x)$$

$$u^{\varepsilon}\ (t,x) = 0\ , \quad x \in \partial D\ ,$$

where the coefficients $(a_{ij}(y))$, $y = x/\varepsilon$, satisfy†

$$\alpha|\xi|^2 \leq \sum_{i,j=1}^{n} a_{ij}(y)\ \xi_i\xi_j \leq \beta|\xi|^2\ , \tag{2}$$

$a_{ij}(y)$ are smooth†† periodic functions of period one in $y_1, y_1, \ldots, y_n$. (3)

We wish to find the limit of $u^{\varepsilon}\ (t,x)$ as $\varepsilon \to 0$ with $x \varepsilon D$ and $0 \leq t \leq t_T < \infty$. The type of convergence (the topology) differs depending on hypotheses and the nature of the approximation. First we shall describe the results.

Let $\chi^k(y)$, $k = 1,2,\ldots,n$ be defined (up to an additive constant) as the smooth periodic (of period one in all variables) solutions of

$$\sum_{i,j=1}^{n} \frac{\partial}{\partial y_i}\left[a_{ij}\frac{\partial}{\partial y_i}\left(\chi^k + y_k\right)\right] = 0\ , \tag{4}$$

$$0 \leq y_i \leq 1\ , \quad i = 1,\ldots,n.$$

Note that these functions are solutions of a <u>cell problem</u> i.e., a local problem involving one individual period cell. Define

$$\bar{a}_{ij} = \int_T \sum_{k=1}^{n} a_{ik}(y)\frac{\partial}{\partial y_k}\left(\chi^j(y) + y_i\right) dy\ . \tag{5}$$

---

† $\quad |\xi|^2 = \sum_{i=1}^{n} \xi_i^2\ .$

††This assumption is made for simplicity only (cf. [3] and references therein).

Let $\bar{u}(t,x)$ be the solution of

$$\frac{\partial \bar{u}(t,x)}{\partial t} = \sum_{i,j=1}^{n} \bar{a}_{ij} \frac{\partial^2 \bar{u}(t,x)}{\partial x_i \partial x_j} , \tag{6}$$

$$x \in D \quad , \quad t > 0 ,$$

$$\bar{u}(0,x) = f(x)$$

$$\bar{u}(t,x) = 0 , \qquad x \in \partial D .$$

Theorem

Under the above assumptions we have

$$\sup_{0 \leq t \leq t_T} \sup_{x \in D} |u^{\varepsilon}(t,x) - \bar{u}(t,x)| \leq C\varepsilon , \tag{7}$$

where C is a constant dependent of ε.

Remark 1

Much weaker smoothness conditions and a correspondingly weaker statement than (7) can be shown to be adequate. Such results are important in (i) nonlinear problems and (ii) in physical situations where the coefficients a_{ij} do change discontinuously (cf. [3]).

Remark 2

The physical significance of the result is as follows. The coefficients of diffusivity in (1) are periodically varying in space which means that the conducting medium is a composite structure that repeats periodically; for example a crystalline solid, a fibre reinforced composite, etc. Because f scaling in (1) the region of interest D contains many of the basic cells that make up the composite. We expect on physical grounds that the problem (1) should be well approximated by a constant coefficient problem (6). The effective diffusivity ($\bar{a}_{ij}$) is given by (5) and involves the solution of the cell problem (4). Note in particular that $\bar{a}_{ij}$ is <u>not</u> simply the average of $a_{ij}(y)$ over a cell.

ASYMPTOTIC EXPANSION

The idea is to let $y = x/\varepsilon$ and introduce a function $v^{\varepsilon}(t,x,y)$ such that $u^{\varepsilon}(t,x) = v\ (t,x,x/\varepsilon)$. Then we expand $v^{\varepsilon}(t,x,y)$ by the usual methods of asymptotic expansion. The function $v^{\varepsilon}(t,x,y)$ clearly satisfies†

$$\frac{1}{\varepsilon^2} \sum_{i,j=1}^{n} \frac{\partial}{\partial y_i} \left(a_{ij}(y)\frac{\partial v^{\varepsilon}}{\partial y_i}\right) \tag{8}$$

$$+ \frac{1}{\varepsilon} \sum_{i,j=1}^{n} \frac{\partial a_{ij}(y)}{\partial y_i} \frac{\partial v^{\varepsilon}}{\partial x_j}$$

$$+ \sum_{i,j=1}^{n} a_{ij}(y) \frac{\partial^2 v^{\varepsilon}}{\partial x_i \partial x_j} - \frac{\partial v^{\varepsilon}}{\partial t} = 0 .$$

Next we write

$$v^{\varepsilon}(t,x,y) = v_0(t,x,y) + \varepsilon v_1(t,x,y) + \varepsilon^2 v_2(t,x,y) + \dots , \tag{9}$$

and find from (8) that

$$\sum_{i,j=1}^{n} \frac{\partial}{\partial y_i} \left(a_{ij}\ (y) \frac{\partial v_0}{\partial y_i}\right) = 0 , \tag{10}$$

$$\sum_{i,j=1}^{n} \frac{\partial}{\partial y_i} \left(a_i(y) \frac{\partial v_1}{\partial y_i}\right) + \sum_{i,j=1}^{n} \frac{\partial a_{ij}(y)}{\partial y_i} \frac{\partial v_0}{\partial x_j} = 0 \tag{11}$$

---

†We use the chain rule so that

$$\frac{\partial}{\partial x_i} \to \frac{\partial}{\partial x_i} + \frac{1}{\varepsilon}\frac{\partial}{\partial y_i} .$$

$$\sum_{i,j=1}^{n} \frac{\partial}{\partial y_1}\left(a_{ij}(y)\frac{\partial v_2}{\partial y_j}\right) \tag{12}$$

$$+ \sum_{i,j=1}^{n} \frac{\partial a_{ij}(y)}{\partial y_j}\frac{\partial v_1}{\partial x_j}$$

$$+ \sum_{i,j=1}^{n} a_{ij}(y)\frac{\partial^2 v_o}{\partial x_i \partial x_j} - \frac{\partial v_o}{\partial t} = 0 ,$$

$$\cdots\cdots\cdots\cdots\cdots\cdots\cdots\cdots\cdots .$$

From (10), the requirement that v_k, $k = 0,1,2,\ldots$ be periodic in y and (2) it follows that

$$v_o = v_o(t,x) ,$$

i.e., v_o is not a function of y. At this stage we do not know what $v_o(t,x)$ should be, however.

We pass to (11). Since (10) has a nontrivial solution, the inhomogeneous term in (11) must integrate to zero over a cell (Fredholm alternative). It does, so we continue with (12). Applying the solvability condition to this equation (i.e. the Fredholm alternative) leads to a differential equation for $v_o(t,x)$. It is easy to see that, in fact, $v_o(t,x) = \bar{u}(t,x)$ as defined by (6) and that the recipe for computing $\bar{a}_{ij}$ (i.e. (4), (5)) follows by simple computations.

Assuming enough regularity a proof can now be given for the theorem above. Since v_1 and v_2 are bounded and smooth it suffices† to show that

$$u^{\varepsilon}(t,x) - (\bar{u}(t,x) + \varepsilon v_1(t,x,x/\varepsilon) + \varepsilon^2 v_2(t,x,x/\varepsilon))$$

---

†Note that $|u^{\varepsilon}-\bar{u}| \leq |u^{\varepsilon}-\bar{u}-\varepsilon v_1-\varepsilon^2 v_2| + \varepsilon|v_1+\varepsilon v_2|$.

is $O(\varepsilon)$. This is true by construction and from the maximum principle for (1).

SOUND PROPAGATION

Consider a fluid (very viscous) that has a spatially periodic structure so that its density is $\rho(x/\varepsilon)$ with $\rho(y)$ periodic and its bulk modulus is $\mu(x/\varepsilon)$ with $\mu(y)$ periodic. Let $p^\varepsilon(t,x)$ be the excess (negative) pressure and $v^\varepsilon(t,x)$ the (small amplitude) velocity deviation. We have the equation

$$\rho(\frac{x}{\varepsilon}) \frac{\partial v^\varepsilon}{\partial t} = \nabla p^\varepsilon \tag{13}$$

$$\frac{\partial p^\varepsilon}{\partial t} = \mu(\frac{x}{\varepsilon}) \nabla . v^\varepsilon$$

which are valid for all $x \in R^3$ with given initial data.

The problem is to find, when ε is small, the effective density $\bar{\rho}$ and the effective bulk modulus $\bar{\mu}$ so that the solution of (13) is approximated well by the solution† of

$$\bar{\rho} \frac{\partial \bar{v}}{\partial t} = \nabla \bar{p} \quad , \tag{14}$$

$$\frac{\partial \bar{p}}{\partial t} = \bar{\mu} \nabla . \bar{v} \ .$$

Following a formal procedure identical to the one of the section above leads to the following formulas for $\bar{\mu}$ and $\bar{\rho}$.

First we solve for the functions $\chi^k(y)$, $k = 1,2,3$ which are periodic and satisfy the cell problem (cf. (4)).

$$\nabla_y . \left\{\frac{1}{\rho(y)} \nabla_y \left\{\chi^k(y)+y_k\right\} \right\} = 0 \ . \tag{15}$$

† $\bar{\rho}$ will turn out to be a tensor (3x3 matrix).

Then we define the tensor $(\bar{\rho}_{ij})$ as the inverse of the tensor (matrix).

$$\left(\int \frac{1}{\rho(y)} \frac{\partial}{\partial y_i} (\chi^j(y)+y_j)dy\right) . \tag{16}$$

This is the desired effective 'density' $\bar{\rho}$.

The effective bulk modulus is given by

$$\bar{\mu} = \left(\int \frac{1}{\mu(y)} dy\right)^{-1} . \tag{17}$$

Note the following important fact. Problem (13) with the periodic spatial inhomogeneities is isotropic. The homogenized problem (14) (with constant coefficients) is anisotropic since the effective material density is a tensor Of course, density is a scalar quantity under any circumstances. What we mean here is that in the homogenized problem the local stress tensor on the fluid element is not simply the gradient of the (negative) pressure but a fixed tensor times the pressure gradient. This fixed tensor is

$$\left(\frac{1}{\int \rho(y)dy} \cdot \int \frac{1}{\rho(y)} \frac{\partial}{\partial y_i} \left\{\chi^j(y)+y_j\right\} dy \right) . \tag{18}$$

The significance of (18) is that it gives the precise amplification factor for the effective stress based on the single cell configurations.

MAXWELL'S EQUATIONS

We repeat the problem of the above section in the electromagnetic case. The small parameter ε will be denoted by γ in this section. The tensor ε will be the dielectric tensor.

Let $\varepsilon(x/\gamma)$ and $\mu(x/\gamma)$ be symmetric 3x3 positive definite matrix functions of y such that $\varepsilon(y)$ and $\mu(y)$ are periodic of period 1 in all variables. The parameter $\gamma > 0$ is small. Let $E^\gamma(t,x)$ and $B^\gamma(t,x)$ be the electric and magnetic field respectively and let $D^{\gamma}(t,x)$ and $H^{\gamma}(t,x)$ be the dielectric displacement and the magnetic induction respectively. Maxwell's equations are

$$\varepsilon(\frac{x}{\gamma})\ \frac{\partial E^{\gamma}(t,x)}{\partial t} = \nabla \times H^{\gamma}(t,x)\ , \tag{19}$$

$$\mu(\frac{x}{\gamma})\ \frac{\partial H^{\gamma}(t,x)}{t} = -\nabla \times E^{\gamma}(t,x)\ ,$$

in all of space, say, and with given initial conditions. The constitutive relations

$$D = \varepsilon E\ , \qquad B = \mu H\ , \tag{20}$$

relate quantities as usual and the conditions

$$\nabla.(\varepsilon E) = 0\ , \qquad \nabla.(\mu H) = 0\ , \tag{21}$$

hold for all time if they hold initially (as we assume).

We want to find the effective dielectric tensor $\bar{\varepsilon}$ and the effective magnetic permeability tensor $\bar{\mu}$. For this purpose we solve first two cell problems (cf. (4)) as follows

$$\nabla_y.(\varepsilon(y)\ \nabla_y(\chi_1(y)+y)) = 0\ , \tag{22}$$

$$\chi_1(y)\ \text{(a 3-vector) periodic}\ ,$$

$$\nabla_y.(\mu(y)\ \nabla_y(\chi_2(y)+y)) = 0\ , \tag{23}$$

$$\chi_2(y)\ \text{(a 3-vector) periodic.}$$

Then we define

$$\bar{\varepsilon}_{ij} = \int \sum_{k=1}^{3} \varepsilon_{ik}(y)\ \frac{\partial}{\partial y_k} \left\{\chi_1^j(y)+y_j\right\}\ dy\ , \tag{24}$$

$$\bar{\mu}_{ij} = \int \sum_{k=1}^{3} \mu_{ik}(y)\ \frac{\partial}{\partial y_k} \left\{\chi_2^j(y)+y_j\right\}\ dy\ , \tag{25}$$

$$i,j = 1,2,3\ .$$

Following the procedure of asymptotic expansion we find that E^γ and H^γ are well approximated when γ is small by the solution of

$$\bar{\varepsilon}\,\frac{\partial \bar{E}}{\partial t} = \nabla \times \bar{H}\ , \tag{26}$$

$$\bar{\mu}\,\frac{\partial \bar{H}}{\partial t} = -\nabla \times \bar{E}\ ,$$

with initial conditions as in (19). One expects, specifically, that for any bounded region D of R^3 the energy

$$\int_D \sum_{i,j=1}^{3} \left[\varepsilon_{ij}\left(\frac{x}{\gamma}\right)E_i^\gamma\,(t,x)E_j^\gamma(t,x) + \mu_{ij}\left(\frac{x}{\gamma}\right)H_i^\gamma(t,x)H_j^\gamma(t,x)\right] dx, \tag{27}$$

converges as $\gamma \to 0$ to the energy associated with problem (26).

STOCHASTIC PROBLEMS

We merely mention that problems where the constitutive parameters are random functions are extremely common and there is a voluminous literature dealing with such problems. We do not know, at present, how to obtain stochastic analogs of the above. We know how to obtain (but not how to prove) results relating to wave propagation. We refer to [4] for a sample and for references to other work.

REFERENCES

1 A. Bensoussan, J.L. Lions and G.C. Papanicolaou, Book in preparation.

2 I. Babuska, Several reports from the University of Maryland.

3 A. Bensoussan, J.L. Lions and G.C. Papanicolaou, C.R. Acad. Sc. Paris, 281(A) (July 1975), pp. 281 and 317. Also 282(A) (January 1976), p. 143.

4 G.C. Papanicolaou and R. Burridge, J. Math. Phys., 16 (1975), p. 2074.

Professor A. Bensoussan, IRIA - Laboria, Domaine de Voluceau, Rocqueucourt, 78150 Le Chesnay, France.

Professor J.L. Lions, IRIA - Laboria, Domaine de Voluceau, Rocqueucourt, 78150 Le Chesnay, France.

Professor G. Papanicolaou, Courant Institute of Mathematical Sciences, New York University, 251 Mercer Street, NY 10012, U.S.A.

T NAKAMIZO and M OSHIRO

On stochastic singular problem of linear dynamical system

INTRODUCTION

A time-invariant linear system with additive Gaussian noise at the input and output is considered here, and thus can be described by the differential equation of the form

$$\dot{x}(t) = Ax(t) + Bu(t) + Gw(t) \tag{1}$$

$$y(t) = Cx(t) + v(t) \tag{2}$$

where x is an nx1 state vector, u is a kx1 control input vector, and y is an mx1 output vector. The constant matrices A, B, C and G are of appropriate dimensions. In equations (1) and (2), w(t) and v(t) are independent zero mean Gaussian white noise with

$$E\{w(t)\ w'(\tau)\} = \Sigma_1 \delta(t-\tau) \tag{3}$$

$$E\{v(t)\ v'(\tau)\} = \Sigma_2 \delta(t-\tau) \tag{4}$$

Both Σ_1 and Σ_2 are non-negative definite symmetric matrices. Further the controlled output vector can be expressed as

$$z(t) = Dx(t) \tag{5}$$

where D is a constant matrix. It is assumed that the system is completely controllable and observable. Consider also the quadratic performance index

$$J = E\ \{z'(t)R_1\ z(t) + u'(t)R_2\ u(t)\} \tag{6}$$

The steady state regulator problem is that of finding u(t) such that J is minimized.

In order for the minimization problem to be well defined, it is usually assumed that (a) the noise covariance Σ_2 is non-singular, which implies that no component of the output can be measured exactly, and (b) the weighting matrix R_2 is non-singular, which implies the limitation of the input. The question now arises as to what happens to the controller and the performance as Σ_2 or R_2 approaches a singular matrix. This is investigated as the main material of this paper. To facilitate the development however, a steady state version of the problem formulation and solution will be given for a time-invariant linear system.

The problem of determining the optimal controller when R_2 or Σ_2 is zero matrix is considered by Friedland [1] and Nakamizo and Oshiro [2]. Moreover some related discussions on this problem are given also by Butman [3], Ho [4] and Kwakernaak [5]. The discussion on the lower bound on the performance is presented by Kwakernaak and Sivan [8] for the deterministic linear system with the perfect observation. The structure of this paper, continuing the investigation of reference [2], is as follows. The standard regulator problem is quickly reviewed for the usual case where both R_2 and Σ_2 are positive definite matrices, and then the basic equations used in computation of a reduced order optimal controller and of the performance for the singular case are described. The stability analysis is performed for the interconnected closed loop system of the plant and the controller and the conditions under which the lower bound on the performance can be achieved are discussed. In the last section the singular observation problem is considered, and it is shown that this problem parallels the singular stochastic control problem in the dual sense.

REVIEW OF REGULATOR PROBLEM

For clarity and for development it is convenient to begin with the quick review of the well known solution to the standard linear regulator problem which is now being applied in various fields of engineering. Consider a linear system governed by equations (1) and (2). The problem is to find the optimal input u(t) in such a way that the quadratic performance index J given by equation (6) is minimized, subjected to equations (1) and (2).

In the non-singular case where both R_2 and Σ_2 are positive definite matrices, the solution to this problem is well developed. It follows from the well known separation theorem [6] that the optimal control input u(t) is given by

$$u(t) = - L\hat{x}(t) \tag{7}$$

with

$$L = R_2^{-1}B'P \tag{8}$$

where P is the unique, positive definite solution of the steady state matrix Riccati equation

$$PA + A'P - PBR_2^{-1}B'P + D'R_1D = 0\,. \tag{9}$$

The estimate $\hat{x}(t)$ is defined as the conditional expectation of x(t) given $y(\tau)$ for $\tau < t$, and is generated by the so called Kalman filter

$$\dot{\hat{x}}(t) = A\hat{x}(t) + Bu(t) + K\left[y(t) - C\hat{x}(t)\right] \tag{10}$$

where the filter gain K is known to be

$$K = \tilde{X}C'\Sigma_2^{-1} \tag{11}$$

The covariance matrix of the estimation error $\tilde{X} = E\{(x-\hat{x})(x-\hat{x})'\}$ is evaluated as the solution of

$$A\tilde{X} + \tilde{X}A' - \tilde{X}C'\Sigma_2^{-1}C\tilde{X} + G\Sigma_1G' = 0\,. \tag{12}$$

When the optimal control law (7) is realized, the closed loop system is stable, and the optimal performance index is given by

$$J^* = \text{trace}\left[D'R_1D\tilde{X} + PK\Sigma_2K'\right]$$

$$= \text{trace}\left[L'R_2L\tilde{X} + PG\Sigma_1G'\right] . \tag{13}$$

The performance has two terms for cost owing to the error in the estimate of the state and the error in control.

In the present paper, main interest is in the case where R_2 or Σ_2 becomes zero. It follows from equation (13) that

$$J_{min} = \lim_{R_2 \to 0} J^* \geq \text{trace}\,(Q\tilde{X}), \qquad Q = D'R_1D \tag{14}$$

$$J_{min} = \lim_{\Sigma_2 \to 0} J^* \geq \text{trace}\,(PW), \qquad W = G\Sigma_1G' . \tag{15}$$

This means that the performance index cannot be less than trace$(Q\tilde{X})$, because of the inevitable accuracy in estimating the state from the noisy observation. Similarly the performance index for the case of no observation noise cannot be less than trace(PW) which exactly corresponds to the case of the perfect observation. Obviously trace$(Q\tilde{X})$ and trace(PW) provide the lower bound on the performance for the respective singular case.

STOCHASTIC SINGULAR CONTROL PROBLEM

In this section, an optimal control problem for a linear system with noisy observation is discussed for the singular case, namely the case where a performance index is the mean square value of the state vector, but does not explicitly depend on the control input vector. Thus the performance index is taken here to be

$$J = E\,\{z'(t)R_1z(t)\}\ , \quad R_1 > 0 . \tag{16}$$

This is a generalization of the mean square error criterion which is widely used in the design of servomechanisms. The problem is that of finding the optimal control u(t) such that the control-free performance index J given by equation (16) is minimized, subject to equations (1) and (2). This is referred to here as the stochastic singular control problem.

Assume that the control weighting matrix in (6) is of the form

$$R_2 = \alpha N \tag{17}$$

with N positive definite and α a positive value. It is known [8] that the limit

$$\bar{P} = \lim_{\alpha \to 0} P \tag{18}$$

exists. If B'QB is non-singular, then J_{min} can be calculated by

$$J_{min} = \text{trace}\ (Q\tilde{X} + \bar{P}K\Sigma_2 K') \tag{19}$$

where nxn matrix $\bar{P}$ is the solution of the algebraic Riccati equation [1, 2]

$$\bar{P}AF + F'A'\bar{P} - \bar{P}AB(B'QB)^{-1}B'A'\bar{P} + QF = 0 \tag{20}$$

and $\bar{P}$ satisfies the constraint $\bar{P}B = 0$. In equation (20)

$$F = I - B(B'QB)^{-1}B'Q\ . \tag{21}$$

These results immediately follow from the results of Friedland [1], so we shall not elaborate on them. It is known also that the sufficient condition for the solution $\bar{P}$ to exist and for $\bar{P} \geq 0$ are that (A,B) be completely controllable and $QF \geq 0$.

Suppose that $\bar{P} = 0$. Then equation (20) gives QF = 0. The necessary condition for QF = 0 to hold is that rank D $\leq$ rank B, and the necessary condition for B'QB to be positive definite is that rank D $\geq$ rank B. Hence the following fact can be observed:

[A] If rank D > rank B, then $J_{min} > \text{trace}(Q\tilde{X})$.

It can be shown that the Riccati equation (20) for nxn matrix $\bar{P}$ is reduced in order. Since the nxk matrix B is assumed to have full rank, there exists an nx(n-k) matrix T such that

$$\hat{B} = \left[\underset{k}{B} \;\vdots\; \underset{n-k}{T} \right] n \tag{22}$$

is non-singular. It is convenient to define the notation partitioned as

$$\hat{B}^{-1} = \underset{n}{\begin{bmatrix} L_1 \\ \hline L_2 \end{bmatrix}} \begin{matrix} k \\ n-k \end{matrix} . \tag{23}$$

Premultiplying $\bar{P}B = 0$ by $\hat{B}'$ yields

$$\hat{B}'\bar{P}\hat{B}\hat{B}^{-1}B = 0 . \tag{24}$$

Define $\Pi = \hat{B}'P\hat{B}$ and note that

$$\hat{B}^{-1}B = \begin{bmatrix} L_1B \\ \hline L_2B \end{bmatrix} = \begin{bmatrix} I_k \\ \hline 0 \end{bmatrix} \tag{25}$$

to obtain

$$\hat{B}'\bar{P}B = \Pi\hat{B}^{-1}B = \begin{bmatrix} \Pi_{11} & \Pi_{12} \\ \hline \Pi'_{12} & \Pi_{22} \end{bmatrix} \begin{bmatrix} I_k \\ \hline 0 \end{bmatrix} = \begin{bmatrix} \Pi_{11} \\ \hline \Pi'_{12} \end{bmatrix} = 0 \tag{26}$$

from which $\Pi_{11} = 0$, $\Pi_{12} = 0$. Hence only Π_{22} remains to be determined. Premultiplying and postmultiplying equation (20) by $\hat{B}'$ and $\hat{B}$ gives

$$\Pi_{22}L_2AFT + T'F'A'L_2'\Pi_{22} + T'QFT$$

$$- \Pi_{22}L_2AB(B'QB)^{-1}B'A'L_2'\Pi_{22} = 0 \tag{27}$$

which is the reduced order Riccati equation to give $\bar{P} = L_2'\Pi_{22}L_2$. This achieves a substantial simplification in the machine computation, since the complexity and cost for solving the equation increase with its dimension.

It can be shown that as $\alpha \to 0$ the stochastic optimal controller reduces to

$$u(t) = -B^{\dagger} \left[\hat{A}(I-BB^{\dagger}) T\eta(t) + Ky(t)\right] \tag{28}$$

$$\dot{\eta}(t) = L_2 \left[\hat{A}(I-BB^{\dagger}) T\eta(t) + Ky(t)\right] \tag{29}$$

where

$$B^{\dagger} = (B'QB)^{-1}B'(A'L_2'\Pi_{22}L_2 + Q) \tag{30}$$

$$\hat{A} = A - KC, \quad \dim(\eta) = n-k \ . \tag{31}$$

This solution follows from the results reported previously by the present authors [2], in which the solution has been obtained directly rather than by a series expansion and limiting approach. The controller (28) and (29) may be expressed in terms of Laplace transforms. After Laplace transformation and some manipulations one obtains

$$U(s) = -\left[B^{\dagger}(sI-\hat{A})^{-1}B\right]^{-1}B^{\dagger}(sI-\hat{A})^{-1}KY(s) \ . \tag{32}$$

This expression is valid provided $\left[B^{\dagger}(sI-\hat{A})^{-1}B\right]^{-1}$ exists. For the particular case where $C = D$ and $\det(CB) \neq 0$, where the output variable is precisely the controlled variable, the expression (32) reduces to

$$U(s) = -\left[C(sI-A)^{-1}B\right]^{-1}C(sI-A)^{-1}KY(s) \ . \tag{33}$$

It immediately follows that the inverse exists provided $\det\left[C(sI-A)^{-1}B\right]$ does not vanish identically in s. Note that $C(sI-A)^{-1}B$ is the open loop transfer function.

It is pertinent here to make three remarks.

(i) The nxn matrix $I-BB^{\dagger}$ has n-k independent columns since $I-BB^{\dagger}$ is of rank n-k. A specific nx(n-k) matrix T can be formed from any n-k linearly independent columns of $I-BB^{\dagger}$. Then it is readily verified that $L_1=B^{\dagger}$ and $B^{\dagger}T=0$. Thus equations (28) and (29) become simply

$$u(t) = -L_1\left[\hat{A}T\ \eta(t) + Ky(t)\right]$$

$$\dot{\eta}(t) = L_2\left[\hat{A}T\ \eta(t) + Ky(t)\right]$$

which is in agreement with the limiting form derived by Friedland [1].

(ii) It should be noted that the matrix $B^{\dagger}$ defined by equation (30) is a left inverse of the input matrix B, i.e. $B^{\dagger}B = I$. For a special case when B is invertible, obviously $B^{\dagger} = B^{-1}$. Thus equation (28) is

$$u(t) = -B^{-1}Ky(t) \tag{34}$$

and the following result can be readily obtained:

[B] If rank B = n, then $J_{min} = \mathrm{trace}(Q\tilde{X})$.

This means that an instantaneous output feedback control given by equation (34) achieves the lower bound of the performance irrespective of rank D, provided B^{-1} exists.

(iii) The optimal controller derived here is of lower order, but the matrix Π_{22} to obtain $B^{\dagger}$ has to be computed by solving the reduced order algebraic Riccati equation which may possess more than one non-negative definite solution. Therefore one has to select the solution that makes the closed loop system stable. This requires the stability analysis.

STABILITY CONSIDERATION

In this section, the stability properties of the closed loop system is analyzed. For the feedback arrangement resulted from interconnecting the plant and the reduced order controller, one can obtain

$$\dot{x} = (A-BB^{\dagger}KC)x-BB^{\dagger}\hat{A}(I-BB^{\dagger})T\eta + Gw-BB^{\dagger}Kv \tag{35}$$

$$\dot{\eta} = L_2KCx+L_2\hat{A}(I-BB^{\dagger})T\eta + L_2Kv \ . \tag{36}$$

Defining

$$e = x - (I-BB^{\dagger})T\eta \tag{37}$$

it is easily obtained from equation (35) and (36) that the augmented vector satisfies the equation

$$\begin{bmatrix} \dot{e} \\ \hline \dot{\eta} \end{bmatrix} = \left[\begin{array}{c|c} \hat{A} & 0 \\ \hline L_2KC & L_2A(I-BB^{\dagger})T \end{array}\right] \begin{bmatrix} e \\ \hline \eta \end{bmatrix} + \left[\begin{array}{c|c} G & -K \\ \hline 0 & L_2K \end{array}\right] \begin{bmatrix} w \\ \hline v \end{bmatrix} . \tag{38}$$

Then the characteristic polynomial is given by

$$\Delta (s) = \det \left[\begin{array}{c|c} sI-(A-KC) & 0 \\ \hline -L_2KC & sI-L_2A(I-BB^{\dagger})T \end{array}\right]$$

$$= \det(sI-A+KC)\det\left[sI-L_2A(I-BB^{\dagger})T\right] . \tag{39}$$

The characteristic values of the interconnected system consist of the characteristic values of (A-KC) and those of $L_2A(I-BB^{\dagger})T$. Since (A-KC) is stable by hypothesis, the closed loop stability depends on the zeroes of

$$\phi(s) = \det\left[sI-L_2A(I-BB^{\dagger})T\right] = 0. \tag{40}$$

For evaluation of $\phi(s)$, it is helpful to consider the characteristic polynomial of $A(I-BB^{\dagger})$:

$$\Delta_2(s) = \det\left[sI-A(I-BB^{\dagger})\right]$$

$$= \det\left[sI-\hat{B}^{-1}A(I-BB^{\dagger})\hat{B}\right]. \tag{41}$$

Using the relation

$$\hat{B}^{-1}A(I-BB^{\dagger})\hat{B} = \left[\begin{array}{c|c} 0 & L_1A(I-BB^{\dagger})T \\ \hline 0 & L_2A(I-BB^{\dagger})T \end{array}\right] \tag{42}$$

then equation (41) becomes

$$\Delta_2(s) = \det \left[\begin{array}{c|c} sI & -L_1A(I-BB^{\dagger})T \\ \hline 0 & sI-L_2A(I-BB^{\dagger})T \end{array}\right]$$

$$= s^k\det\left[sI-L_2A(I-BB^{\dagger})T\right]$$

$$= s^k\phi(s). \tag{43}$$

On the other hand, the alternate expression of $\Delta_2(s)$ can be obtained by the string of equations as

$$\begin{aligned}\Delta_2(s) &= \det\left[sI-A(I-BB^{\dagger})\right] \\ &= \det(sI-A)\det\left[I+(sI-A)^{-1}ABB^{\dagger}\right] \\ &= \det(sI-A)\det\left[B^{\dagger}(I+(sI-A)^{-1}A)B\right] \\ &= \det(sI-A)\det\left[B^{\dagger}(sI-A)^{-1}(sI-A+A)B\right] \\ &= \det(sI-A)\det\left[sB^{\dagger}(sI-A)^{-1}B\right] \\ &= s^k \det(sI-A)\det\left[B^{\dagger}(sI-A)^{-1}B\right]. \end{aligned} \tag{44}$$

It follows from equations (43) and (44) that

$$\phi(s) = \det(sI-A)\det\left[B^{\dagger}(sI-A)^{-1}B\right]. \tag{45}$$

This shows that $\phi(s)$ is exactly the numerator polynomial of the transfer function matrix

$$G_p(s) = B^{\dagger}(sI-A)^{-1}B \tag{46}$$

i.e.,

$$\det\left[G_p(s)\right] = \frac{\phi(s)}{\det(sI-A)} . \tag{47}$$

Note that the degree of $\phi(s)$ is exactly n-k. Thus one arrives at the following fact:

[C] The closed loop system is stable, if the transfer function matrix $B^{\dagger}(sI-A)^{-1}B$ has all of its zeroes in the left half plane of s.

This explicit criterion may not be useful in the practical situation because it depends on the matrix $B^{\dagger}$. However if u(t) and z(t) have the same dimensions, this leads to an interesting result. If rank D = rank B and $\det(DB) \neq 0$, then obviously QF = 0 in equation (27). Hence a possible solution to equation (27) is $\Pi_{22} = 0$. In such a case, one can get

$$B^{\dagger} = (DB)^{-1}D .$$

Hence the fact [C] can be rephrased as

[D] Let rank D = rank B and $\det(DB) \neq 0$. Then the closed loop system is stable, if the open loop transfer function $D(sI-A)^{-1}B$ has all of its zeroes in the left half plane.

Turning to the control problem, this fact shows:

[E] Let rank D = rank B. If $\det(DB) \neq 0$, and the open loop transfer function $D(sI-A)^{-1}B$ has all of its zeroes in the left half plane, then the maximum accuracy can be achieved, i.e., $J_{min} = \text{trace}(Q\tilde{X})$, by the reduced order controller.

It should be noted that the condition $\det(DB) \neq 0$ assures that no pure differentiators appear in the controller. The above fact [E] states that one may obtain the lower bound on the performance for the case where there are as many control input variables as controlled variables. Therefore the following fact is the immediate consequence from the result of Kwakernaak and Sivan.

[F] For the case where rank D < rank B, if there exists a kxn matrix M such that

(i) $\det(DBM) \neq 0$

(ii) $D(sI-A)^{-1}BM$ has minimum phase property then the maximum accuracy can be achieved, i.e., $J_{min} = \text{trace}(Q\tilde{X})$.

This can be obtained only by replacing the control input u(t) with another control Mu(t), the dimension of which is the same as that of controlled variables. The facts described here are closely related to the results of Kwakernaak and Sivan which are derived by a different method for the deterministic problem with the perfect state observation.

To conclude this section, main results developed here are now summarized: The solution to the stochastic singular problem described in the third section is given as follows. Assume that B'QB is non-singular. Then the optimal controller is expressed by

$$u(t) = -B^{\dagger}[\hat{A}(I-BB^{\dagger})T\eta(t) + Ky(t)]$$

$$\dot{\eta}(t) = L_2[\hat{A}(I-BB^{\dagger})T\eta(t) + Ky(t)]$$

where

$$\hat{A} = A - KC$$

$$K = \tilde{X}C'\Sigma_2^{-1}.$$

The matrices T, L_1, L_2 and $B^{\dagger}$ are defined as

$$\hat{B} = [B \mid T]$$

$$\hat{B}^{-1} = \begin{bmatrix} L_1 \\ \hline L_2 \end{bmatrix}$$

$$B^{\dagger} = (B'QB)^{-1}B'(A'L_2'\Pi_{22}L_2+Q)$$

where the matrix T can be found arbitrarily such that the partitioned matrix $\hat{B}$ is non-singular. The (n-k) × (n-k) matrix Π_{22} is the solution of the reduced order matrix Riccati equation (27):

$$\Pi_{22}L_2AFT + T'F'A'L_2'\Pi_{22} + T'QFT$$
$$-\Pi_{22}L_2AB(B'QB)^{-1}B'A'L_2'\Pi_{22} = 0\,.$$

If equation (27) possesses more than one non-negative definite solution, one has to select the solution that makes the closed loop system stable. The associated performance index is given by

$$J^* = \text{trace}\ (Q\tilde{X} + L_2'\Pi_{22}L_2K\Sigma_2K')\,.$$

In particular, when z(t) and u(t) have the same dimensions, the lower bound of the performance can be achieved, i.e.,

$$J^* = \text{trace}\ (Q\tilde{X})$$

provided that the open loop system

$$G(s) = D(sI-A)^{-1}B$$

has minimum phase properties. It should be noticed that, when the matrix B'QB is singular, the optimal controller requires pure differentiators. Such a problem has little practical significance. However if the dimension of u(t) is greater than that of z(t), the problem may be modified by decreasing the number of control inputs.

SINGULAR OBSERVATION PROBLEM

The rest of the paper is devoted to the discussion on the singular observation problem, namely the case where the observation noise covariance Σ_2 vanishes. Such a problem is exactly dual to the singular control problem discussed in previous sections.

Consider a time-invariant linear system described by

$$\dot{x}(t) = Ax(t) + Bu(t) + Gw(t) \tag{48}$$

$$y(t) = Cx(t) \tag{49}$$

$$z(t) = Dx \ . \tag{50}$$

If the system be controlled such that for $R_2 > 0$

$$J = E\{z'(t)R_1z(t) + u'(t)R_2u(t)\} \tag{51}$$

is minimized, then the optimal control law is given by

$$u(t) = - L\hat{x}(t) \tag{52}$$

with

$$L = R_2^{-1}B'P \tag{53}$$

$$A'P + PA - PBR_2^{-1}B'P + Q = 0 \ . \tag{54}$$

The estimate $\hat{x}(t)$ is generated by the reduced order filter of Bryson and Johansen [9]. The condition that this filter does not contain differentiators is $\det(CWC') \neq 0$.

Assume that the covariance matrix is of the form

$$\Sigma_2 = \alpha N \tag{55}$$

with N positive definite and α a positive scalar. Then the limit

$$\bar{X} = \lim_{\alpha \to 0} \tilde{X} \tag{56}$$

exists. Here $\tilde{X}$ is the solution of equation (12). It can be shown that the limit $\bar{X}$ satisfies

$$\bar{X}A'S + SA\bar{X} - \bar{X}A'C'(CWC')^{-1} CA\bar{X} + SW = 0 \tag{57}$$

$$S = I - WC'(CWC')^{-1}C$$

provided that CWC' is non-singular. By reducing in order, the Bryson-Johansen filter can be obtained. Since the matrix C can be assumed to have full rank, it is always possible to find an nxn non-singular matrix $\hat{C}$ partitioned as

$$\hat{C} = \underset{n}{\begin{bmatrix} C \\ \hline H \end{bmatrix}} \begin{matrix} m \\ n-m \end{matrix} . \tag{58}$$

It is convenient to define the following partition

$$\hat{C}^{-1} = \begin{bmatrix} \underset{m}{M_1} & \vdots & \underset{n-m}{M_2} \end{bmatrix} n . \tag{59}$$

By quite a similar way as in the third section, equation (57) reduces to

$$HSAM_2\Gamma_{22} + \Gamma_{22}M_2'A'S'H' + HSWH' - \Gamma_{22}M_2'A'C'(CWC')^{-1}CAM_2\Gamma_{22} = 0 \tag{60}$$

to obtain $\bar{X} = M_2\Gamma_{22}M_2'$. Then the reduced order filter can be expressed as

$$\hat{x}(t) = M_2\mu(t) + C^{\dagger}y(t) \tag{61}$$

$$\dot{\mu}(t) = H(I-C^{\dagger}C)\tilde{A}\left[M_2\mu(t) + C^{\dagger}y(t)\right] \tag{62}$$

where

$$\tilde{A} = A - BL, \qquad \dim(\mu) = n-m \tag{63}$$

$$C^{\dagger} = (M_2\Gamma_{22}M_2'A' + W)C'(CWC')^{-1} . \quad (64)$$

In terms of Laplace transform, the optimal control can be expressed as

$$U(s) = -L(sI-\tilde{A})^{-1}C^{\dagger}\left[C(sI-\tilde{A})^{-1}C^{\dagger}\right]^{-1}Y(s) . \quad (65)$$

The associated optimal performance index is given by

$$J^* = \text{trace}\left[PW + \bar{X}LR_2L'\right] . \quad (66)$$

The closed loop system that results from interconnecting the plant with the controller can be described as

$$\begin{bmatrix} \dot{x} \\ \dot{e} \end{bmatrix} = \begin{bmatrix} \tilde{A} & BLM_2 \\ 0 & H(I-C^{\dagger}C)AM_2 \end{bmatrix}\begin{bmatrix} x \\ e \end{bmatrix} + \begin{bmatrix} G \\ H(I-C^{\dagger}C)G \end{bmatrix} w .$$

The characteristic polynomial is given by

$$\Delta(s) = \det(sI-\tilde{A})\det\left[sI-H(I-C^{\dagger}C)AM_2\right] . \quad (67)$$

Hence the stability of the closed loop system depends on the zeroes of

$$\psi(s) = \det\left[sI-H(I-C^{\dagger}C)AM_2\right] = 0 . \quad (68)$$

It can be easily shown that $\psi(s)$ is the numerator polynomial of the transfer function matrix

$$C(sI-A)^{-1}C^{\dagger} .$$

Consequently the following facts can be derived:

[G] The closed loop system is stable, if the transfer function matrix $C(sI-A)^{-1}C^{\dagger}$ has all of its zeroes with negative real parts.

[H] If rank C < rank G, then $J_{min} >$ trace (PW) .

[I] Let rank C = rank G. Then $J_{min} =$ trace (PW) if

(i) $\det(CG) \neq 0$

(ii) $C(sI-A)^{-1}G$ is minimum phase.

[J] Let rank C > rank G. If there exists a matrix Ω such that

(i) $\det(\Omega CG) \neq 0$

(ii) $\Omega C(sI-A)^{-1}G$ is minimum phase, then J_{min} = trace (PW).

It turns out that the singular observation problem is dual with the singular control problem. To check this more concretely, let

$$A \to A', \quad B \to C', \quad C \to B', \quad W \to Q$$

It is easily observed that

$$M_1 \to L_1', \quad M_2 \to L_2', \quad H \to T', \quad C \to B'$$

$$\Pi_{22} \to \Gamma_{22}, \quad C^{\dagger} \to (B^{\dagger})', \quad \tilde{A} \to A' .$$

ILLUSTRATIVE EXAMPLE

Now as a simple example to illustrate the general procedure for the stochastic singular control problem, consider the time-invariant second order linear system

$$\dot{x}(t) = \begin{bmatrix} 0 & 1 \\ 0 & -1 \end{bmatrix} x(t) + Bu(t) + \begin{bmatrix} 0 \\ 1 \end{bmatrix} w(t)$$

$$y(t) = \begin{bmatrix} 1 & 0 \end{bmatrix} x(t) + v(t)$$

$$z(t) = Dx(t)$$

where

$$E\{w(t)w(\tau)\} = \sigma_w^2 \delta(t-\tau)$$

$$E\{v(t)v(\tau)\} = \sigma_v^2 \delta(t-\tau) .$$

The performance index of interest is

$$J = E\left\{z'(t) \begin{bmatrix} 1 & 0 \\ 0 & 1 \end{bmatrix} z(t)\right\} .$$

For the present case, the error covariance is given by

$$\tilde{X} = \sigma_v^2 \begin{bmatrix} -1+\sqrt{1+2\gamma} & 1+\gamma-\sqrt{1+2\gamma} \\ 1+\gamma-\sqrt{1+2\gamma} & -1-2\gamma+(1+\gamma)\sqrt{1+2\gamma} \end{bmatrix}$$

where

$$\gamma = \sigma_w/\sigma_v .$$

It follows that the optimal filter gain is given by

$$K = \begin{bmatrix} k_1 \\ k_2 \end{bmatrix} = \begin{bmatrix} -1+\sqrt{1+2\gamma} \\ 1+\gamma-\sqrt{1+2\gamma} \end{bmatrix} .$$

[Example 1]: Consider the case where

$$B = \begin{bmatrix} 0 \\ 1 \end{bmatrix} \qquad D = \begin{bmatrix} 1 & 0 \\ 0 & 1 \end{bmatrix} .$$

The matrices T, L_1 and L_2 can be taken as the simplest form

$$T = \begin{bmatrix} 1 \\ 0 \end{bmatrix} , \quad L_1 = \begin{bmatrix} 0 & 1 \end{bmatrix} , \quad L_2 = \begin{bmatrix} 1 & 0 \end{bmatrix} .$$

The reduced order Riccati Equation (27) for Π_{22} takes the form

$$-\Pi_{22}^2 + 1 = 0$$

yielding $\Pi_{22} = 1$. Thus from equation (30)

$$B^{\dagger} = \begin{bmatrix} 1 & 1 \end{bmatrix} .$$

Hence the optimal controller can be determined as

$$U(s) = - G_c(s)Y(s)$$

with

$$G_c(s) = \frac{s+1}{s + \sqrt{1+2\gamma}} .$$

The associated performance index is given by

$$J_{min} = \sigma_v^2 \gamma\sqrt{1+2\gamma} .$$

[Example 2]: Consider the case where

$$B = \begin{bmatrix} 1 \\ 1 \end{bmatrix} \qquad D = \begin{bmatrix} 1 & 0 \end{bmatrix} .$$

Since rank B = rank D = 1, det(DB) $\neq$ 0 and

$$D(sI-A)^{-1}B = \frac{s + 2}{s(s + 1)}$$

is minimum phase, then the maximum accuracy can be achieved. The optimal controller is found to be

$$G_c(s) = \frac{k_1 s+k_1+k_2}{s + 2} .$$

The performance index becomes

$$J_{min} = (\sqrt{1+2\gamma} - 1)\sigma_v{}^2 .$$

[Example 3]: Consider the case where

$$B = \begin{bmatrix} 0 \\ 1 \end{bmatrix} \qquad D = \begin{bmatrix} 1 & 0 \end{bmatrix} .$$

Obviously rank B = rank D, and

$$D(sI-A)^{-1}B = \frac{1}{s(s+1)}$$

is minimum phase. But det(DB) = 0, which indicates that the controller requires the pure differentiator. In fact, the optimal controller to achieve the maximum accuracy becomes

$$G_c(s) = k_1 s + k_1 + k_2 .$$

[Example 4]: Consider the case where

$$B = \begin{bmatrix} -1 \\ 2 \end{bmatrix} \qquad D = \begin{bmatrix} 1 & 0 \end{bmatrix} .$$

In this case, the open loop system with the transfer function

$$D(sI-A)^{-1}B = \frac{s-1}{s(s+1)}$$

is non-minimum phase, while rank B = rank D and det(DB) $\neq$ 0. This means that it is impossible to achieve the maximum accuracy. By the procedure given in this paper however, the optimal controller is obtained as

$$G_c(s) = \frac{\gamma(s+1)}{s + \gamma + \sqrt{1+2}} .$$

CONCLUSION

An optimal stochastic control problem has been discussed for the singular case where the control input weighting matrix is the zero matrix. The optimal control law can be realized by a dynamical linear system which is of lower order than the controlled plant. The present technique requires to find only (n-k)x(n-k) matrix as the solution of Riccati equation. Through the stability analysis for the closed loop system, the conditions under which the lower bound on the performance is achieved are established. The maximum accuracy is obtained when the control input variables are as many as the controlled variables. The extra requirement is the minimum phase property of the open loop transfer function which ensures that the closed loop system is stable. These explicit conditions for the maximum accuracy are closely related to the results of Kwakernaak and Sivan for the deterministic problem with the perfect state observation. We have also briefly discussed the singular observation problem which turns out to be dual with the singular control problem. We have directed our attention to the totally singular cases in which R_2 or Σ_2 is zero matrix, but the results presented here can also be easily extended to the partially singular case, namely the case where R_2 or Σ_2 is not zero but singular matrix.

REFERENCES

1 B. Friedland, Limiting form of optimal stochastic linear systems with noisy observation. ASME J.Dyn. Sys., Meas. and Contr. 93, 134 (1971).

2 T. Nakamizo and M. Oshiro, Minimum mean square error control of linear stochastic systems with noisy observation. IFAC Symp. on Stoch. Contr., Budapest (1973).

3 S. Butman, A method for optimizing control-free costs in systems with linear controllers. IEEE Trans. Autom. Contr. 13, 554 (1968).

4 Y.C. Ho, Linear stochastic singular control problems. J. Opt. Theory and Appl. 9, 24 (1972).

5 H. Kwakernaak, Optimal low sensitivity linear feedback systems. Automatica 5, 279 (1969).

6 W.H. Wonham, On the separation theorem of stochastic control. SIAM J. Control 6, 312 (1968).

7 H. Kwakernaak and R. Sivan, Linear Optimal Control Systems. Wiley-Interscience (1972).

8 H. Kwakernaak and R. Sivan, The maximally achievable accuracy of linear optimal regulators and linear optimal filters. IEEE Trans. Autom. Contr. 17, 79 (1972)

9 A.E. Bryson and D.E. Johansen, Linear filtering for time-varying systems using measurements containing colored noise. IEEE Trans. Autom. Contr. 10, 4 (1965).

Professor T. Nakamizo, Department of Mechanical Engineering, National Defense Academy, Yokosuka, Japan.

Dr. M. Oshiro, Department of Mechanical Engineering, National Defense Academy, Yokosuka, Japan.

DISCUSSION

In the first part of the discussion (with Drenick, Willems) it was emphasized that the stability of the closed loop system depends not on the poles but on the zeroes of the transfer function of the open loop system (results [C], [D]). More exactly the open-loop transfer function for the present case is is expressed by

$$G_p(s) = B^{\dagger} (sI - A)^{-1}B$$

which is referred to as the modified open-loop transfer function in this paper. For the special case where rank D = rank B and det(DB) $\neq$ 0, the poles of the closed loop transfer function corresponds exactly to the zeroes of the truly open loop transfer function $D(sI - A)^{-1}B$.

In a comment Professor Willems pointed out that the covariance matrix Σ_1 does not affect the solution of the problem. Thus, since only steady state errors are considered, all stabilizing feedback will give zero steady state error if the covariance matrix Σ_2 becomes O, i.e. if one has undisturbed observations. In the singular observation problem, if we consider the particular case where the weighting matrix R_1 is equal to zero, our problem may be interpreted as a stabilization problem, and there seems to exist an infinite number of optimal solutions such that the closed loop system is stable. However since the state x(t) does not affect the performance index J, the trivial solution u(t) = O is still optimal, but in general this will be impractical for the case when the matrix A is not a stable matrix. Considering the steady state version of the problem formulation and solution requires the stability of the resultant system. Consequently R_1 = O makes the problem degenerated, and R_1 should be selected to ensure that every non-zero trajectory produces non-zero cost.

A final question (Hammond) dealt with the possibility of getting the solution of the singular control problem using some reduced order Luenberger observer with the singular observation problem and dualizing the results. Professor Nakamizo explained that he had just intended to work out this duality between the two singular problems by means of the approach used in the paper.

Y SUNAHARA, T ASAKURA and Y MORITA

On the asymptotic behaviour of nonlinear stochastic dynamical systems considering the initial states

INTRODUCTION

Many dynamical systems always exhibit various kinds of nonlinearities and undergo random changes with time and environment and may result in unwanted vibrations. For example, measurements of ship motion or an aircraft flying through turbulent air reveal that motions with random perturbation may be discussed only from stochastic viewpoints [1].

For linear stochastic systems, the procedure for examining the system response including the examination of system stability has become well known and it has received application to numerous engineering problems.

A current problem of great importance in the theory of stochastic dynamical systems is the asymptotic behaviour of nonlinear stochastic systems dependent on their initial states which is an inherent characteristic due to the existence of nonlinearities.

In general, there seems to be a fundamental question regarding the asymptotic behaviour mentioned above, because there are as yet no convincing evidence that the more sophisticated approaches provide useful improvements relative to the widely applied methods that are well reviewed by Kozin [2].

Consequently, this paper is concerned with developing a realizable approach to solve stochastical asymptotic stability for nonlinear systems with (1) a random parameter modelled by white Gaussian random process and (2) two random parameters modelled by a white Gaussian and a finite state Markov chain processes respectively.

The process obtained by measurements is subjected to disturbance due to many individually negligible causes. The classical central limit theorem shows us that such disturbance can be considered as a Gaussian random variable with zero mean and variance chosen properly. On the other hand, for a random parameter modelled by a Markov chain process, it cannot be regarded as a diffusion process but as a pure jump process. It is, thus, reasonable to treat dynamical systems with random perturbations, both acting continuously in time and involving random shocks at random moments of time.

Although stability analysis has been investigated by many investigators [3]-[6], a stochastic Lyapunov function approach to explore the asymptotic stability is demonstrated in the following two sections, taking into account the influence of the initial conditions. In later sections, a general class of nonlinear dynamical systems with two random parameters is considered. For the purpose of examining the asymptotic behaviour, the concept of random evolution [7]-[12] is introduced. The final section is devoted to demonstrate illustrative examples.

Figure 1 illustrates the analytical procedure which will be developed in the sequel. As shown in Figure 1, we shall consider two kinds of dynamical systems actually encountered in the practice: one is a system whose random parameter is considered to be a white Gaussian process $\dot{\xi}(t)$ and the other a system with two parameters modelled respectively by a white Gaussian process and a Markov chain process $\alpha(t)$ with the finite state. In the former case, for the purpose of finding the differential generator of the r(t)-process, the averaging principle by Khasminskii is applied to the Kolmogorov equation derived by using the polar coordinate transformation. On the other hand, in the latter case, the concept of random evolutions is introduced instead of the averaging principle because of the existence of the parameter of Markov chain type. The system stability is finally examined by using the differential generator in terms of stochastic behaviours of the r(t)-process.

DIFFERENTIAL GENERATOR ASSOCIATED WITH BASIC EQUATION

We shall consider a nonlinear dynamical system modelled by

$$\ddot{x} + \omega^2 x + \varepsilon g(x,\dot{x}) = \delta h(x,\dot{x})\dot{\xi}(t) \tag{1}$$

with the given initial values $x(0) = x_0$ and $\dot{x}(0) = \dot{x}_0$, where ε and δ are small constants, g and h nonlinear functions respectively and $\dot{\xi}(t)$ a white Gaussian noise and where "." expresses the differentiation with respect to time t. Equation (1) may be considered as a generalization of mathematical models of dynamical systems such that the system is lightly damped, weakly nonlinear and that the system response is related to a random excitation with relatively small magnitude [13],[14].

With $x = x_1$, $\dot{x} = x_2$, Equation (1) is expressed by the following stochastic differential equation of Itô-type [15],

$$dx_1 = x_2 dt \tag{2a}$$

$$dx_2 = -\{\omega^2 x_1 + \varepsilon g(x_1,x_2)\}\ dt + \delta h(x_1,x_2)\ dw(t) \tag{2b}$$

where the w(t)-process is the Brownian motion process related to the $\dot{\xi}(t)$-process with relation [16],

$$w(t) = \int_0^t \dot{\xi}(s)ds \tag{3}$$

Furthermore, the following properties are well-known:

$$E_b\ \{dw(t)\} = 0,\ E_b\ \{(dw(t))^2\} = \sigma^2 dt \tag{4}$$

where σ is a constant and the symbol E_b denotes the mathematical expectation.

It can easily be expected that the two-dimensional dynamical system given by (2) is converted into the one-dimensional system along the relation,

$$x_1 = \frac{r}{\omega}\sin\theta, \qquad x_2 = r\cos\theta. \tag{5}$$

Naturally, the converted one-dimensional process r(t) is

$$r^2(t) = \omega^2 x_1^{\,2} + x_2^{\,2} \tag{6}$$

After somewhat tedious calculations using the averaging principle [17], it may be found that the r(t)-process is Markovian with the differential generator (see Appendix A),

$$L_r = U^2(r)\,\frac{d^2}{dr^2} + V(r)\,\frac{d}{dr}\ , \tag{7}$$

where

$$U^2(r) = \frac{\sigma^2}{4\pi}\int_0^{2\pi} h^2\left(\frac{r}{\omega}\sin\theta,\ r\cos\theta\right)\cos^2\theta\,d\theta \tag{8}$$

and

$$V(r) = \frac{\sigma^2}{4\pi}\int_0^{2\pi} h^2\left(\frac{r}{\omega}\sin\theta, r\cos\theta\right)\frac{\sin^2\theta}{r}\,d\theta - \frac{1}{2\pi}\int_0^{2\pi} g\left(\frac{r}{\omega}\sin\theta, r\cos\theta\right)\cos\theta\,d\theta. \tag{9}$$

A NEW LYAPUNOV FUNCTION

We need the following lemma associated with the asymptotic stability criteria of the r(t)-process.

<u>Lemma</u> For a fixed m, assume the following conditions (A.1) to (A.3):

(A.1) $W_L(r)$ is a non-negative and continuous in the open set $Q_m \triangleq \{r;\ W_L(r) < m\}$.

(A.2) r(t) is a right continuous strong Markov process with the weak infinitesimal operator $\tilde{A}_m$ defined in Q_m.

(A.3) $\tilde{A}_m W_L(r) = -\,k(r) \le 0$.

Letting $R_m = Q_m \cap \{r;\ k(r) = 0\}$, then

$$P_0\left\{\lim_{t\to\infty} r(t)\ \varepsilon R_m\right\} \ge 1 - \frac{W_L(r_0)}{m}\ , \tag{10}$$

where $P_0\{\cdot\}$ is the probability of $\{\cdot\}$, provided that the r(t)-process starts with $r(0) = r_0$ at the initial time t = 0.

Furthermore, assume that the assumptions (A.1) to (A.3) hold for $m > 0$. Letting $R \triangleq \bigcup_{m=1}^{\infty} R_m$, then

$$P_0 \{\lim_{t\to\infty} r(t) \varepsilon R\} = 1 \tag{11}$$

Since the proof is straightforward by using the supermartingale property of $W_L(r)$, we shall omit to write here.

The following theorem gives sufficient conditions of the asymptotic stability with probability one.

<u>Theorem 1</u>

For an arbitrarily fixed initial value $r(0) = r_0$, assume that the coefficients of L_r in(7) satisfy the following conditions:

(C.1) $U^2(r) = 0$ if and only if $r = 0$, and $V(0) = 0$.

(C.2) $\lim_{r\to 0} \left| \frac{V(r)}{U^2(r)} \right| \exp\{ \int_r^{r_0} \frac{V(\zeta)}{U^2(\zeta)} d\zeta < \infty$.

(C.3) $\int_{r_0}^{\infty} \frac{V(\zeta)}{U^2(\zeta)} d\zeta < \infty$.

Then, for any initial value $r_0 \; \varepsilon [0, \infty)$, the following equality holds:

$$P_0 \{\lim_{t\to\infty} r(t) = 0 \} = 1 . \tag{12}$$

<u>Proof</u>: Let $\Psi(r)$ be an arbitrary positive smooth function such that

$$\int_0^{\infty} \Psi(r) dr < \infty . \tag{13}$$

Define $W_L(r)$ by

$$W_L(r) \triangleq \int_0^r \exp\{\int_\eta^{r_0} (\frac{V(\zeta)}{U^2(\zeta)} + \Psi(\zeta)) d\zeta \} d\eta . \tag{14}$$

From (14), it follows that

$$\frac{dW_L(r)}{dr} = \exp\{\int_r^{r_0} (\frac{V(\zeta)}{U^2(\zeta)} + \Psi(\zeta)) d\zeta \} \tag{15}$$

and

$$\frac{d^2W_L(r)}{dr^2} = -\left[\frac{V(r)}{U^2(r)} + \Psi(r)\right] \exp\left\{\int_r^{r_0}\left(\frac{V(\zeta)}{U^2(\zeta)} + \Psi(\zeta)\right) d\zeta\right\}. \tag{16}$$

Hence, it can easily be examined that

$$L_r W_L(r) = -U^2(r)\Psi(r)\exp\left\{\int_r^{r_0}\left(\frac{V(\zeta)}{U^2(\zeta)} + \Psi(\zeta)\right) d\zeta\right\} \leq 0. \tag{17}$$

The assumptions (A.1) to (A.3) are thus satisfied and $W_L(r)$ becomes a Lyapunov function, if the following conditions (C.i) and (C.ii) are satisfied:

(C.i) $W_L(r) \to \infty$ as $r \to \infty$.

(C.ii) For any bounded r,

$$W_L(r) < \infty, \quad \frac{dW_L(r)}{dr} < \infty, \quad \left|\frac{d^2W_L(r)}{dr^2}\right| < \infty.$$

With the definition (14), the first condition (C.i) is

$$\lim_{r\to\infty} \exp\left\{\int_r^{r_0}\left(\frac{V(\zeta)}{U^2(\zeta)} + \Psi(\zeta)\right) d\zeta\right\} \nrightarrow 0 \tag{18}$$

or equivalently

$$\int_{r_0}^{\infty}\left(\frac{V(\zeta)}{U^2(\zeta)} + \Psi(\zeta)\right) d\zeta < \infty . \tag{19}$$

From (C.3), (13) and (19) the condition (C.i) holds.

Since it is apparent that, if $|d^2W_L(r)/dr^2| < \infty$, then $W_L(r) < \infty$ and $dW_L(r)/dr < \infty$, it is sufficient to show that the third inequality of the condition (C.ii) holds. From (C.1), the origin r=0 is the only one singular point. Consequently, with the conditions (C.1) and (C.2) and the inequality (13) we have,

$$\lim_{r\to\infty}\left|\frac{V(r)}{U^2(r)} + \Psi(r)\right| \exp\left\{\int_r^{r_0}\left(\frac{V(\zeta)}{U^2(\zeta)} + \Psi(\zeta)\right) d\zeta\right\} < \infty, \tag{20}$$

from which the condition (C.ii) holds. Thus, the proof has been completed.

We shall proceed to state sufficient conditions of the asymptotic stability with the probability appraisal.

Theorem 2

Assume that the following condition holds together with the conditions (C.1) and (C.2).

(C.4) There exists a positive constant M such that, for any $r \varepsilon (0, M)$, the drift term V(r) given by (9) is negative.

Then we have

$$P_0 \{ \lim_{t\to\infty} r(t) = 0 \} \geq 1 - \frac{W_L(r_0)}{W_L(M)} \quad . \tag{21}$$

Proof: Define

$$Q_m' \stackrel{\Delta}{=} \{r;\ r \varepsilon [0, M)\} \equiv \{r;\ W_L(r) < m\}, \tag{22}$$

where $m = W_L(M)$. Furthermore, for $r \varepsilon Q_m'$, let $\Phi(r)$ be an arbitrary smooth and positive function related to the function $\Psi(r)$ by

$$\Psi(r) \stackrel{\Delta}{=} - V(r)\Phi(r), \tag{23}$$

where

$$\int_0^M \{- V(r)\Phi(r)\}dr < \infty \ . \tag{24}$$

From (14) and (17) the Lyapunov function $W_L(r)$ and its differential generator are respectively expressed by

$$W_L(r) = \int_0^r \exp\{ \int_\eta^{r_0} (\frac{V(\zeta)}{U^2(\zeta)} - V(\zeta)\Phi(\zeta))\ d\zeta \}\ d\eta \tag{25}$$

and

$$L_r W_L(r) = V(r)U^2(r)\Phi(r) \exp\{ \int_r^{r_0} (\frac{V(\zeta)}{U^2(\zeta)} - V(\zeta)\Phi(\zeta))\ d\zeta \} \leq 0 \tag{26}$$

for $r, r_0 \varepsilon Q_m'$.

We shall examine the conditions (C.i) and (C.ii) in the proof of Theorem 1. By the conditions (C.1) and (C.2), it is obvious that the condition (C.ii) holds. Furthermore, it is a direct consequence from (25) that $W_L(r)$ is monotone increasing with respect to r in Q_m'. Consequently, using the inequality (10), the asymptotic stability is concluded with appraisal $1 - W_L(r_0)/W_L(M)$.

EXTENSION TO DYNAMICAL SYSTEMS WITH RANDOM COEFFICIENTS

In this section, an extension of the results obtained in the previous section is demonstrated to the dynamical system modelled by

$$\ddot{x} + \omega^2 x + \varepsilon g(x,\dot{x}, \alpha(t)) = \delta h(x,\dot{x})\dot{\xi}(t) \tag{27}$$

with the given initial conditions, $x(0) = x_0$ and $\dot{x}(0) = \dot{x}_0$, where $\alpha(t)$ is a parametric noise process expressed mathematically by an ergodic Markov chain with finite n-stages and $(0) = \alpha_i$ $(i=1,2,\ldots,n)$. Equation (27) may be considered as a mathematical model of a class of lightly damped nonlinear dynamical systems excited by a random input, whose parameter changes with time taking n modes according to a continuous-time Markov chain with the infinitesimal generator Q.

The stochastic differential equation of Itô-type associated with Equation (27) is easily derived as

$$dx_1 = x_2 dt \tag{28a}$$

$$dx_2 = - \{\omega^2 x_1 + \varepsilon g(x_1,x_2,\alpha(t))\}\, dt + \sqrt{\varepsilon} h(x_1,x_2)\, dw(t) \tag{28b}$$

where, for convenience of discussions, we set as $\delta = \sqrt{\varepsilon}$.

Noting that a Markov chain process may be regarded as a special class of Poisson processes, it can easily be understood that the joint process $(x_1,x_2,\alpha(t))$ is a pair Markov process perturbed randomly by both the Brownian motion and Poisson processes [18], [19]. Hence, defining the probability density of a transition from the state $X_i = (x_1, x_2, \alpha_i)$ to another state $Y_j = (y_1,y_2,\alpha_j)$ by $p_i = p_i(X_i;t;Y_j)$, the probability density p_i for the fixed Y_j satisfies

$$\frac{\partial p_i}{\partial t} = - x_2 \frac{\partial p_i}{\partial x_1} - \{\omega^2 x_1 + \varepsilon g(x_1,x_2,\alpha_i)\} \frac{\partial p_i}{\partial x_2} + \frac{\varepsilon^2}{2^\sigma} h^2(x_1,x_2) \frac{\partial^2 p_i}{\partial x_2^2} + \sum_{k=1}^{n} q_{ik}\, p_k \tag{29}$$

with the initial condition $p_i(X_i;0;Y_j) = \delta_{ij}\delta(x_1-y_1,x_2-y_2)$, where q_{ik} is the (i,k)th element of an $n \times n$ matrix Q, δ_{ij} the Kronecker delta and δ the Dirac delta function. It is well-known that

$$\lim_{\Delta t \to 0} \Pr \{\alpha(t + \Delta t) = \alpha_k | \alpha(t) = \alpha_i\} = q_{ik}\Delta t + o(\Delta t) \tag{30a}$$

and

$$\lim_{\Delta t \to 0} \Pr \{\alpha(t + \Delta t) = \alpha_i | \alpha(t) = \alpha_i\} = 1 + q_{ii}\Delta t + o(\Delta t) \tag{30b}$$

where $q_{ik} \geq 0$ for $i \neq k$, $q_{ii} \leq 0$ and $\sum_{k=1}^{n} q_{ik} = 0$ for $i=1,2,\ldots,n$.

We shall convert the $(x_1,x_2,\alpha(t))$-process into the $(r,\theta,\alpha(t))$-process along the relation,

$$x_1 = \frac{r}{\omega} \sin \theta, \qquad x_2 = r \cos \theta \tag{31a}$$

and

$$r^2 = \omega^2 x_1^{\,2} + x_2^{\,2} \quad . \tag{31b}$$

Noting that the zero solution $x_1 = x_2 = 0$ to Equation (27) implies $r=0$, which is a reflecting barrier, the $r(t)$-process may be considered within the semi-interval $r \varepsilon [0,\infty)$. With the relation (31), we write $v(r,\theta,\alpha_i;t;r_1,\theta_1,\alpha_j)$ for $p_i(x_1,x_2,\alpha_i;t;y_1,y_2,\alpha_j)$ and abbreviate it by $v_i(r,\theta,t)$ for a set of fixed values, r_1,θ_1 and α_j. Letting $\theta = \phi - \omega t$, then, after somewhat tedious calculations, we have

$$\begin{aligned}\frac{\partial v_i}{\partial t} = \; & \varepsilon\{\frac{\sigma^2}{2} h^2 \Big[\cos^2(\phi-\omega t) \frac{\partial^2 v_i}{\partial r^2} - \frac{\sin 2(\phi-\omega t)}{r} \frac{\partial^2 v_i}{\partial r \partial \phi} \\ & + \frac{\sin^2 (\phi-\omega t)}{r^2} \frac{\partial^2 v_i}{\partial \phi^2} + \frac{\sin 2(\phi-\omega t)}{r^2} \frac{\partial v_i}{\partial \phi} \\ & + \frac{\sin^2 (\phi-\omega t)}{r} \frac{\partial v_i}{\partial r} \Big] + g(\alpha_i) \frac{\sin (\phi-\omega t)}{r} \frac{\partial v_i}{\partial \phi} \\ & - g(\alpha_i) \cos (\phi-\omega t) \frac{\partial v_i}{\partial r}\} + \sum_{k=1}^{n} q_{ik} v_k, \quad (i=1,2,\ldots,n)\end{aligned} \tag{32}$$

with the initial condition $v_i(r,\phi,0) = \delta_{ij} (r-r_1,\phi-\phi_1)$, where $g(\alpha_i) = g\{r \sin(\phi-\omega t)/\omega, r \cos(\phi-\omega t), \alpha_i\}$ and $\phi_1 = \theta_1$. The limiting behaviour of $v_i(t)$ is investigated by tending ε to zero and t to infinitive under the condition that εt is constant. To do this, changing the time scale t for τ and writing $v_i^{(\varepsilon)}(\tau)$ for $v_i(\tau/\varepsilon)$, where $\tau = \varepsilon t$, Equation (32) is written by

$$\begin{aligned}\frac{\partial v_i^{(\varepsilon)}}{\partial \tau} = & \frac{\sigma^2}{2} h^2 \Big[\cos^2 (\phi - \frac{\omega\tau}{\varepsilon}) \frac{\partial^2 v_i^{(\varepsilon)}}{\partial r^2} - \frac{\sin 2(\phi - \frac{\omega\tau}{\varepsilon})}{r} \frac{\partial^2 v_i^{(\varepsilon)}}{\partial r \partial \phi} \\ & + \frac{\sin^2(\phi - \frac{\omega\tau}{\varepsilon})}{r^2} \frac{\partial^2 v_i^{(\varepsilon)}}{\partial \phi^2} + \frac{\sin 2(\phi - \frac{\omega\tau}{\varepsilon})}{r^2} \frac{\partial v_i^{(\varepsilon)}}{\partial \phi} \\ & + \frac{\sin^2(\phi - \frac{\omega\tau}{\varepsilon})}{r} \frac{\partial v_i^{(\varepsilon)}}{\partial r}\Big] + g(\alpha_i) \frac{\sin (\phi - \frac{\omega\tau}{\varepsilon})}{r} \frac{\partial v_i^{(\varepsilon)}}{\partial \phi} \\ & - g(\alpha_i) \cos (\phi - \frac{\omega\tau}{\varepsilon}) \frac{\partial v_i^{(\varepsilon)}}{\partial r} \\ & + \frac{1}{\varepsilon} \sum_{k=1}^{n} q_{ik} v_k^{(\varepsilon)}, \qquad (i = 1,2,\dots, n). \end{aligned} \tag{33}$$

Let $A_i(r,\phi - \frac{\omega\tau}{\varepsilon})$ be the differential generator of the right hand side in Equation (33). Equation (33) is written by

$$\frac{\partial v_i^{(\varepsilon)}}{\partial \tau} = A_i v_i^{(\varepsilon)} + \frac{1}{\varepsilon} \sum_{k=1}^{n} q_{ik} v_k^{(\varepsilon)}, \quad (i = 1,2,\dots, n). \tag{34}$$

Assume that the value of ε is sufficiently small and, for any fixed $\tau > 0$, define $L_{\alpha i}$ by

$$L_{\alpha i} \triangleq \lim_{T\to\infty} \frac{1}{T} \int_0^T A_i\ (r,\phi - \frac{\omega\tau}{\varepsilon})\ d(\frac{\tau}{\varepsilon}) \tag{35}$$

Noting that $A_i\ (r,\phi - \frac{\omega\tau}{\varepsilon})$ is periodic with respect to $\phi - (\omega\tau/\varepsilon) = \psi$, then, from (35), $L_{\alpha i}$ is computed to be

$$\begin{aligned} L_{\alpha i} = & \frac{1}{2\pi} \int_0^{2\pi} A_i(r,\psi) d\psi = \frac{\sigma^2}{4\pi} \Big[(\int_0^{2\pi} h^2 \cos^2\psi d\psi) \frac{\partial^2}{\partial r^2} \\ & - (\int_0^{2\pi} h^2 \frac{\sin 2\psi}{r} d\psi) \frac{\partial^2}{\partial r \partial \phi} + (\int_0^{2\pi} h^2 \frac{\sin^2\psi}{r^2} d\psi) \frac{\partial^2}{\partial \phi^2} \\ & + (\int_0^{2\pi} h^2 \frac{\sin 2\psi}{r^2} d\psi) \frac{\partial}{\partial \phi} + (\int_0^{2\pi} h^2 \frac{\sin^2\psi}{r} d\psi) \frac{\partial}{\partial r} \Big] \end{aligned}$$

$$+ \left(\frac{1}{2\pi}\int_0^{2\pi} g(\alpha_i)\,\frac{\sin\psi}{r}\,d\psi\right)\frac{\partial}{\partial\phi} - \left(\frac{1}{2\pi}\int_0^{2\pi} g(\alpha_i)\,\cos\psi\,d\psi\right)\frac{\partial}{\partial r}\,. \tag{36}$$

Although Equation (36) plays a basic role to explore stochastic behaviours of the nonlinear dynamical system (27), our attention is focussed on the asymptotic aspect of Equation (28) with an averaged differential generator rather than $v_i^{(\varepsilon)}$ themselves.

ASYMPTOTIC THEOREMS FOR RANDOM EVOLUTIONS

Under an assumption that a random evolution $M^{(\varepsilon)}(\sigma,\tau)$ exists almost surely, we shall now define $M^{(\varepsilon)}(\sigma,\tau)$ as the unique solution of the stochastic initial value problem, (for more details, see Appendix B).

$$\frac{\partial M^{(\varepsilon)}(\sigma,\tau)}{\partial\tau} = M^{(\varepsilon)}(\sigma,\tau)\; L_{\alpha(\tau/\varepsilon)}, \quad \sigma < \tau\,, \tag{37a}$$

$$M^{(\varepsilon)}(\sigma,\sigma) = I\,, \tag{37b}$$

where $L_{\alpha i}$ is the generator of a semigroup of operators on a Banach space $\mathcal{B}$. We need the following theorem.

<u>Theorem 3</u> (Feynman-Kac) [9]

Let f be a $\mathcal{B}$-valued function. If

$$u_i^{(\varepsilon)}(\tau) = E_i\,\{M^{(\varepsilon)}(0,\tau)\; f\,[\alpha(\tfrac{\tau}{\varepsilon})]\} \tag{38}$$

exists as a strongly continuous $\mathcal{B}$-valued function of τ, then the function $u_i^{(\varepsilon)}(\tau)$ satisfies the following deterministic abstract Cauchy problem.

$$\frac{du_i^{(\varepsilon)}(\tau)}{d\tau} = L_{\alpha i}\, u_i^{(\varepsilon)} + \frac{1}{\varepsilon}\sum_{j=1}^{n} q_{ij}\, u_j^{(\varepsilon)}\,, \tag{39a}$$

$$u_i^{(\varepsilon)}(0) = f\,, \tag{39b}$$

where E_i denotes the mathematical expectation with respect to the probability measure generated by $\alpha(0) = \alpha_i$. It should be noted that Equation (34) coincides with Equation (39a) and that, in the sequel, the function $v_i^{(\varepsilon)}(\tau)$ may be expressed by $u_i^{(\varepsilon)}(\tau)$.

In what follows, we shall investigate the asymptotic behaviour of $u_i^{(\varepsilon)}$ as ε tends to zero. To do this, let $\bar{p}_i$ be

$$\bar{p}_i = \lim_{t\to\infty} \Pr\{\alpha(t) = \alpha_i\} \tag{40}$$

for any $\alpha(0) = \alpha_j$, where $\Sigma_i \bar{p}_i = 1$. Furthermore, define

$$\bar{\alpha} = \sum_{i=1}^{n} \bar{p}_i \ \alpha_i \tag{41}$$

and

$$\bar{L} = \sum_{i=1}^{n} \bar{p}_i \ L_{\alpha i} \ . \tag{42}$$

We are now ready to describe the following theorem.

Theorem 4

If $\bar{L}$ generates a strongly continuous semigroup on $\mathcal{B}$, $\mathcal{D}$ is dense in $\mathcal{B}$ and Q is ergodic, then

$$\lim_{\varepsilon\to 0} E_i \{M^{(\varepsilon)}(\ ,\tau)\} = \exp(\tau\bar{L}), \tag{43}$$

where $\mathcal{D} = \bigcap_{i,j} \{$ domain of $L_{\alpha i} \ L_{\alpha j}\}$.

Since the essential portion of the proof of Theorem 4 follows that of Theorem 1 in Reference [9], the whole aspect of the proof is included in Appendix C.

From Theorem 4, it may be concluded that the solution $u_i^{(\varepsilon)}(\tau)$ to Equation (39a) converges to $u^{(o)}(\tau)$ satisfying

$$\frac{\partial u^{(o)}(\tau)}{\partial \tau} = \bar{L}u^{(o)}(\tau) \tag{44a}$$

$$u^{(o)}(0) = \sum_{i=1}^{n} \bar{p}_i \; f_i \; , \tag{44b}$$

where, from (36), we obtain

$$\begin{aligned}\bar{L} = & \left(\frac{\sigma^2}{4\pi} \int_0^{2\pi} h^2 \cos^2 \psi d\psi\right) \frac{\sigma^2}{\partial r^2} - \left(\frac{\sigma^2}{4\pi} \int_0^{2\pi} h^2 \frac{\sin 2\psi}{r} d\psi\right) \frac{\partial^2}{\partial r \partial \phi} \\ & + \left(\frac{\sigma^2}{4\pi} \int_0^{2\pi} h^2 \frac{\sin^2 \psi}{r^2} d\psi\right) \frac{\partial^2}{\partial \phi^2} + \left(\frac{\sigma^2}{4\pi} \int_0^{2\pi} h^2 \frac{\sin 2\psi}{r^2} d\psi \right. \\ & + \frac{1}{2\pi} \int_0^{2\pi} g(\bar{\alpha}) \frac{\sin \psi}{r} d\psi) \frac{\partial}{\partial \phi} + \left(\frac{\sigma^2}{4\pi} \int_0^{2\pi} h^2 \frac{\sin^2 \psi}{r} d\psi \right. \\ & - \left.\frac{1}{2\pi} \int_0^{2\pi} g(\bar{\alpha}) \cos \psi d\psi\right) \frac{\partial}{\partial r} \quad . \end{aligned} \tag{45}$$

ASYMPTOTIC STABILITY THEOREMS

Asymptotic behaviours of the stochastic nonlinear system given by Equation (27) is, in this section, examined. Our main concern is the asymptotic behaviour of the zero solution $x_1 = x_2 = 0$ which implies $r = 0$. Our attention is thus directed to the r(t)-process whose differential generator is given by (45). However, if there exists a stationary density $p(r, \phi)$ for the process described by (45), then it will apparently be independent of ϕ. This fact allows us to write (45) in a simpler form as

$$\bar{L}_r = U^2 \frac{\partial^2}{\partial r^2} + V_r(r,\bar{\alpha}) \frac{\partial}{\partial r} \quad , \tag{46}$$

where

$$\begin{aligned} V_r(r,\bar{\alpha}) = & \frac{\partial^2}{4\pi} \int_0^{2\pi} h^2\left(\frac{r}{\omega} \sin \psi, r \cos \psi\right) \frac{\sin^2 \psi}{r} d\psi \\ & - \frac{1}{2\pi} \int_0^{2\pi} g\left(\frac{r}{\omega} \sin \psi, \; r \cos \psi, \; \bar{\alpha}\right) \cos \psi d\psi. \end{aligned} \tag{47}$$

It can thus be understood that theoretical considerations run on the same line as described in the third section. The following theorem gives sufficient conditions of the asymptotic stability in the large.

Theorem 5

Assume that, for any fixed initial value $r(0) = r_0$, the coefficients $U^2(r)$ and $V_r(r,\bar{\alpha})$ in (46) satisfy the following conditions:

(C.5) $U^2(r) = 0$, if and only if $r = 0$ and $V_r(0,\bar{\alpha}) = 0$

(C.6) $$\lim_{r \to 0} \left| \frac{V_r(r,\bar{\alpha})}{U^2(r)} \right| \exp \left\{ \int_r^{r_0} \frac{V_r(\zeta,\bar{\alpha})}{U^2(\zeta)} d\zeta \right\} < \infty .$$

(C.7) $$\int_r^{\infty} \frac{V_r(\zeta,\bar{\alpha})}{U^2(\zeta)} d\zeta < \infty.$$

Then, for any initial values $r_0 \varepsilon [0,\infty)$ and α_i $(i=1,2,\ldots,n)$, we have

$$\Pr \{\lim_{t\to\infty} r(t) = 0 \mid r(0) = r_0, \ \alpha(0) = \alpha_i\} = 1 \tag{48}$$

The following theorem gives also sufficient conditions with the probability appraisal.

Theorem 6

Assume that the following conditions are satisfied together with the conditions (C.5) and (C.6):

(C.8) There exists a positive constant M such that, for any $r\varepsilon(0,M)$, the drift term $\bar{V}(r,\bar{\alpha})$ in (46) is negative.

(C.9) The initial value $\alpha(0) = \alpha_i$ satisfies that, for any fixed r, $V_r(r,\bar{\alpha}) > V_r(r,\alpha_i)$.

Then we have

$$\Pr \{\lim_{t\to\infty} r(t) = 0 \mid r(0) = r_0 < M, \ \alpha(0) = \alpha_i\} \geq 1 - \frac{W_\alpha(r_0)}{W_\alpha(M)} \tag{49}$$

where

$$W_\alpha(r) = \int_0^r \exp \left\{ \int_\eta^{r_0} \left(\frac{V_r(\zeta,\bar{\alpha})}{U^2(\zeta)} - V_r(\zeta,\bar{\alpha})\Phi(\zeta)\right) d\zeta \right\} d\eta . \tag{50}$$

Since proofs of Theorems 5 and 6 are essentially same as those of Theorems 1 and 2 except for the condition (C.9), descriptions are omitted here. For the condition (C.9), in the neighbourhood of $t = 0$, it is necessary to assume that the $r(t)$-process does not go out of the domain $0 \leq r < M$. To do this, bearing in mind the fact that the initial value $\alpha(0) = \alpha_i$ is considered so as to satisfy

$$u^{(o)}(0) = \sum_{i=1}^{n} \bar{p}_i f(\alpha_i) \tag{51}$$

where

$$f(\alpha_i) = u_i(r, \theta, \alpha_i; 0, r_1, \theta_1; \alpha_j) \quad , \tag{52}$$

the initial value $\alpha(0) = \alpha_i$ lies on the domain such that the inequality $V_r(r,\bar{\alpha}) > V_r(r,\alpha_i)$ holds for any fixed r.

ILLUSTRATIVE EXAMPLES

Example 1

We shall consider a nonlinear dynamical system given by

$$\ddot{x} + \{2\varepsilon\beta - \delta\dot{\xi}(t)\}\,\dot{x} + \omega^2 x + \varepsilon x^3 = 0 \; . \tag{53}$$

Equation (53) may be considered as a mathematical model of dynamical systems whose damping coefficient is white Gaussian and restoring force is a nonlinearity of the cubic order. From Equation (53), both the nonlinear functions g and h are respectively identified by

$$g(x,\dot{x}) = g(x_1,x_2) = x_1^3 + 2\beta x_2 \tag{54}$$

$$h(x,\dot{x}) = h(x_1,x_2) = x_2 \quad . \tag{55}$$

The precise interpretation of Equation (53) is made by the following stochastic differential equation of Itô-type:

$$dx_1 = x_2 dt \tag{56a}$$

$$dx_2 = -\{\omega^2 x_1 + \varepsilon(2\beta x_2 + x_1^3)\}\; dt + \delta x_2 dw(t) \; . \tag{56b}$$

From (8) and (9), the diffusion and drift coefficients are respectively computed to be

$$U^2(r) = \frac{3\sigma^2 r^2}{16} \quad (57)$$

and

$$V(r) = (\frac{\sigma^2}{16} - \beta) r. \quad (58)$$

Using (57) and (58), the conditions (C.1) to (C.3) in Theorem 1, are examined as follows:

(1) The condition (C.1) is obviously satisfied.

(2) The condition (C.2) holds for $\beta > \sigma^2/4$.

(3) The condition (C.3) holds for $\beta > \sigma^2/16$.

Consequently, we may conclude that the origin of the system (56) is asymptotically stable in the large under the condition

$$\beta > \frac{\sigma^2}{4} \quad (59)$$

On the other hand, choose the function $W_L(x_1, x_2)$ as [20]

$$W_L(x_1, x_2) = x_2^2 + 2 \int_0^{x_1} (\omega^2 y + \varepsilon y^3)\, dy . \quad (60)$$

Let L be the differential generator of Equation (58). Then, from the relation,

$$LW_L = x_2 \frac{\partial W_L}{\partial x_1} - \{\omega^2 x_1 + \varepsilon(2\beta x_2 + x_1^3)\} \frac{\partial W_L}{\partial x_2} + \frac{\delta^2 \sigma^2 x_2^2}{2} \frac{\partial^2 W_L}{\partial x_2^2}$$

$$= - x_2^2 (4\varepsilon\beta - \delta^2 \sigma^2) \quad (61)$$

and the assumption, $\delta^2 =$, it follows that, for $\beta > \sigma^2/4$,

$$LW_L \leq 0. \quad (62)$$

It is obvious that the function W_L is the Lyapunov function and that

$$\Pr \{\lim_{t\to\infty} x_2(t) = 0\} = 1 . \quad (63)$$

Since the result (63) brings $\Pr\{\lim_{t\to\infty} x_1(t) = 0\} = 1$ [20], for any initial value r_0, we have the summarized result:

$$\Pr \{\lim_{t\to\infty} x_1(t) = x_2(t) = 0\} = 1 \quad (64)$$

under the condition given by (59).

Example 2

We shall consider a nonlinear dynamical system with a random coefficient given by

$$\ddot{x} + \{2\varepsilon\alpha(t) - \delta\dot{\xi}(t)\}\,\dot{x} + \omega^2 x + \varepsilon x^3 = 0 \tag{65}$$

where the parameter $\alpha(t)$ is considered to be a Markov chain discussed in the previous section. From Equation (65), the stochastic differential equation of Itô-type becomes

$$dx_1 = x_2 dt \tag{66a}$$

$$dx_2 = -\left[\omega^2 x_1 + \varepsilon\{2\alpha(t)x_2 + x_1^3\}\right] dt + \delta x_2\, dw(t) . \tag{66b}$$

Bearing the relation (6) in mind and using (8) and (47), it follows that

$$U^2(r) = \frac{3}{16}\sigma^2 r^2 \tag{67}$$

and

$$V_r(r,\bar{\alpha}) = (\frac{\sigma^2}{16} - \bar{\alpha})r . \tag{68}$$

From (67) and (68), the conditions (C.5) to (C.7) in Theorem 5 are examined as follows:

(i) The condition (C.5) is satisfied.

(ii) It can easily be seen that the condition (C.6) holds for $\bar{\alpha} > \sigma^2/4$.

(iii) For $\bar{\alpha} > \sigma^2/16$, the condition (C.7) is satisfied.

Hence, from Theorem 5, the equality (48) holds for any initial values r_0 and $\alpha(0) = \alpha_i$ under the condition $\bar{\alpha} > \sigma^2/4$.

Example 3

Consider a dynamical system modelled by the nonlinear differential equation of Van der Pol type:

$$\ddot{x} + x + \varepsilon(1 - x^2)\,\dot{x} = \delta x\dot{\xi}(t) , \tag{69}$$

where $x(0) = x_0$ and $\dot{x}(0) = \dot{x}_0$ as usual. It is well known that, if $\delta = 0$, then the system exhibits an unstable limit cycle and is asymptotically stable with respect to the origin.

Converting equation (69) into

$$dx_1 = x_2 dt \tag{70a}$$

$$dx_2 = -\{x_1 + \varepsilon(1 - x_1^2)\, x_2\}\, dt + \delta x_1 dw(t) , \tag{70b}$$

and letting $r^2 = x_1^2 + x_2^2$, the r(t)-process is the scalar Markov process whose diffusion and drift coefficients are respectively computed to be

$$U^2(r) = \frac{\sigma^2}{16} r^2 \tag{71}$$

and

$$V(r) = \frac{3\sigma^2}{16} r - \frac{r}{4}\left(2 - \frac{r^2}{2}\right). \tag{72}$$

By using (71) and (72), the conditions (C.1), (C.2) and (C.4) in Theorem 1 are examined as follows:

(i) The condition (C.1) holds.

(ii) The condition (C.2) holds, provided that $\sigma^2 < 2$.

(iii) From (72), since V(r) is negative for $r < \sqrt{4 - 3\sigma^2/2}$, the domain satisfying the condition (C.4) is

$$Q_m' = \{r;\ r < M = \sqrt{4 - \frac{3\sigma^2}{2}}\ \} . \tag{73}$$

Thus, from Theorem 2, it may be concluded that, for the r(t)-process initiating at $r_0 \varepsilon Q_m'$,

$$P_0 \{\lim_{t\to\infty} r(t) = 0\} \geq 1 - \frac{W_L(r_0)}{W_L(\sqrt{4 - \frac{3\sigma^2}{2}})} \tag{74}$$

Example 4

The same system as in Example 3 is considered, besides the system parameter is modelled by a Markov chain, i.e.,

$$\ddot{x} + x + \varepsilon\{1 - \alpha(t)x^2\}\,\dot{x} = \delta x \dot{\xi}(t) , \tag{75}$$

where $\alpha(0) = \alpha_i$ (i = 1,..., n). Equation (75) is converted into

$$dx_1 = x_2 dt \tag{76a}$$

$$dx_2 = - \left[x_1 + \varepsilon\{1 - \alpha(t)x_1^{\,2}\}\, x_2\right] dt + \delta x_1\, dw(t) \ . \tag{76b}$$

Hence, from (47) we have

$$V_r(r,\bar{\alpha}) = \frac{\bar{\alpha}}{8}\, r \left(r^2 - \frac{8 - 3\sigma^2}{2\bar{\alpha}}\right) . \tag{77}$$

By examining the conditions (C.5), (C.6), (C.8) and (C.9) and using Theorem 6 the sufficient conditions for the asymptotic stability is found to be $\sigma^2 < 2$. Furthermore, it can easily be found that

$$Q_m{}' = \{r;\ r < M = \sqrt{\frac{8 - 3\sigma^2}{2\bar{\alpha}}}\ \} . \tag{78}$$

where we assumed that $\bar{\alpha} > 0$, because the system is easily shown to be asymptotically stable in the large, if $\bar{\alpha} < 0$.

The condition (C.9) shows

$$V_r(r,\bar{\alpha}) > V_r(r,\alpha_i) \ , \tag{79}$$

that is, in this example, $\bar{\alpha} > \alpha_i$. Hence, for the r(t)-process starting at the initial value $r_0 < \sqrt{(8 - 3\ ^{2})/2\bar{\alpha}}$ and $\alpha_i < \bar{\alpha}$, it may be concluded that

$$\Pr\{\lim_{t\to\infty} r(t) = 0 \,|\, r(0) = r_0,\ \alpha(0) = \alpha_i\} \geq 1 - \frac{W_\alpha(r_0)}{W_\alpha\left(\sqrt{\frac{8 - 3\sigma^2}{2\,\bar{\alpha}}}\right)} \ . \tag{80}$$

Equation (75) was simulated on a digital computer where realizations of the random parameter $\alpha(t)$ are piece-wise constant functions with the four stages $\alpha_1 = 0, 0.5, 1.5$ and 2, i.e., $\bar{\alpha} = 1.0$ and changes uniformly with the probability 1/4 at intervals of 1/2 sec. Other parameters in Equation (75) were set as $\varepsilon = 0.1$, $\delta = \sqrt{0.1}$, $\sigma^2 = 1$, and $\alpha(0) = \alpha_2 = 0.5$. A variety of sample runs was simulated. The results presented below are representative of the simulation experiments. Two sample runs are comparatively shown in Figure 2. The initial values of the solid run were $x_1(0) = 0.9$, i.e., $r(0) =$

$r_0 = 1.27$, while those of the dotted run $x_1(0) = 1.1$, $x_2(0) = 1.0$, i.e., $r_0 = 1.49$. The probability appraisal is evaluated by using (50) and the right hand side of the inequality (80). This was 0.530 for the solid run and 0.187 for the dotted run respectively. Furthermore, the value of M is computed to be M = 1.6, which gives us the upper bound of the asymptotic stability with respect to the initial value r_0. For the purpose of comparative observations, the deterministic process where $\dot{\xi}(t) = 0$ and $\alpha(t)$ = constant = 1 in Equation (75) is also plotted in Figure 2.

CONCLUSIONS

In this paper, a new approach has been developed to analyse the asymptotic stability of nonlinear dynamical systems with a random parameter behaving as a white Gaussian process. The basic notion presented here is a choice of the stochastic Lyapunov function with an advantage that influences of initial values of the system states come out.

Introducing the concept of random evolution associated with the limit theorem, the stability analysis is extended to a general class of nonlinear dynamical systems involving two kinds of random parameters modelled by a white Gaussian and a Markov chain process respectively.

Throughout this paper, the relation between the asymptotic behaviour of nonlinear stochastic systems and the domain of their initial values are examined by using the useful Theorems giving sufficient conditions for the asymptotic stability with the probability appraisal.

Y. Sunahara, Faculty of Polytechnic Science, Kyoto Institute of Technology, Matsugasaki, Sakyo-ku, Kyoto 606, Japan.

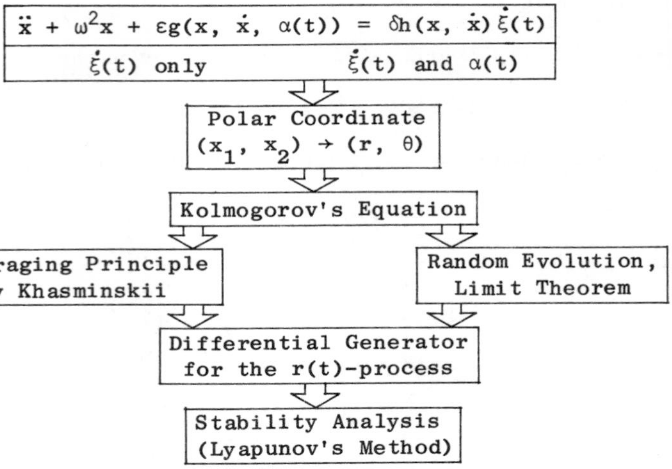

Figure 1

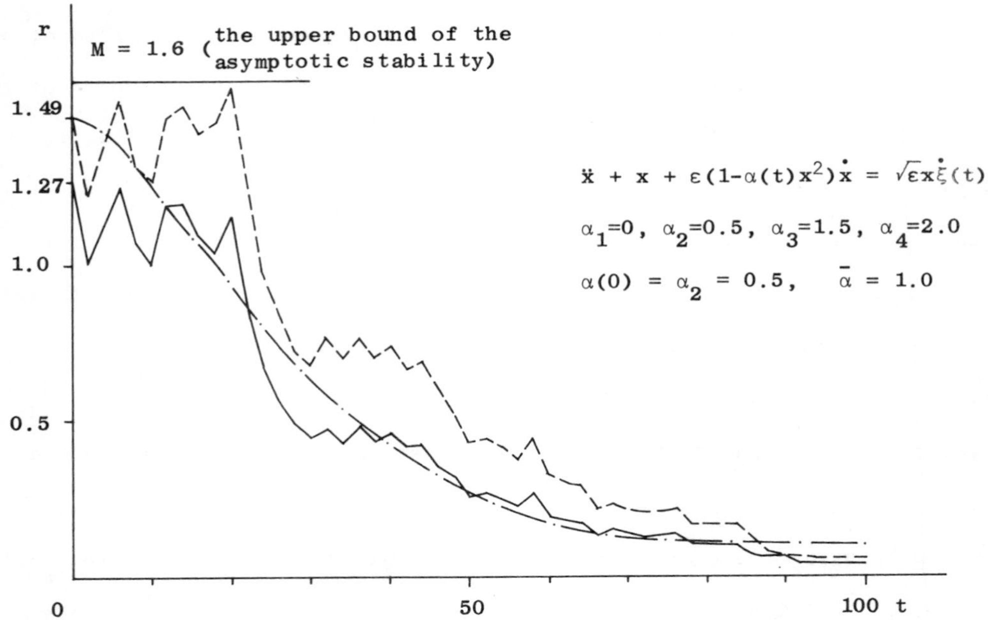

—·— Deterministic process ($\dot{\xi}(t) = 0$).

– – – – The r(t)-run which is asymptotically stable with the probability appraisal 0.187.

——— The r(t)-run which is asymptotically stable with the probability appraisal 0.530.

Figure 2

REFERENCES

1 Y. Sawaragi et al. Statistical Studies on Nonlinear Control Systems. Osaka, Nippon Printing and Publishing. (1962).

2 F. Kozin A survey of stability of stochastic systems. Automatica Journal of the International Federation of Automatic Control, 5, p. 95-112. (1969).

3 A. Dold and B. Eckmann Stability of Stochastic Dynamical Systems. Berlin, Springer-Verlag. (1972).

4 I.I. Gihman and A.V. Skorohod Stochastic Differential Equations. Berlin, Springer-Verlag. (1972).

5 H.J. Kushner On the stability of stochastic dynamical systems. Proc. National Academy of Sciences, 53, p. 8-12. (1965).

6 W.M. Wonham Lyapunov criteria for weak stochastic stability. Journal of Differential Equation. 2, p. 195-207. (1966).

7 R. Griego and R. Hersh Random evolution, Markov chain and systems of partial differential equations. Proc. National Academy of Sciences, 62, p. 305-308. (1969).

8 R. Griego and R. Hersh Theory of random evolutions with applications to partial differential equations. Trans. American Mathematical Society, 156, p. 405-418. (1971).

9 R. Hersh and G.C. Papanicolaou Noncommuting random evolutions and an operator-valued Feynman-Kac formula. Communications on Pure and Applied Mathematics, 25, p. 337-366. (1972).

10 R. Hersh and A.M. Pinsky Random evolutions are asymptotically Gaussian. Communications on Pure and Applied Mathematics, 25, p. 33-44. (1972).

11 T.Z. Kurtz A limit theorem for perturbed operator semigroups with applications to random evolutions. Journal of Functional Analysis, 12, p. 55-67. (1973).

12 G.C. Papanicolaou and J.B. Keller Stochastic differential equations with applications to random harmonic oscillators and wave propagation in random media. SIAM Journal on Applied Mathematics, 21, p. 287-305. (1971).

13 Y. Sunahara et al. General conditions for noise stabilization of non-linear dynamical systems. Transactions of the Japan Society of Mechanical Engineers, 40, p. 752-763. (1974). (in Japanese).

14 Y. Sunahara et al. On the noise stabilization of nonlinear dynamical systems. Transactions of the Japan Society of Mechanical Engineers, 40, p. 2852-2861. (1974). (in Japanese).

15 K. Itô On stochastic differential equations. Memoirs of the American Mathematical Society, 4. (1961).

16 Y. Yaglom An Introduction to the Theory of Stationary Random Functions. Englewood Cliffs, New Jersey. Prentice-Hall. (1962).

17 R.Z. Khasminskii Principle of averaging for parabolic and elliptic differential equations and for Markov processes with small diffusion. Theory of Probability and Applications, 8, p. 1-21. (1963).

18 W.M. Wonham Probabilistic Methods in Applied Mathematics, 2. (edited by Bharucha-Reid). New York, Academic Press. (1970).

19 I.I. Gihman and A.Ya. Dorogovcev On stability of solutions of stochastic differential equations. Ukrainian Mathematical Journal USSR, 17. (1965).

20 H.J. Kushner Stochastic Stability and Control. New York, Academic Press. (1967).

Professor Y. Sunahara, Faculty of Polytechnic Science, Kyoto Institute of Technology, Matsugasaki, Sakyo-ku, Kyoto 606, Japan.

T. Asakura, Research Assistant, Kyoto Institute of Techology, Kyoto.

Y. Morita, Graduate Student, Osaka University, Osaka.

APPENDIX A APPLICATION OF THE AVERAGING PRINCIPLE

It is well-known that the solution process $(x_1(t), x_2(t))$ to Equation (2) is a time uniform Markov process in the phase space (x_1, x_2). Let $p(x_1, x_2; t; y_1, y_2)$ be the probability density of a transition from the state (x_1, x_2) to (y_1, y_2) within time t. For fixed y_1 and y_2, the probability density satisfies

$$\frac{\partial p}{\partial t} = x_2 \frac{\partial p}{\partial x_1} - \{\omega^2 x_1 + \varepsilon g(x_1, x_2)\} \frac{\partial p}{\partial x_2} + \frac{\varepsilon\sigma^2}{2} h^2(x_1, x_2) \frac{\partial^2 p}{\partial x_2^2} \tag{A.1}$$

with the initial condition $p(x_1, x_2; 0; y_1, y_2) = \delta(x_1 - y_1, x_2 - y_2)$.

Along the relation (5), letting

$$p(\frac{r}{\omega} \sin\theta, r\cos\theta; \frac{r_1}{\omega} \sin\theta_1, r_1 \cos\theta_1) = v(r, \theta; r_1, \theta_1) \tag{A.2}$$

and $\theta \equiv \phi - \omega t$, then, after somewhat tedious calculations, we have

$$\begin{aligned} \frac{\partial v}{\partial t} = \varepsilon\{ & \frac{\sigma^2}{2} h^2 \Big[\cos^2(\phi - \omega t) \frac{\partial^2 v}{\partial r^2} - \frac{\sin 2(\phi - \omega t)}{r} \frac{\partial^2 v}{\partial r \partial \phi} \\ & + \frac{\sin^2(\phi - \omega t)}{r^2} \frac{\partial^2 v}{\partial \phi^2} + \frac{\sin 2(\phi - \omega t)}{r^2} \frac{\partial v}{\partial \phi} + \frac{\sin^2(\phi - \omega t)}{r} \frac{\partial v}{\partial r}\Big] \\ & + g \frac{\sin(\phi - \omega t)}{r} \frac{\partial v}{\partial \phi} - g\cos(\phi - \omega t) \frac{\partial v}{\partial r} \} . \end{aligned} \tag{A.3}$$

Changing the time variable t for τ with the relation $\tau = \varepsilon t$ and tending ε to zero in such a way that $\tau = \varepsilon t$ is constant, the averaging principle can be applied to Equation (A.3).

The averaging principle is roughly stated as follows: Let x be a point of the n-dimensional Euclidean space. Consider the partial differential equation $\partial v/\partial t = \varepsilon L(t,x)v$, where L is an elliptic or a parabolic second-order differential operator. As ε goes to zero, the solution v is uniformly approximated over a time interval $0 \leq t < (T/\varepsilon)$ by the solution to the equation $\partial v_0/\partial t = \varepsilon L_0(x) v_0$, where

$$L_0(x) = \lim_{T\to\infty} \frac{1}{T} \int_0^T L(t,x)\, dt \tag{A.4}$$

The solution v to Equation (A.3) can thus be uniformly approximated by the solution to the partial differential equation,

$$\frac{\partial v_0}{\partial t} = \frac{\sigma^2}{2\pi} \{ (\frac{1}{2\pi} \int_0^{2\pi} h^2 \cos^2 \psi d\psi) \frac{\partial^2 v_0}{\partial r^2} - (\frac{1}{2\pi} \int_0^{2\pi} \frac{h^2 \sin 2\psi}{r} d\psi) \frac{\partial^2 v_0}{\partial r \partial \phi}$$

$$+ (\frac{1}{2\pi} \int_0^{2\pi} h^2 \frac{\sin^2 \psi}{r^2} d\psi) \frac{\partial^2 v_0}{\partial \phi^2} + (\frac{1}{2\pi} \int_0^{2\pi} h^2 \frac{\sin 2\psi}{r^2} d\psi) \frac{\partial v_0}{\partial \phi}$$

$$+ (\frac{1}{2\pi} \int_0^{2\pi} h^2 \frac{\sin^2 \psi}{r} d\psi) \frac{\partial v_0}{\partial r} + (\frac{1}{2\pi} \int_0^{2\pi} g \frac{\sin \psi}{r} d\psi) \frac{\partial v_0}{\partial \phi}$$

$$- (\frac{1}{2\pi} \int_0^{2\pi} g \cos \psi d\psi) \frac{\partial v_0}{\partial r} , \qquad (A.5)$$

where $\psi = \phi - (\omega\tau/\varepsilon)$. Since we assume that ε is sufficiently small in Equation (1), ϕ may be considered to be uniformly distributed. Consequently, v_0 will evidently be independent of ϕ and, from (A.5), we obtain the differential generator (7).

APPENDIX B RANDOM EVOLUTIONS

The equation presented here is of the form,

$$\frac{dy}{dt} = \varepsilon L_{\alpha(t)} y, \quad y(0) = f, \tag{B.1}$$

where y and f are elements in a Banach space $\mathcal{B}$. Assume that there exists a solution to Equation (B.1) of the form,

$$y(s) = M(s,t)y(t), \tag{B.2}$$

where M(s,t) is an operator satisfying the differential equation

$$\frac{dM(s,t)}{dt} = \varepsilon M(s,t)\, L_{\alpha(t)}, \quad t \geq s \tag{B.3}$$

and

$$M(s,s) = I . \tag{B.4}$$

Note that the operator M(s,t) defined by (B.3) is for the backward equation $dy/ds = -\varepsilon L_{\alpha(t)} y$, where $s \leq t$, $y \in \mathcal{B}$. Letting $t' = t - s$, then M(s,t) becomes M(t-t',t) associated with the solution to the forward equation $dy/dt' = \varepsilon L_{\alpha(t')} y$, $t' \geq 0$.

From Equation (B.3), it follows that, for $t_1 \leq t_2 \leq t_3$,

$$M(t_1,t_2)\, M(t_2,t_3) = M(t_1,t_3), \tag{B.5}$$

Changing the time scales s and t for σ and τ respectively along the relation $\sigma = \varepsilon s$ and $\tau = \varepsilon t$, and writing $M(\sigma/\varepsilon, \tau/\varepsilon)$ for $M^{(\varepsilon)}(\sigma,\tau)$, Equations (B.3) and (B.4) are expressed by

$$\frac{d}{d\tau} M^{(\varepsilon)}(\sigma,\tau) = M^{(\varepsilon)}(\sigma,\tau)\, L_{\alpha(\tau/\varepsilon)}, \quad \sigma \leq \tau \tag{B.6}$$

and $M^{(\varepsilon)}(\sigma,\sigma) = I$. We call $M^{(\varepsilon)}(\sigma,\tau)$ the random evolution.

APPENDIX C PROOF OF THEOREM 4

Following symbolic conventions in Reference [9], let σ_j be the epoch of the j-th return to the state $\{\alpha_1\}$. The successive return times $\tau_j = \sigma_{j+1} - \sigma_j$ are mutually independent and identically distributed random variables with respect to j, where $\tau_0 = \sigma_1$.

By integrating and iterating Equation (37a), we have

$$M^{(\varepsilon)}(\sigma,\tau) = I + \int_\sigma^\tau L(\frac{\rho}{\varepsilon})\, d\rho + \int_\sigma^\tau \int_\sigma^{\rho_2} M^{(\varepsilon)}(\sigma,\rho_1)\, L(\frac{\rho_1}{\varepsilon})\, L(\frac{\rho_2}{\varepsilon})\, d\rho_1\, d\rho_2 .^{*} \quad (C.1)$$

By setting $\sigma = 0$ and $\tau = \varepsilon\tau_1$ in (C.1) and operating E_1 to both sides, it follows that

$$S_\varepsilon = I + S_1 + S_2 , \quad (C.2)$$

where

$$S_\varepsilon = E_1 \{M^{(\varepsilon)}(0,\varepsilon\tau_1) \} , \quad (C.3)$$

$$S_1 = E_1 \{ \int_0^{\varepsilon\tau_1} L(\frac{\rho}{\varepsilon})\, d\rho \} \quad (C.4)$$

and

$$S_2 = E_1 \{ \int_0^{\varepsilon\tau_1} \int_0^{\rho_2} M^{(\varepsilon)}(0,\rho_1)\, L(\frac{\rho_1}{\varepsilon})\, L(\frac{\rho_2}{\varepsilon})\, d\rho_1\, d\rho_2 \} . \quad (C.5)$$

We need the following lemma.

Lemma For any $f \in \mathcal{D}$, it follows that

$$\overline{L}f = \frac{S_1 f}{\varepsilon\mu} \quad (C.6)$$

and

$$S_2 f \to O(\varepsilon^2) , \quad (C.7)$$

where μ is the mean value of τ_k.

---

* We abbreviate $L_{\alpha(\rho)}$ to $L(\rho)$ in this Appendix.

Proof: Note that, in view of the time homogeneity of $\alpha(t)$,

$$S_1 f = \varepsilon E_1 \{ \int_0^{\tau_1} L(\rho)d\rho \} f = \varepsilon E \{ \int_{\sigma_j}^{\sigma_{j+1}} L(\rho)d\rho \} f, \tag{C.8}$$

where E is the mathematical expectation with respect to the probability measure on the $\alpha(t)$-process whose initial point is distributed with $\bar{p}_1$. For any T, we have

$$\int_0^T L(\rho)fd\rho = \int_0^{\sigma_1} L(\rho)fd\rho + \sum_{j=1}^{N(T)} \int_{\sigma_j}^{\sigma_{j+1}} L(\rho)fd\rho - \int_{\sigma_{N(T)-1}}^{\sigma_{N(T)+1}} L(\rho)fd\rho + \int_{\sigma_{N(T)-1}}^{T} L(\rho)fd\rho \tag{C.9}$$

where

$$N(T) = \min \{ j | \sigma_j \geq T \}. \tag{C.10}$$

From the property of τ_j, it is obvious that

$$E_1 \{ \int_0^{\tau_1} L(\rho)fd\rho \} = \frac{1}{\varepsilon} S_1 f. \tag{C.11}$$

Furthermore

$$E \{N(T)\} = \frac{T}{\mu} + o(T) . \tag{C.12}$$

By using Wald's Theorem, the mathematical expectation of the second term of the right-hand side in (C.9) becomes

$$E \{ \sum_{j=1}^{N(T)} \int_{\sigma_j}^{\sigma_{j+1}} L(\rho)fd\rho \} = \frac{T}{\mu} . \frac{S_1 f}{\varepsilon} + o(T) . \tag{C.13}$$

With the help of bounded properties such that

$$||E \{ \int_0^{\sigma_1} L(\rho)fd\rho|| \leq E \{\sigma_1\} \sup_{\alpha_i} || L_{\alpha i} f || , \tag{C.14}$$

$$||E \{ \int_{\sigma_{N(T)-1}}^{\sigma_{N(T)+1}} L(\rho)fd\rho|| \leq E\{\sigma_{N(T)+1} - \sigma_{N(T)-1}\} \sup_{\alpha_i} || L_{\alpha i} f||$$
$$\leq 2\mu \sup_{\alpha_i} || L_{\alpha i} f|| \tag{C.15}$$

and

$$E\{ \int_{\sigma_{N(T)-1}}^{T} L(\rho) f d\rho \} \leq E\{T - \sigma_{N(T)-1}\} \sup_{\alpha_i} || L_{\alpha i} f || \leq \mu \sup_{\alpha_i} || L_{\alpha i} f ||, \tag{C.16}$$

it follows that

$$\frac{1}{T} E\{ \int_0^T L(\rho) f d\rho \} = \frac{S_1 f}{\mu\varepsilon} + \frac{1}{T} o(T) , \tag{C.17}$$

where o(T) in (C.17) is the collection of terms which are independent of T.

Letting $T \to \infty$, the left-hand side of (C.17) becomes

$$\lim_{T\to\infty} \frac{1}{T} E\{ \int_0^T L(\rho) f d\rho \} = \lim_{T\to\infty} \frac{1}{T} \{ \int_0^T E(L(\rho) f) d\rho \}$$

$$= \lim_{T\to\infty} \frac{1}{T} \{ \int_0^T \sum_{i=1}^{n} \bar{p}_i L_{\alpha i} f d\rho \} = \sum_{i=1}^{n} \bar{p}_i L_{\alpha i} f = \bar{L} f. \tag{C.18}$$

Hence, it is a simple exercise to show that (C.17) yields (C.6).

From (C.5), it can be seen that

$$||S_2 f|| \triangleq ||E_1\{ \int_0^{\varepsilon\tau_1} \int_0^{\rho_2} M^{(\varepsilon)}(0,\rho_1) L(\frac{\rho_1}{\varepsilon}) L(\frac{\rho_2}{\varepsilon}) d\rho_1 d\rho_2 \} f ||$$

$$\leq \varepsilon^2 E_1\{\tau_1^2\} \sup_{\alpha_i, \alpha_j} || L_{\alpha i} L_{\alpha j} f || . \tag{C.19}$$

This shows that (C.7) holds.

Thus, from (C.6) and (C.7), (C.2) is expressed by

$$S_\varepsilon = I + \varepsilon\mu \bar{L} + O(\varepsilon^2) . \tag{C.20}$$

Keeping $\tau/\varepsilon\mu$ integer valued, we obtain

$$(S_\varepsilon)^{\tau/\varepsilon\mu} \to e^{\tau\bar{L}} . \tag{C.21}$$

Combining this with (C.3), it may be concluded that (43) holds.

DISCUSSION

Professor Ariaratnam mentioned that in the stochastic averaging procedures the convergence to a Markov process is weak and that therefore no statements on probability of convergence can strictly be made. However, on the assumption that this technical inconsistency is unimportant, Professor Ariaratnam then went on to present an alternative way of dealing with certain similar problems. In particular, in the case of example 1, equation (53), the differential generator corresponds to an Itô process.

$$dr = (\frac{\sigma^2}{16} - \beta)\ rdt + \sqrt{2}\ .\ \frac{\sqrt{3}\sigma r}{4}\ dB,$$

for which sample stability can be obtained directly by defining $\rho = \log r$ and integrating the transformed equation to obtain

$$\log \rho - \log \rho_o = \int \phi dt,$$

where ρ_o is the initial condition and ϕ (= constant in this case) is the ϕ function used in Kozin's work (q.v.). Stability is thus ensured if $\phi < 0$, a condition which corresponds to $\rho_o > \sigma^2/4$ as found in the present paper. The case of a random coefficient can also be treated in this manner and it was suggested that the method might lead to the choice of a simple Liapunov function (for example $v = r^2$) and that the standard Theorems of Kushner [20] could then be used in the nonlinear case.

In reply to a question on the possible non-equivalence between stability conditions for the original and averaged equations (Willems) it was noted that the averaged equation was used because it made the choice of Liapunov function simpler. It was mentioned (Kozin) that it would indeed be interesting to obtain a time-invariant Liapunov function from the averaged equation and then to apply it to the original time-varying equation but that then (Ariaratnam) the Liapunov function itself might become time varying and hence the computations complex. Such an approach might however (Willems) enable one to obtain stability criteria for the original equation with greater confidence.

Replying to a query on the assumption of Gaussian White Noise (Bendat) Professor Sunahara confirmed that the present application was limited to that case and that it could unfortunately not be extended to cover physical noise. It was noted (Kozin) that white noise is indeed a fiction but its use is encouraged by the existence of Itô's calculus, which makes its analysis relatively simple, and that it does appear from simulation results that the white noise assumption is not unreasonable. It was also mentioned (Lin) that the development of the stochastic averaging or Khasminskii limit Theorem and the consequent view of white noise as a limiting process has added considerably to the acceptance and use of Itô's calculus by practising engineers.

In response to a further question (Lin) it was confirmed that the α-process analysed was a continuous Markov chain.

Finally it was stressed that the averaging approach applied to ODE's with non-white coefficients by workers such as Ariaratnam is quite distinct from the averaging technique, also due to Khasminskii, applied in the present case to a PDE (equation (29)) with time-varying coefficients. In the former case the averaging is necessary in order to approximate a non-white process by a Markov process in the limit; in the second case, one starts with an Itô equation with white noise terms and can thus write down the differential generator almost directly (equations (2)-(9)). The time varying terms in the latter must then be averaged.

WRITTEN DISCUSSION by Professor J.L. Willems

Expressions (57) and (58) in the paper show that the process r(t) of Example 1 is governed by the Itô differential equation.

$$dr(t) = (\frac{\sigma^2}{16} - \beta)r(t)dt + \frac{\sqrt{3}}{4}\sigma r(t)dw(t) \qquad (1)$$

A direct stability analysis of this linear first order equation is possible; it yields:

(i) the inequality $\beta > \sigma^2/16$ is a necessary and sufficient condition for the asymptotic stability of the first moment $E(r)$;

(ii) the inequality $\beta > \sigma^2/4$ is a necessary and sufficient condition for the asymptotic stability of the second moment $E(r^2)$;

(iii) the inequality $\beta > -\sigma^2/8$ is a necessary and sufficient condition for almost sure asymptotic stability with probability one.

This shows that the criterion obtained in the paper is rather conservative, which of course is often the case for stability results obtained by means of Lyapunov's direct method. The criterion presented in the paper is not only sufficient for almost sure asymptotic stability, but it is even sufficient for mean square asymptotic stability, which is a much stronger stability property. This is confirmed by the fact that Lyapunov function (60) also yields a mean square stability criterion.

Example 2 can be discussed in a similar way; as a matter of fact the results and conclusions concerning Example 1 are valid here, provided β is replaced by $\bar{\alpha}$. This simple analysis cannot be used for Examples 3 and 4 since the Itô equations obtained by means of averaging are nonlinear. However we have constructed a Lyapunov function which for Examples 1 and 2 leads to necessary and sufficient stability conditions; in other cases it yields less conservative criteria than the Lyapunov function proposed by the authors.

Indeed consider the Lyapunov function $W(r) = r^q$ where q is a positive constant. From (7) one readily obtains $LW(r) = q\, r^{q-1} V(r) + q(q-1) r^{q-2} U^2(r)$. If $V(0) = 0$, $U^2(0) = 0$, then the following stability results are derived from this Lyapunov function :

(i) the null solution is q-th mean asymptotically stable (i.e. $E(r(t)^q) \to 0$ as $t \to \infty$ for small $r(0)$), if

$$rV(r) + (q-1)U^2(r) < 0 \tag{2}$$

for small positive r;

(ii) the null solution is globally g-th mean asymptotically stable if (2) holds for all positive r;

(iii) the null solution is almost surely asymptotically stable if

$$rV(r) < U^2(r) \tag{3}$$

for small positive r;

(iv) the null solution is globally almost surely asymptotically stable if (3) holds for all positive r.

In case (iii) an appraisal of the probability of the convergence of the solution generated by an initial condition can also be derived from the Lyapunov function. If (2) holds for $0 < r < R_q$, then the estimate is

$$\text{Prob}\left[x(t) \to 0 \quad \text{as} \quad t \to \infty\right] > 1 - \operatorname*{Inf}_q \left(\frac{r_o}{R_q}\right)^q \quad \text{where } r_o = ||x(0)||.$$

It is easily checked that the above criteria yield the necessary and sufficient stability results for Examples 1 and 2 of the paper. Also for Examples 3 and 4 criteria are obtained which are less conservative than those derived in the paper. The condition for asymptotic stability with probability one is $\sigma^2 < 4$ in both cases; here again the condition of the paper is sufficient for mean square asymptotic stability.

Finally we want to point out that in Examples 1 and 2 the system can be stabilized by the introduction of the noise. This however is due to the fact that ideal white noise was considered. If it is assumed that the noise $\dot{\xi}(t)$ in (1) is very wide band noise, then the Stratonovitch correction term should be introduced in the Itô equation (2); for Example 1 this would mean that $2\varepsilon\beta$ should be replaced by $(2\varepsilon\beta - \frac{1}{2}\sigma^2 \delta^2)$. The necessary and sufficient condition for asymptotic stability with probability one then becomes

$$\beta > + \sigma^2/8$$

which is clearly more acceptable. Similarly the other cases can be dealt with.

AUTHORS COMMENTS on the written contribution from Professor J. Willems

1 In Example 1 of the paper, the condition $\beta > \sigma^2/4$ becomes necessary and sufficient for the mean square asymptotic stability for the first order equation, which is not "sufficient for the mean square asymptotic stability" as you said. The condition $\beta > \sigma^2/4$ differs from the necessary and sufficient condition $\beta > -\sigma^2/8$ for the almost sure asymptotic stability. (Please see 4 also).

2 In Examples 3 and 4, although our Theorems 1 and 6 give only sufficient conditions for the almost sure asymptotic stability, they allow us to determine the domain of the initial values based on the condition for the almost sure asymptotic stability. Our Lyapunov function is compared with $W(r) = r^q$ in your letter. Consider Equation (2) in your letter with a fixed q = 2 (the mean square sense). The same values of V(r) and $U^2(r)$ are set as in Example 3. Then Equation (2) is written by

$$r\left\{\frac{3\sigma^2}{16} r - \frac{r}{4}\left(2 - \frac{r^2}{2}\right)\right\} + \frac{\sigma^2 r^2}{16} < 0$$

It is easily examined that, in your case, the domain of the initial values for assuring the asymptotic stability becomes $Q_m' = \{r;\ r < M = \sqrt{4 - 2\sigma^2}$, and that the condition for the asymptotic stability is $\sigma^2 < 2$, where M corresponds to R_q in your letter. Our Theorem 1 shows that the domain Q_m' becomes $Q_m' = \{r;\ r < M = \sqrt{4 - \frac{3\sigma^2}{2}}\}$, and that the condition for the asymptotic stability is $\sigma^2 < 2$ in Equation (73). Hence, although the condition $\sigma^2 < 2$ obtained above is the same as that in your case, a wider domain can be obtained than your Lyapunov function, where q = 2. Therefore, by using M, the probability appraisal $1 - \frac{W(r_0)}{W(M)}$,
may be pleasant for the mean square asymptotic stability. However,

our Lyapunov function cannot give such a condition (w.p.1) given by Equation (3) in your letter. (See Remark 4 also).

3 As you pointed out in your letter, since a state-dependent white Gaussian noise process is introduced, a correction term should be taken into consideration. However, our discussions did not touch with the version of the correction term. Further discussions about this problem will be reported in the subsequent paper.

4 Finally, our remark is directed to the consideration that our method does not give the necessary and sufficient conditions for the almost sure asymptotic stability for the first order equation. A general class of the second order nonlinear equation is considered. Therefore, when the Lyapunov function is applied to the first order equation derived from the second order equation, the differentiability at the origin $r = 0$ (i.e. $x_1 = x_2 = 0$) is set, which weakens the condition for the asymptotic stability. If we only treat the first order equation, more strict conditions can be obtained for the almost sure asymptotic stability by excluding differentiability at $r = 0$. In order to show this, consider the first order Itô-equation:

(*) $dr(t) = ar(t)dt + r(t)dw(t)$.

The differential generator L is

$$L = ar \frac{\partial}{\partial r} + \frac{\sigma^2 r^2}{2} \frac{\partial^2}{\partial r^2} \equiv V(r) \frac{\partial}{\partial r} + U^2(r) \frac{\partial^2}{\partial r^2} .$$

In this case, our Lyapunov function $W_L(r)$ becomes

$$\begin{cases} W_L(r) = \displaystyle\int_0^r \exp \left\{ \int_\eta^{r_0} \left(\frac{V(\zeta)}{U^2(\zeta)} + \Psi(\zeta) \right) d\zeta \right\} d\eta \\ L\, W_L(r) = - U^2(r)\Psi(r) \exp \left\{ \displaystyle\int_r^{r_0} \left(\frac{V(\zeta)}{U^2(\zeta)} + \Psi(\zeta) \right) d\zeta \right\} \leq 0 \end{cases}$$

As we are not concerned with the differentiability at $r = 0$, the boundedness of $W_L(r)$ should be examined:

$$W_L(r) = \int_0^r \exp\{\int_\eta^{r_0} \frac{2a}{\sigma^2 \zeta} d\zeta\} \exp\{\int_\eta^{r_0} \Psi(\zeta)\, d\zeta\} d\eta$$

$$= \int_0^r (\frac{r_0}{\eta})^{2a/\sigma^2} \exp\{\int_\eta^{r_0} \Psi(\zeta)\, d\zeta\} d\eta \quad .$$

Noting that $\Psi(r)$ is assumed to be positive, sufficiently smooth and integrable, we easily have the following condition for the boundedness of $W_L(r)$;

$$\int_0^r (\frac{r_0}{\eta})^{2a/\sigma^2} d\eta < \infty \quad \text{for any bounded} \quad r > 0,$$

that is, $-\ 2a/\sigma^2 + 1 > 0$ or $a < \sigma^2/2$.

This is the necessary and sufficient condition for the almost sure asymptotic stability of Equation (*). A similar calculation is performed in Example 3, which shows the condition $\sigma^2 < 4$; the condition for the asymptotic stability with probability one (in your letter).

K PISZCZEK

Influence of random disturbances on determined nonlinear vibrations

INTRODUCTION

It is considered in the theory of mechanical vibrations, and specially in machine dynamics that factors causing vibrations are deterministic magnitudes, thus, deterministic functions of time, or they represent deterministic parameters of the vibrating system.

In the recent couple of years a method of analysis of nonlinear vibrating systems has been developed in which the factors inducing vibrations represent random magnitudes regarded as stochastic processes, stationary or nonstationary. The achievements in this field are considerable, especially for stationary excitation and the literature concerning this subject is rich. A number of methods have been elaborated, of which some are exact and others approximate ones [1].

An important group of problems in the nonlinear theory of vibrations is constituted by those in which two or more mechanisms exciting vibrations act simultaneously, one of them being deterministic and the others random. Out of the problems of this type we are interested in the influence of random perturbations on the amplitude, and frequency of vibrations or resonance regions known from deterministic mechanisms of excitation. The behaviour of the vibrating system can be analysed with respect to the mean value of the amplitude, mean square value of the amplitude, dispersion, spectral density and other stochastic characteristics. Subsequently a short review will be given of the groups of problems in which two vibration excitation mechnisms are involved and to which, as it is known, the super-position principle does not apply. Confining to one degree of freedom systems and neglecting investigations of the possible influence of these disturbances on stability of the system, five groups of problems can be distinguished which can be presented by means of known differential equations as representatives of a specified group.

(a) In the first group, those systems will be included in which excitation is harmonically changed with time lapse at a concomitant action of load in form of a stochastic process of external character. For this group, the Duffing's equation with random disturbance $\eta(t)$ added on the right side of the equality sign is representative. As the assumption $\eta(t) = 0$ the solution of the above equation in the first approximation is a harmonic function of a deterministic amplitude, frequency and phase. In case of a simultaneous action of two mechanisms of excitation the influence of the random excitation on the above mentioned parameters of the deterministic solution will be investigated.

(b) The other group of problems are self-excited systems, perturbated by the action of random factors. The representative of this group of systems is the system described by the Van der Pol's differential equation with random excitation $\eta(t)$ on the right side of the equality sign.

(c) The third group of problems covers parametric vibrations, nonlinearity of the system being considered. Two sub-groups can be distinguished, i.e.:

(α) the parametric action is a stochastic process at a simultaneous action of a determined external load.

(β) the external force constitutes the stochastic process and the variable parameter of the system is a deterministic function of time. The representative of this group of problems is the Mathieu equation with a suitably assumed function on the right side of the sign of an equality.

(d) The other group will include such systems in which the mechanism of self-excitation and of parametric excitation regarded as a stochastic process will occur simultaneously. An inverse case is possible i.e. a case of random self-excitation mechanism and a deterministic parametric excitation.

(e) The last group presents nonlinear problems with parametric excitation consisting of a deterministic and random part.

It is quite obvious that the division presented does not include all systems without any exception and a specific vibration system can be given which cannot be included into any of these groups.

The examples given above refer to systems of one degree of freedom. It is quite clear that the problems discussed also occur in systems of a finite and infinite number of degrees of freedom. Thus it is necessary to elaborate methods of analysis of this type of problem.

In the following we shall deal with some of the previously mentioned groups of problems, attention being drawn to the method of their analysis and to the results obtained.

SELF EXCITED VIBRATIONS

The method of analysis of the influence of small stationary random disturbances on the amplitude and frequency of self-excited vibrations of the system of one degree of freedom was given by the author [2]. As a matter of fact, it consists in applying to a respective differential equation harmonic and statistic linearization successively.

Let equation

$$L\left[\ddot{x},\ \dot{x},\ x;\ p\right] \equiv \ddot{x} + f_1(x,\dot{x}) + \omega_0^2\, x - p(t) = 0 \tag{1}$$

describe the vibrations of the system of one degree of freedom, where the function $f_1(x,\dot{x})$ has the property that at $p(t)\equiv 0$ stable self-excited vibrations occur. The function p(t) presents small perturbations in form of a stationary, normal stochastic process of mean value equal to zero.

Let ξ_1 represent random vibrations excited by external load p(t) and ξ_2 - stationary self-excited vibrations which, in general, can be written in form of Fourier's series

$$\xi_2(t) = \xi_{02} + A \sin \omega t + a_2 \sin (2\omega t + \phi_2) + \dots \tag{2}$$

In the above relation ξ_{02} = const. represents the displacement of the vibration centre, A is in the first approximation the amplitude of vibrations, ω on the other hand represents the mean angular frequency. The other expressions in the relation (2) give the harmonics of a higher order. The magnitudes A, ω, a_2 ϕ_2,... are constant values.

Similarly as in the case of a deterministic problem concerning the influence of external excitation on self-excited vibrations the solution of equation (1) will be in form:

$$x = \xi_1 + \xi_2 \ . \tag{3}$$

This assumption represents the influence of random excitation ξ_1 on the final form of the solution x, which is approximate to solution ξ_2 at small random perturbations.

Substituting relations (2) and (3) into equation (1) and presenting the expression obtained in such a way in form of Fourier's series we obtain

$$0 \equiv \Psi_0 + \Psi_1 \cos \omega t + X_1 \sin \omega t + \Psi_2 \cos 2\omega t + X_2 \sin 2\omega t \ . \tag{4}$$

With respect to the fact that the above given equality is to satisfy every t, all the coefficients of the expression (4) should equal zero. Confining to the first approximation i.e. maintaining in the relation (2) the two first expressions and in consequence - in the relation (4) assuming the first three coefficients to be zero we get:

$$\Psi_0 \; (\ddot{\xi}_1, \; \dot{\xi}_1, \; \xi_1; \; A,\omega) = 0 \tag{5}$$

$$(\dot{\xi}_1, \; \xi_1; \; A,\omega) = 0$$

$$(\dot{\xi}_1, \; \xi_1; \; A,\omega) = 0 \; .$$

Considering the expressions giving the Fourier's coefficients, the relations (5) will be written in form:

$$\ddot{\xi}_1 + \omega_0^2 \, \xi_1 + \omega_0^2 \, \xi_{02} + \frac{1}{T}\int_0^T f_1 \; (\xi_1 + \xi_{02} + A \sin \omega t,$$

$$\dot{\xi}_1 + A \, \dot{\omega} \cos \omega) \, dt - p = 0 \; ;$$

$$\int_0^T f_1 \; (\xi_1 + \xi_{02} + A \sin \omega t, \; \dot{\xi}_1 + A \, \omega \cos \omega t) \cos \omega t \, dt = 0 \tag{6}$$

$$A(\omega_0^2 - \omega^2) + \frac{2}{T}\int_0^T f_1 \; (\xi_1 + \xi_{02} + A \sin \omega t, \; \dot{\xi}_1 + A \, \omega \cos \omega t)$$

$$\times \sin \omega t \, dt = 0 \; .$$

From the last two equations of the system (6) A and ω can be calculated as functions of ξ_1, $\dot{\xi}_1$, ξ_{02}. The exact form of these functions depends on the type of nonlinearity occurring in equation (1). In certain cases, which were analysed in detail in papers [2] and [3] the last two equations of the system (6) can be given in form

$$\begin{aligned} \omega^2 \, A^2 &= \Phi_1^0 + \Phi_1 \; (\xi_1, \; \dot{\xi}_1) \\ \omega^2 &= \Phi_2^0 + \Phi_2 \; (\xi_1, \; \dot{\xi}_1), \end{aligned} \tag{7}$$

where

$$\begin{aligned} \Phi_1 \; (0,0) &= 0 \; , \qquad & \Phi_2 \; (0,0) &= 0 \\ \Phi_1^0 &= \bar{\omega}^2 \, \bar{A}^2 \; , \qquad & \Phi_2^0 &= \bar{\omega}^2 \; , \end{aligned} \tag{8}$$

where $\bar{\omega}$ and $\bar{A}$ are the frequency and amplitude of self-excited vibrations in case of absence of disturbances.

Taking into consideration (7) in the first equation of the system (6) and calculating the constant value ξ_{02} we finally obtain for discussion equation

$$\ddot{\xi}_1 + f\ (\xi_1,\ \dot{\xi}_1) = p(t)\ . \tag{9}$$

The problem of perturbation will be analysed with respect to the mean square value of the amplitude. The expression we are interested in will be obtained from (7).

$$\begin{aligned} <A^2> &= \Phi_1^0 + <\Phi_1\ (\xi_1,\ \dot{\xi}_1)> \frac{1}{<\omega^2>} \\ <\omega^2> &= \Phi_1^0 + <\Phi_2\ (\xi_1,\ \dot{\xi}_1)>\ , \end{aligned} \tag{10}$$

where it was assumed that values A^2 and ω^2 are statistically independent.

In order to obtain the mean square value of the displacement ξ_1, the statistical linearization technique will be applied in equation (9). For this purpose the following relations will be taken:

$$\xi_1 = <\xi_1> + x_1 = m + x_1 \tag{11}$$

$$k = \frac{< f(m + x_1, x_1)x_1 >}{< \dot{x}_1^2 >}$$

$$\Omega^2 = \frac{< f(m + x_1, \dot{x}_1)x_1 >}{< x_1^2 >}\ ,$$

on the strength of which the equation (9) will take the form:

$$\ddot{x}_1 + k\ \dot{x}_1 + \Omega^2\ x_1 = p(t)\ . \tag{12}$$

Moreover the relation

$$< f(m + x_1,\ \dot{x}_1)> = 0 \tag{13}$$

is satisfied, since $<\ddot{\xi}_1> = <p> = 0$.

The solution of the equation (12) will be taken in the form

$$x_1 = a \sin (\Omega t + \Phi) = a \sin \theta \ , \qquad (14)$$

assuming the probability density function of the amplitude in the form of Rayleigh's formula

$$(a) = \frac{a}{<x_1^2>} \exp \left[- \frac{a^2}{2<x_1^2>} \right] . \qquad (15)$$

The relations (11) and (13) take, thus, the form

$$k = \frac{1}{2\pi<x_1^2>} \int_0^{2\pi} \int_0^{\infty} f \ (m + a \sin \theta \ ,$$

$$a \ \Omega \cos \theta) \ a \ \Omega \cos \theta \ (a) \ da \ d\theta$$

$$\Omega^2 = \frac{1}{2\pi<x_1^2>} \int_0^{2\pi} \int_0^{\infty} f \ (m + a \sin \theta \ , \qquad (16)$$

$$a \ \Omega \cos \theta) \ a \sin \theta \ (a) \ da \ d\theta$$

$$\int_0^{2\pi} \int_0^{\infty} f \ (m + a \sin \theta, \ a \ \Omega \cos \theta) \ (a) \ da \ d\theta = 0$$

From the above relations and from the formula

$$<\dot{x}_1^2> = \Omega^2 <x_1^2> \qquad (17)$$

k and Ω^2 will be expressed as functions of $<x_1^2>$ and parameters of the vibrating system, included in the function $f(\xi_1, \xi_1)$.

The mean square value of the displacement x_1 will be calculated from the relation

$$<x_1^2> = \int_{-\infty}^{+\infty} \left| H \ (j\omega) \right|^2 Sp \ (\omega) \ d\omega \ , \qquad (18)$$

where

$$H(j\omega) = \frac{1}{\Omega^2 - \omega^2 + j\,k\,\omega} \tag{19}$$

and Sp (ω) is the spectral density of excitation p. In the papers [2] and [3] a number of examples of self-excited vibrations were analysed. And so, for example, in the case of Van der Pol's equation

$$\ddot{x} - \varepsilon\,(1 - \gamma\,x^2)\,\dot{x} + \omega_0^2\,x = p(t) \tag{20}$$

with excitation in form of 'white noise' we have

$$\langle A^2 \rangle = \frac{2}{\gamma}\left[1 + \sqrt{1 - \frac{4\pi\,S0\,\gamma}{\omega_0^2\,\varepsilon}}\right] \tag{21}$$

$$\langle \omega^2 \rangle = \omega_0^2 ,$$

which relation is identical with the one obtained previously on another way in paper [4].

Analysis was also made of equations in which nonlinearity was expressed with following relations:

$$\begin{aligned}
f_1(x,\dot{x}) &= -(a_1^2 + \tfrac{1}{2}\beta\,\dot{x} - \tfrac{1}{3}\gamma^2\,\dot{x}^2)\dot{x}\\
f_1(x,\dot{x}) &= -\,\varepsilon(1 - x^2 - \mu\,x^4)\dot{x}\\
f_1(x,\dot{x}) &= -\,\varepsilon(1 - x^2)\,\dot{x} + \mu\,x^3\\
f_1(x,\dot{x}) &= -\,\varepsilon(1 - \tfrac{1}{3}\dot{x}^2)\,\dot{x} + \mu\,x^3
\end{aligned} \tag{22}$$

In all cases it was found that (a) the mean square value of the amplitude decreases with the increase of random disturbance, (b) the influence of non-linear members on the value of the amplitude is different and depends on the type of nonlinearity, (c) the mean square value of frequency does not depend on the external random perturbations.

It should be added that in paper [5] the author analysed the influence of random perturbations on uni-frequency self-excited vibrations of a material continuum applying the Krylov-Bogolubov-Mitropolsky method and the diffusion equation. As a criterion of the influence of random disturbances he assumed the most probable amplitude value, from which it follows that the increase of random perturbation causes also an increase of its value.

VIBRATIONS EXCITED BY HARMONIC FORCE

The method presented in the foregoing paragraph was applied for the case of vibrations with a nonlinear elastic force given by the equation

$$L\left[x,\dot{x},\ddot{x};t\right] \equiv \ddot{x} + 2\,\delta\dot{x} + f_1(x) - B_0 - B\cos\nu t = 0 . \tag{23}$$

As it is known an approximate solution describing harmonic vibrations is

$$\tilde{x} = a_0 + \tilde{a}\sin\nu t + \tilde{b}\cos\nu t \tag{24}$$

Applying harmonic linearization, for calculation of the constants a_0, $\tilde{a}$, $\tilde{b}$ the following system of equations will be obtained:

$$B_0 - \frac{1}{T}\int_0^T f_1(\tilde{x})\,dt = 0 \tag{25}$$

$$\tilde{a}\,\nu^2 + 2\tilde{b}\,\delta\,\nu - \frac{2}{T}\int_0^T f_1(\tilde{a})\sin\nu t\,dt = 0$$

$$\tilde{b}\,\nu^2 - 2\tilde{a}\,\delta\,\nu + B - \frac{2}{T}\int_0^T f_1(\tilde{x})\cos\nu t\,dt = 0$$

Assume that, apart from the exciting force in form of $B_0 + B\cos\nu t$ a random excitation p(t) also acts which represents small perturbations of harmonic excitation as a stationary, normal, stochastic process with mean value equal to zero.

The solution of the equation (23) completed with the function p(t) will be searched for in the form (3) where at present

$$\xi_2 = \tilde{x}_1 = a_0 + a \sin \nu t + b \cos \nu t \; . \tag{26}$$

From the assumption made it follows that $a \to \tilde{a}$, $b \to \tilde{b}$, $\xi_1 \to 0$ when $p \to 0$.

The equation system corresponding to (6) take now the form

$$\ddot{\xi}_1 + 2\,\delta\,\dot{\xi}_1 + \frac{1}{T}\int_0^T f_1\;(\xi_1 + \tilde{x}_1)\;dt - B_0 - p = 0 \tag{27}$$

$$b\,\nu^2 + 2b\,\delta\,\nu + B - \frac{2}{T}\int_0^T f_1\;(\xi_1 + \tilde{x}_1)\,\cos \nu t\;dt = 0$$

$$a\,\nu^2 + 2b\,\delta\,\nu - \frac{2}{T}\int_0^T f_1\;(\xi_1 + \tilde{x}_1)\,\sin \nu t\;dt = 0\;.$$

From the last two equations we shall calculate

$$a = a\;(\xi_1;\;a_0)$$

$$b = b\;(\xi_1;\;a_0)\,, \tag{28}$$

which expressions after substituting into the first equation of the system (27) give equation (9) for a respective function $f(\xi_1,\;\dot{\xi}_1)$ which does not contain any more the component of a constant value.

An exact calculation of coefficients (28) meets, in a general case, great difficulties. These values will be, thus, calculated in an approximate way. Considering them as analytical functions of the variable ξ_1 they are presented in form of series

$$a = \alpha_0 + \alpha_1\,\xi_1 + \alpha_2\,\xi_1^2 + \ldots.$$

$$b = \beta_0 + \beta_1\,\xi_1 + \beta_2\,\xi_1^2 + \ldots.\,,$$

where $\quad \alpha_0 = a\;(0) = \tilde{a}, \quad \beta_0 = b\;(0) = \tilde{b}$,

whereas $\quad \alpha_i,\;\beta_i,\quad i = 1,2,\ldots.$ (29)

should be calculated. Substituting (29) into (26) we obtain

$$f_1(x) = f_1(\xi_1 + \tilde{x}_1) = \sum_{j=0}^{\infty} f_{ij}\, \xi_1^j \,. \tag{30}$$

Some of the first coefficients of the series (30) take the form

$$\begin{aligned}
f_{10} &= f_1(a_0 + \alpha_0 \sin \nu t + \beta_0 \cos \nu t) \\
f_{11} &= \frac{df_1}{dx} \cdot \frac{dx}{d\xi_1} \Bigg|_{\xi_1 = 0} \\
f_{12} &= \frac{1}{2!} \left[\frac{d^2 f_1}{d\, x^2} \left(\frac{dx}{d\xi_1}\right) + \frac{df_1}{dx} \frac{d^2\, x}{d\, \xi_1^2} \right]_{\xi_1 = 0} \\
f_{13} &= \frac{1}{3!} \left[\frac{d^3 f_1}{d\, x^3} \left(\frac{dx}{d\xi_1}\right)^3 + 3 \frac{d^2 f_1}{d\, x^2} \frac{dx}{d\xi_1} \frac{d^2 x}{d\xi_1^2} + \frac{df_1}{dx} \frac{d^3 x}{d\xi_1^3} \right]_{\xi_1 = 0}
\end{aligned} \tag{31}$$

and

$$x = \xi_1 + a_0 + \sum_{i=0}^{\infty} (\alpha_i \sin \nu t + \beta_i \cos \nu t)\, \xi_1^i \,. \tag{32}$$

From (26) and (29) the square of the amplitude is obtained

$$A^2 = a^2 + b^2 = \sum_{j=0}^{\infty} A_j\, \xi_1^j \,, \tag{33}$$

where

$$\begin{aligned}
A_0 &= \alpha_0^2 + \beta_0^2 \,, \quad A_1 = 2(\alpha_0\, \alpha_1 + \beta_0\, \beta_1) \\
A_2 &= \alpha_1^2 + \beta_1^2 + 2\,(\alpha_0\, \alpha_2 + \beta_0\, \beta_2) \\
A_3 &= 2\,(\alpha_1\, \alpha_2 + \beta_1\, \beta_2 + \ldots.) \\
A_4 &= \alpha_2^2 + \beta_2^2 + \ldots. \,,
\end{aligned} \tag{34}$$

whereas the mean square value of the amplitude is

$$\langle A^2 \rangle = \sum_{j=0}^{\infty} A_j \langle \xi_1^j \rangle \,. \tag{35}$$

After substituting (30) into the two last equations of the equation system (27) and taking into consideration the relation (29) and assuming the value of coefficients occurring at the same powers of the variable ξ_1 as zero a recurrent equation system is obtained for calculating α_j, β_j.

$$\nu^2 \alpha_j + 2 \delta\nu\beta j - \frac{2}{T}\int_0^T f_{1j} \sin \nu t \, dt = 0$$

$$\nu^2 \beta j - 2 \delta\nu\alpha j + B \delta_{oj} - \frac{2}{T}\int_0^T f_{ij} \cos \nu t \, dt = 0 \tag{36}$$

$$j = 0,1,2,\ldots ,$$

where

$$\delta_{oj} = \begin{cases} 1, & j = 0 \\ 0, & j = 1 . \end{cases} \tag{37}$$

The calculated coefficients of development (29) should be considered in the first equation of the system (27) where the constant $a_0 = \bar{a}_0$ is calculated from the relation

$$\frac{1}{T}\int_0^T f_{10} \, dt = B_0 \tag{38}$$

The function $f(\xi_1, \dot{\xi}_1)$ occurring in the general equation (9) has in the considered case the final form

$$f(\xi_1, \dot{\xi}_1) = 2 \delta \dot{\xi}_1 + \sum_{j=0}^{\infty} \left(\frac{1}{T}\int_0^T f_{ij}\Big|_{a_0 = \bar{a}_0} dt\right) \xi_1^j \tag{39}$$

In detailed calculations we confine ourselves to calculations of some first coefficients of development (29).

In the paper [6] the problem concerning Duffing's equation with nonsymetric characteristic was considered.

$$f_1(x) = \gamma_1 x + \gamma_2 x^2 + \gamma_3 x^3, \ \gamma_1 > 0, \ \gamma_3 \geq 0 \tag{40}$$

Now the results concerning a particular case will be presented for which $\gamma_2 = 0$, $B_0 = 0$.

After considering the relations (29) in the first equation system (27) it will take the form

$$\ddot{\xi}_1 + 2 \delta \dot{\xi}_1 + \tilde{\gamma}_1 \xi_1 + \tilde{\gamma}_3 \xi_1^3 = p(t) \tag{41}$$

where for the assumptions made

$$\begin{aligned} \tilde{\gamma}_1 &= \gamma_1 + \frac{3}{2} \gamma_3 (\alpha_0^2 + \beta_0^2) \\ \tilde{\gamma}_3 &= \gamma_3 \left[1 + 3 (\alpha_0 \alpha_2 + \beta_0 \beta_2) \right] . \end{aligned} \tag{42}$$

The magnitudes α_0, β_0, α_2, β_2, will be calculated from the equation system (36).

CALCULATION OF PROBABILISTIC CHARACTERISTICS

In a general case the random process determined by the relation (3) considering (26) is a nonstationary process. With respect to the fact that the functions sin νt and cos νt are functions of the second group, the subsequently given theory will be used for calculating the probabilistic characteristics.

The mean value of the second kind of the nonstationary process (3) will be calculated. We have

$$\boxed{x} = \lim_{T\to\infty} \frac{1}{2T} \int_{-T}^{T} \langle x \rangle \, dt = \langle \xi_1 \rangle + a_0 \ , \tag{43}$$

where it was assumed that the process ξ_1 is a stationary process which causes the fact that $\langle \xi_1 \rangle$ and $\langle a \rangle$ and $\langle b \rangle$ are not time dependent.

The correlation function of the second kind is calculated from the relation

$$\Phi_x(\tilde{t}) = \lim_{T\to\infty} \frac{1}{2T} \int_{-T}^{T} K\left[t, t+\tau\right] dt \tag{44}$$

where

$$K\left[t, t+\tau\right] = <(x(t) - \left[x\right])(x(t+\tau) - \left[x\right])> \tag{45}$$

Taking into consideration the expressions (3) and (43) we get

$$\Phi_x(\tau) = <\xi_1 \xi_{1\tau}> - <\xi_1>^2 + \frac{1}{2} <a\, a_{\tilde{2}}> \cos \nu\tau$$

$$+ \frac{1}{2} <b\, b_\tau> \sin \nu\tau - \frac{1}{2} <a\, b_\tau> \sin \nu\tau + \frac{1}{2} <b\, b_\tau> \cos \nu\tau \tag{46}$$

where $Z_\tau = S(t + \tau)$ was determined.

The dispersion of the second kind of the random function $x(t)$ is calculated from (46) assuming $\tau = 0$. We thus obtain from (35)

$$D_x = \Phi_x(0) = K_{\xi_1}(0) + \frac{1}{2}(<a^2> + <b^2>) = K_{\xi_1} + \frac{1}{2} <A^2> \tag{47}$$

or

$$\sigma_x^2 = \sigma_{\xi_1}^2 + \frac{1}{2} <A^2>, \tag{48}$$

where

$$\sigma_{\xi_1}^2 = K_{\xi_1}(0) = <\xi_1^2> \tag{49}$$

since from the assumption made we get $<\xi_1> = 0$. The subsequently applied method of statistic linearization as shown earlier leads to equation (18) which in case of 'white noise' takes the form

$$2\,\delta <\xi_1^2> (\tilde{\gamma}_1 + 3\,\tilde{\gamma}_3 <\xi_1^2>) = \pi\, S_0 \tag{50}$$

whose positive root is expressed as follows

$$<\xi_1^2> = (6\,\tilde{\gamma}_3)^{-1} \left(-\tilde{\gamma}_1 + \sqrt{\tilde{\gamma}_1^2 + 6\,\tilde{\gamma}_3 \cdot \frac{\pi S_0}{\delta}} \right) \tag{51}$$

Taking into consideration the relations (34), (35), (48) and (49) and confining to two expressions in (35) we have

$$\sigma_x^2 = \frac{1}{2} (\alpha_0^2 + \beta_0^2) + (1 + \alpha_0\alpha_2 + \beta_0\beta) <\xi_1^2> , \tag{52}$$

where (51) should be considered. It is easy to check that in the case of a linear characteristic ($\tilde{\gamma}_3 \to 0$) the dispersion of displacement is expressed by the relation

$$\bar{\sigma}_x^2 = \bar{\sigma}_{x_0}^2 + \frac{\pi S_0}{2\gamma_1\delta} , \tag{53}$$

where $\bar{\sigma}_{x_0}^2$ is the dispersion of displacement x in the case of deterministic excitation and $\frac{\pi S_0}{2\gamma_1\delta}$ is a displacement dispersion with random excitation. The principle of superposition occurs then, whereas, in expression (52) the influence of deterministic excitation on a part connected with random excitation takes place.

As detailed numerical calculations show the influence of random perturbations is such that they cause a decrease of the amplitude square for a frequency less than resonance frequency (resonance vibrations) and the amplitude square increases for frequencies higher than resonance frequencies (super-resonance vibrations). This causes a decrease of the amplitude jump at its break connected with a change of frequency.

REFERENCES

1 K. Piszczek, Poznań Probabilistic methods in the theory of non-linear vibrations (in Polish) (1974).

2 K. Piszczek, Archive of Applied Mechanics, V 25 No. 5, Influence of random perturbations on self-excited vibrations of a system with one degree of freedom (1973).

3 J. Luczko, Theoretical and Applied Mechanics V 13 No. 3. Influence of additional perturbations on the probabilistic characteristics of classical equations describing self-excited vibrations (in Polish) (1975).

4 T.K. Caughey, I Appl. Mechanics September. Response of Van der Pol's oscillator to random excitation (1959).

5 W.G. Kolomiez, Kiev Random vibrations of nonlinear systems with continuous parameters (in Russian) (1975).

6 K. Piszczek, Problems of Nonlinear Vibrations, PAN Warszawa. Influence of random excitation on harmonic vibrations of a system of one degree of freedom (in press). (1976).

Professor Dr. K. Piszczek, Politechnika Krakowska, Instytut M-1, ul. Warszawska 24, 31 - 155 Kraków, Poland.

DISCUSSION

Professor Ariaratnam presented an alternative analysis which did not depend on the principle of equivalent linearisation which he mentioned that Kozin had cautioned against in certain problems; in particular the negative linear damping of (20) might cause problems.

Consider the Van der Pol equation (20) of the paper:

$$\ddot{x} - \varepsilon(1 - \gamma x^2)\dot{x} + \omega_0^2 x = p(t)$$

where $\langle p(t)\, p(t+\tau)\rangle = \varepsilon S_0\, \delta(\tau)$.

The transformation:

$$x = a(t)\cos\Phi(t), \quad \dot{x} = -a\,\omega_0 \sin\Phi(t), \quad \Phi = \omega_0 t + \phi(t)$$

yields the pair of first-order equations:

$$\dot{a} = \varepsilon(1 - \gamma a^2 \cos^2\Phi)\, a \sin^2\Phi - \frac{1}{\omega_0}\, p(t) \sin\Phi$$

$$a\dot{\phi} = \varepsilon(1 - \gamma a^2\cos^2\Phi)\, a\sin\Phi\cos\Phi - \frac{1}{\omega_0}\, p(t)\cos\Phi.$$

Application of the averaging method of Stratonovich and Khas'minski leads to the following Itô equation for the amplitude a(t);

$$d\,a = \varepsilon\left(\frac{a}{2} - \frac{\gamma a^3}{8} + \frac{\alpha}{2a}\right) dt + (\varepsilon\alpha)^{\frac{1}{2}}\, dW$$

where $\alpha = S_0/(2\omega_0^2)$ and W(t) is the Wiener process of unit intensity.

The Fokker-Planck equation governing the conditional probability density p(a,t) is then

$$\frac{\partial p}{\partial t} = \varepsilon \frac{\partial}{\partial a}\left[-\left(\frac{a}{2} - \frac{\gamma a^3}{8} + \frac{\alpha}{2a}\right) p\right] + \frac{\alpha}{2}\, \frac{\partial^2 p}{\partial a^2},$$

whose stationary solution is

$$p(a) = C\, a \exp\left[\frac{1}{2\alpha}\left(a^2 - \frac{\gamma a^4}{8}\right)\right]$$

where

$$C^{-1} = 2 \left(\frac{\pi\alpha}{\gamma}\right)^{\frac{1}{2}} \exp\left(\frac{1}{\alpha\gamma}\right) \left[1 + \Psi\left(\frac{1}{\sqrt{\alpha\gamma}}\right)\right],$$

$\Psi(x)$ being the error function defined by

$$\Psi(x) = \frac{2}{\sqrt{\pi}} \int_0^x e^{-t^2}\, dt .$$

The mean square amplitude $<a^2>$ is evaluated to be:

$$<a^2> = \frac{4}{\gamma}\left[1 + \frac{\sqrt{\frac{\alpha\gamma}{\pi}} \exp\left(-\frac{1}{\alpha\gamma}\right)}{1 + \Psi\left(\frac{1}{\sqrt{\alpha\gamma}}\right)}\right].$$

In the absence of the stochastic excitation, the limit cycle amplitude is $2/\gamma^{1/2}$. Thus the mean square amplitude is increased as a result of the stochastic forcing. This is at variance with the result of the author (eq. (21)) as well as that of Caughey Ref. [4] but agrees with that found by Khas'minski [7] using a slightly different approach. It may be remarked, however, that the mean amplitude $< a >$ is decreased due to the stochastic excitation (see Stratonovich [8] , p. 127, eq. (5.19)).

REFERENCES

7 R.Z. Khas'minski, The behaviour of a self-oscillating system acted upon by slight noise. P.M.M. (English Transl.), 27, (4), 1963, pp. 1035-1044.

8 R.L. Stratonovich, Topics in the theory of random noise, Vol. II, Gordon and Breach, New York, (1967).

AUTHOR'S REPLY TO COMMENTS BY ARIARATNAM

First, I would like to stress that equivalent linearisation is applied after transformation to equation (9) where $f(\xi_1, \dot{\xi}_1)$ has a coefficient greater than zero and that Professor Ariaratnam's objection might thus be inapplicable.

It is also possible that there are mistakes in the calculations presented by Professor Ariaratnam, which lead to results different from mine. The best way to check theoretical results is by experiment. In the case of the influence of stochastic perturbations on amplitude of the Van der Pol equation, an experiment has confirmed my results (see [2]). Moreover, the linearisation technique was not applied to the original equation but to equation (9) obtained in the proper way as in [2]. As follows from [2], equation (9) clearly has a positive coefficient to the linear damping term.

Besides the present results can be checked in the other way by applying the averaging method to coefficients of Fokker-Planck equation, as in the papers: I.A. Mitropolsky - Averaging Method in Nonlinear Mechanics, Kiev, 1971, (in Russian), K. Piszczek - On certain methods for investigation the influence of random perturbations to deterministic vibrations, (in Polish), 'Wibroakustyka' T.I., Nr 4, 1976, AHG Krakow.

Applying this method for the equation written by Professor Ariaratnam (with $\gamma = 1$) and transforming it in the same way, we obtain a Fokker-Planck equation of the form

$$\frac{d}{da} \{\tfrac{1}{2} a \left[1 - \tfrac{1}{4} a^2\right] w\} = \frac{\sigma^2}{4} \frac{d^2 w}{da^2} ,$$

and its solution as a probability density function is

$$w(a) = C \exp\left[- \frac{a^2}{\sigma^2} \left(\frac{a^2}{8} - 1\right) \right].$$

From the normalization condition follows

$$C = 2 \frac{\exp\left(-\frac{1}{\sigma^2}\right)}{\sqrt{2\pi\sigma}\; D_{-\frac{1}{2}}\left(-\frac{2}{\sigma}\right)} ,$$

where $D_q(z)$ represents the parabolic cylinder function. The mean square value of amplitude takes the form

$$\langle a^2 \rangle = \sigma \frac{D_{-3/2}\left(-\frac{2}{\sigma}\right)}{D_{-\frac{1}{2}}\left(-\frac{2}{\sigma}\right)} ,$$

$$\langle a^2 \rangle = 4 + 2\sigma \frac{D_{\frac{1}{2}}\left(-\frac{2}{\sigma}\right)}{D_{-\frac{1}{2}}\left(-\frac{2}{\sigma}\right)} ,$$

where the relationship

$$\tfrac{1}{2} D_{-3/2}(z) = D_{\frac{1}{2}}(z) - z\, D_{-\frac{1}{2}}(z)$$

was applied.

The function $\langle a^2 \rangle$ versus σ is indicated on Figure 1. It is clear that for small values of σ (as was assumed in my papers) random perturbations cause the decrease of the mean square value of amplitude of vibrations.

In this way the previous results obtained by another method are confirmed.

Note added in Final Preparation

In a recent paper on a related subject (Juan Lin and Peter B. Kahn, Limit Cycles in random environments, S.I.A.M. J.Appl. Math. 32(1), 260 (1977)) it is shown that Limit Cycles occurring in predator-prey models suffer a reduction of amplitude in the presence of random perturbations. The Stratonovich-Khasminskii Limit Theorem is used and it is found that '(a) the radius of the limit cycle decreases as noise increases, (b) If noise dispersion is relatively large, the stationary probability distribution of a small deterministic limit cycle may be difficult to differentiate from the distribution of a stable focus, (c) The dispersion of the angular variable increases linearly with time.' This appears to confirm Ariaratnam's remark that 'the mean amplitude <a> is decreased due to stochastic excitation'. The magnitudes of third and higher moments increase with noise, but in this case the second moment or mean square amplitude appears to be unaffected.

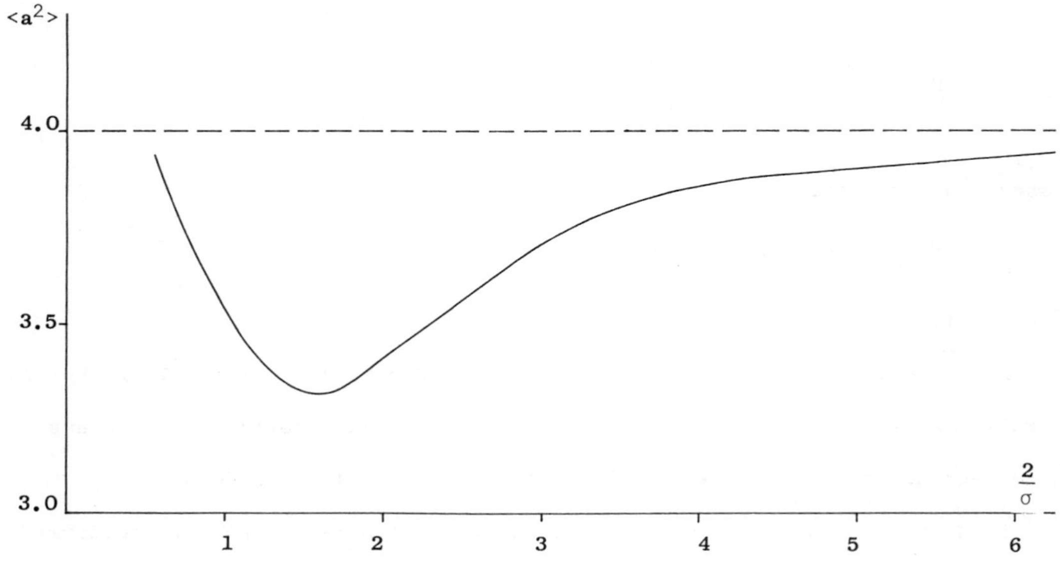

Figure 1 Variation of function $\langle a^2 \rangle$ with σ

G SCHMIDT

Probability densities of parametrically excited random vibrations

INTRODUCTION

The internal resonance of a system under the action of broad band random excitation is investigated. The Fokker Planck Kolmogorov equation method combined with the averaging method yields the two-dimensional stationary probability density. The behaviour of the probability density is discussed.

We investigate the internal resonance of a system of two vibration equations

$$x_i'' + \omega_i^2 x_i = F_i \ (t, x_\nu, x_\nu', x_\nu''), \quad i = 1,2; \ \nu = \{1,2\} \tag{1}$$

where ω and $\omega_2 > \omega_1$ are circular frequencies, F_i are nonlinear functions, and derivatives with respect to the time t are denoted by primes.

Introducing the partial amplitudes a_i and phases $\phi_i = \omega_i t + \theta_i$ by

$$x_i = a_i \cos \phi_i, \ x_i' = -\omega_i \ a_i \sin \phi_i,$$

we can write (1) in the standard form

$$- \omega_i \ a_i' = F_i \sin \phi_i, \quad - \omega_i \ a_i \ \theta_i' = F_i \cos \phi_i \tag{2}$$

In order to investigate internal resonances, for instance of parametrically excited bars and sandwich plates [5] or of systems with autoparametric coupling as autoparametric vibration absorbers [3], [4] we assume the functions F_i in the form

$$F_1 = - \beta_1 x_1' - b \ x_1^2 x_1' + c \ x_1^3 + d \ x_1 x_2 + e \ x_1 x_2''$$
$$+ f \ x_1(x_1'^2 + x_1 x_1'') + \gamma \ \dot{\xi}(t) x_1 \quad ,$$

$$F_2 = - \beta_2 x_2' + g\, x_1^2 + h\,(x_1'^2 + x_1 x_1'') + \varepsilon\, \dot{\xi}(t)$$

with small linear damping terms β_1, β_2, a nonlinear damping term b, a white noise random excitation $\dot{\xi}$ (t) acting as a parametric excitation in the first and as a forced excitation in the second equation with the small coefficients γ and $\varepsilon > 0$, and other nonlinear terms with the coefficients c, d, e, f, g, and h which cause a nonlinear coupling of the two equations. The terms with the coefficients f and h are known as nonlinear inertia terms.

The smallness of the righthand functions F_i is guaranteed without the assumption of small coefficients b, c, d, ... of the nonlinear terms. We have only to consider the solution x_1, x_2 and its derivatives as small, an assumption underlying the equations (1) with these and no higher order nonlinearities.

Having to get rid of the second derivatives in the right-hand functions F_i, we insert from (1)

$$x_i'' = - \omega_i x_i + F_i(t, x_\nu, x_\nu', x_\nu'')$$

and get second derivatives only in nonlinear terms of higher degree not taken into consideration in (1). By repeating this procedure, we could restrict the second derivatives to nonlinear terms of a degree as high as you like. In what follows we confine ourselves to nonlinearities of the degrees considered already in (1).

With such functions F_i, the first order differential equations (2) yield

$$\begin{aligned} - \omega_1 a_1' = {} & \omega_1 a_1 (\beta_1 + b a_1^2 \cos^2\phi_1) \sin^2\phi_1 + c\, a_1^3 \sin\phi_1 \cos^3\phi_1 \\ & + d a_1 a_2 \sin\phi_1 \cos\phi_1 \cos\phi_2 - \omega_2^2\, e\, a_1 a_2 \sin\phi_1 \cos\phi_1 \cos\phi_2 \\ & + \omega_1^2\, f\, a_1^3 (\sin^2\phi_1 - \cos^2\phi_1) \sin\phi_1 \cos\phi_1 + \gamma a_1\, \dot{\xi}(t) \sin\phi_1 \cos\phi_1 \quad , \end{aligned}$$

$$- \omega_2 a_2' = \omega_2 \beta_2 a_2 \sin^2\phi_2 + g\, a_1^2 \cos^2\phi_1 \sin\phi_2$$

$$+ \omega_1^2 h\, a_1^2 (\sin^2\phi_1 - \cos^2\phi_1) \sin\phi_2 + \varepsilon\, \dot{\xi}(t) \sin\phi_2 \quad ,$$

$$- \omega_1 a_1 \theta_1' = \omega_1 a_1 (\beta_1 + b a_1^2 \cos^2\phi_1) \sin\phi_1 \cos\phi_1 + c\, a_1^3 \cos^4\phi_1$$

$$+ d\, a_1 a_2 \cos^2\phi_1 \cos\phi_2 - \omega_2^2 e\, a_1 a_2 \cos^2\phi_1 \cos\phi_2$$

$$+ \omega_1^2 f\, a_1^3 (\sin^2\phi_1 - \cos^2\phi_1) \cos^2\phi_1 + \gamma a_1\, \dot{\xi}(t) \cos^2\phi_1 \quad ,$$

$$- \omega_2 a_2 \theta_2' = \omega_2 \beta_2 a_2 \sin\phi_2 \cos\phi_2 + g\, a_1^2 \cos^2\phi_1 \cos\phi_2$$

$$+ \omega_1^2 h\, a_1^2 (\sin^2\phi_1 - \cos^2\phi_1) \cos\phi_2 + \varepsilon\, \dot{\xi}(t) \cos\phi_2$$

These differential equations are assumed as equations of Stratonovich type. If they are written in the form

$$dy_k = P_k\, dt + Q_k\, d\xi \quad , \qquad k = 1,\ldots,4$$

where

$$y_i = a_i \;,\quad y_{i+2} = \theta_i \;\; (i = 1,2) \quad \text{and} \quad d\xi = \dot{\xi}(t)\, dt \;,$$

the corresponding Itô equations [1] are

$$dy_k = (P_k + \Pi_k)\, dt + Q_k\, d\xi \;, \qquad \Pi_k = \frac{1}{2} \sum_{\ell=1}^{4} \frac{\partial Q_k}{\partial y_\ell} Q_\ell \tag{3}$$

In our case the first two additional Itô terms are

$$\Pi_1 = \frac{\gamma^2 a_1}{2\omega_1^2} \cos^4\phi_1 \quad , \qquad \Pi_2 = \frac{\varepsilon^2}{2\omega_2^2 a_2} \cos^2\phi_2 \;.$$

After trigonometric transformation, the Itô equations (3) can be written

$$da_i = p_i dt + q_i d\xi \quad , \qquad d\theta_i = r_i dt + s_i d\xi$$

with

$$p_1 = -\frac{\beta_1 a_1}{2} + \frac{\beta_1 a_1}{2}\cos 2\phi_1 - \frac{ba_1^3}{8} + \frac{ba_1^3}{8}\cos 4\phi_1$$

$$-\frac{ca_1^3}{4\omega_1}\sin 2\phi_1 - \frac{ca_1^3}{8\omega_1}\sin 4\phi_1$$

$$-\frac{d-\omega_2^2 e}{4\omega_1} a_1 a_2 \sin(2\phi_1+\phi_2) - \frac{d-\omega_2^2 e}{4\omega_1} a_1 a_2 \sin(2\phi_1-\phi_2)$$

$$+\frac{\omega_1 f\, a_1^3}{4}\sin 4\phi_1 + \frac{3\gamma^2 a_1}{16\,\omega_1^2} + \frac{\gamma^2 a_1}{4\omega_1^2}\cos 2\phi_1 + \frac{\gamma^2 a_1}{16\omega_1^2}\cos 4\phi_1 \quad ,$$

$$p_2 = -\frac{\beta_2 a_2}{2} + \frac{\beta_2 a_2}{2}\cos 2\phi_2 - \frac{ga_1^2}{2\omega_2}\sin\phi_2$$

$$-\frac{ga_1^2}{4\omega_2}\sin(2\phi_1+\phi_2) + \frac{ga_1^2}{4\omega_2}\sin(2\phi_1-\phi_2)$$

$$+\frac{\omega_1^2 ha_1^2}{2\omega_2}\sin(2\phi_1+\phi_2) - \frac{\omega_1 ha_1^2}{2\omega_2}\sin(2\phi_1-\phi_2)$$

$$+\frac{\varepsilon^2}{4\omega_2^2 a_2} + \frac{\varepsilon^2}{4\omega_2^2 a_2}\cos 2\phi_2 \quad ,$$

$$q_1 = -\frac{\gamma a_1}{2\omega_1}\sin 2\phi_1 \ , \qquad q_2 = -\frac{\varepsilon}{\omega_2}\sin\phi_2$$

and corresponding expressions for r_i, s_i.

The Fokker Planck Kolmogorov equation determines the probability density as a function of a_i, θ_i, and t. As a first approximation of a successive method [6], [2], we use the averaging method in the coefficients of the FPK equation, that is we omit the oscillating terms containing ϕ_1, ϕ_2. Then the FPK equation for the probability density $v(a_1,a_2,t)$ reads

$$\frac{\partial v}{\partial t} = - \frac{\partial v(\bar{p}_1 v)}{\partial a_1} - \frac{\partial(\bar{p}_1 v)}{\partial a_2} + \frac{1}{2}\frac{\partial^2(\bar{q}_1^2 v)}{\partial a_1^2} + \frac{\partial^2(\overline{q_1 q_2} v)}{\partial a_1\, \partial a_2} + \frac{1}{2}\frac{\partial^2(\bar{q}_2^2 v)}{\partial a_2^2} \qquad (4)$$

where

$$\bar{p}_1 = -\frac{\beta_1 a_1}{2} - \frac{b a_1^3}{8} + \frac{3\gamma^2 a_1}{16\,\omega_1^2}\ , \quad \bar{p}_2 = -\frac{\beta_2 a_2}{2} + \frac{\varepsilon^2}{4\omega_2^2 a_2}\ ,$$

$$\overline{q_1^2} = \frac{\gamma^2 a_1^2}{8\omega_1^2}\ , \quad \overline{q_1 q_2} = \delta_{\omega_2}^{2\omega_1} \frac{\gamma\varepsilon\, a_1}{4\omega_1\omega_2}\ , \quad \bar{q}_2^2 = \frac{\varepsilon^2}{2\omega_2^2}$$

The expression containing the Kronecker symbol $\delta_{\omega_2}^{2\omega_1}$ only appears in the case $\omega_2 = 2\,\omega_1$ of an internal resonance; corresponding with the approximation in hand, in this expression was set $\cos(2\theta_1 - \theta_2) = 1$.

First we neglect the variable x_2 in (1), confining ourselves to the first vibration equation (1) with parametric excitation. Then the FPK equation (4) has a stationary solution ($\frac{\partial v}{\partial t} = 0$) satisfying the boundary conditions $\bar{p}_1 v \to 0$, $\frac{d}{da_1}(\overline{q_1^2} v) \to 0$ for $a \to \infty$.

$$v = C\, e^{\int \frac{2\bar{p}_1 - \frac{d\overline{q_1^2}}{da_1}}{\overline{q_1^2}} da_1} = C\, a_1^{(1 - \frac{8\omega_1^2\beta_1}{\gamma^2})}\, e^{(-\frac{\omega_1^2 b}{\gamma^2} a_1^2)} \qquad (5)$$

The normalisation condition leads to

$$C = \frac{2\left(\frac{\omega_1^2 b}{\gamma^2}\right)^{(1 - \frac{4\omega_1^2\beta_1}{\gamma^2})}}{\Gamma(1 - \frac{4\omega_1\beta_1}{\gamma^2})} \qquad (\gamma^2 > 4\omega_1^2\,\beta_1,\ b > 0)$$

where Γ is the gamma function.

If we take into consideration both the equations (1), the FPK equation (4) yields for the stationary case

$$\frac{\gamma^2 a_1^2}{16\omega_1^2}\frac{\partial^2 v}{\partial a_1^2} + \frac{\varepsilon^2}{4\omega_2^2}\frac{\partial^2 v}{\partial a_2^2} + \delta^{2\omega_1}_{\omega_2}\frac{\gamma\varepsilon a_1}{4\omega_1\omega_2}\frac{\partial^2 v}{\partial a_1 \partial a_2} \tag{6}$$

$$+ \left(\frac{\gamma^2 a_1}{16\omega_1^2} + \frac{\beta_1 a_1}{2} + \frac{b a_1^3}{8}\right)\frac{\partial v}{\partial a_1} + \left(\frac{\beta_2 a_2}{2} + \delta^{2\omega_1}_{\omega_2}\frac{\gamma\varepsilon}{4\omega_1\omega_2} - \frac{\varepsilon^2}{4\omega_2^2 a_2}\right)\frac{\partial v}{\partial a_2}$$

$$+ \left(-\frac{\gamma^2}{16\omega_1^2} + \frac{\beta_1+\beta_2}{2} + \frac{3ba_1^2}{8} + \frac{\varepsilon^2}{4\omega_2^2 a_2^2}\right) v = 0$$

The direct integration of this equation, as in the one-dimensional case, fails. Generalizing (5), we successively seek a solution of (6) in the form

$$v = C\, a_1^i a_2^j \exp \sum_{k+\ell=1}^{4} C_k \; a_1^k a_2^\ell$$

$$(k, \ell \geq 0)$$

Inserting into (6) leads at least to the solution

$$v = C\, a_1^i a_2 e^{(c_{01}a_2 + c_{03}a_2^3 + c_{22}a_1^2 a_2^2 \; c_{04}a_2^4)} \tag{7}$$

where

$$c_{01} = -\;\delta^{2\omega_1}_{\omega_2}\frac{2\gamma}{\varepsilon}(i+1)\;, \qquad c_{03} = \delta^{2\omega_1}_{\omega_2}\frac{16\omega_1^2\beta_2\gamma}{g\varepsilon^3}(i+1)\;,$$

$$c_{22} = -\frac{\omega_2^2 b}{8\varepsilon^2}(i+3)\;, \qquad c_{04} = \delta^{2\omega_1}_{\omega_2}\frac{2\omega_1^2\beta_2\gamma^2}{3\varepsilon^4}(i+1)^2$$

and

$$i = -\frac{4\omega_1^2\beta_1}{\gamma^2} \pm \sqrt{\left(1 - \frac{4\omega_1^2\beta_1}{\gamma^2}\right)^2 - \frac{16\;\omega_1^2\beta_2}{\gamma^2}}$$

In the special case $\beta_2 = 0$ is

$$i_1 = 1 - \frac{8\omega_1^2\beta_1}{\gamma^2}\;, \quad i_2 = -1$$

In the general case $\beta_2 > 0$ the radicand is smaller and, consequently, i_1 is smaller and i_2 greater than for $\beta_2 = 0$. Two real solutions exist as long as

$$\beta_2 \leq \frac{\gamma^2}{16\omega_1^2}\left(1 - \frac{4\omega_1^2\beta_1}{\gamma^2}\right)^2$$

The solution i_1 is non-negative for

$$\beta_1 + 2\beta_2 \leq \frac{\gamma^2}{8\omega_1^2} \quad .$$

In order to find the maximum value of the probability density, we differentiate (7):

$$\frac{\partial v}{\partial a_1} = C\, a_1^{i-1}\, a_2 (i + 2c_{22} a_1^2 a_2^2)\, e^{(c_{01}a_2 + c_{03}a_2^2 + c_{22}a_1^2 a_2^2 + c_{04}a_2^4)}$$

Extreme values demand $i + 2\, c_{22}\, a_1^2\, a_2^2 = 0$, that is

$$a_1 a_2 = E\ , \qquad E = \frac{2\varepsilon}{\omega_2}\sqrt{\frac{i}{b(i+3)}} \tag{8}$$

a hyperbola in the a_1, a_2 plane. For it, (7) yields

$$v_h = C\, E a_1^{i-1}\, e^{\left(c_{22}\, E^2 + \frac{c_{01}E}{a_1} + \frac{c_{03}E^3}{a_1^3} + \frac{c_{04}E^4}{a_1^4}\right)} \tag{9}$$

with the derivative

$$\frac{dv_h}{da_1} = C\, E a_1^{i-2}\left(i - 1 - \frac{c_{01}E}{a_1} - \frac{3c_{03}E^3}{a_1^3} - \frac{4c_{04}E^4}{a_1^4}\right) e^{\cdots}$$

where the exponent of e is the same as in (9). The sign of the expression in parenthesis informs about the behaviour of the probability density on the hyperbola. For $\omega_2 \neq 2\,\omega_1$ it is negative, v_h diminishes when a_1 augments.

In the case $\omega_2 = 2\,\omega_1$ of an internal resonance, we first assume $\beta_2 = 0$. When $a_1 < \frac{c_{01}E}{i-1} = \frac{4\gamma}{\omega_2}\frac{(1+i)}{(1-i)}\sqrt{\frac{i}{b(i+3)}}$ the expression in parenthesis is positive, consequently v_h augments, for larger values of a_1, v_h diminishes. The maximum of the probability density therefore appears for

$$a_1 = \frac{4\gamma(1+i)}{\omega_2(1-i)}\sqrt{\frac{i}{b(i+3)}}, \quad a_2 = \frac{\varepsilon(1-i)}{2\gamma(1+i)} \quad ,$$

the last expression because of (8). If β_2 is not zero, a second maximum can appear. The values $\omega_1 = 1$, $\gamma = 10$, $\varepsilon = 0.1$, $\beta_1 = 0.001$, $b = 1$, and $\beta_2 = 0$ for instance give the only maximum for $a_2 = 1/30$ and $a_1 = \frac{3}{20\,\omega_2}$ (=0.075 for $\omega_2 = 2$) whilst $\beta_1 = 0.0001$, $\omega_2 = 2$ yields two maxima for $a_1 = 0.0748$ and $a_1 = 0.000223$ and a minimum for $a_1 = 0.00408$.

The formulae obtained show that the vibrational behaviour basically depends on the nonlinear damping.

Figure 1 gives an example of the two-dimensional probability density v for the case $\omega_2 = 2\omega_1$ of an internal resonance and $\gamma = 0.1$, $\varepsilon = 0.01$, $\beta_1 = 0.001$, $\beta_2 = 0$, and $b = 1$. The dashed line gives the hyperbola (8) of the maximum values in the a_1a_2-plane, the full-lined hyperbola the corresponding v values. The maximum value v is taken for $a_2 = 0.075$, it is approximately 80. The corresponding density for $\omega_2 = 2.2$ (no internal resonance) is given in Figure 2. Compared with Figure 1 here large values of the probability density are nearer to the a_2 axis; with other words, the probability of large a_1 amplitude values is now smaller, the a_1 vibration is not so dangerous. This behaviour is even more pronounced for the case $\omega_2 = 5$ (Figure 3). In Figure 1, the case of internal resonance, greater probabilities of large a_1 amplitudes correspond to smaller probabilities of large a_2 amplitudes, a sort of absorber effect of the a_2 vibration by the coupled a_1 vibration, in correspondence with a similar result found in [4].

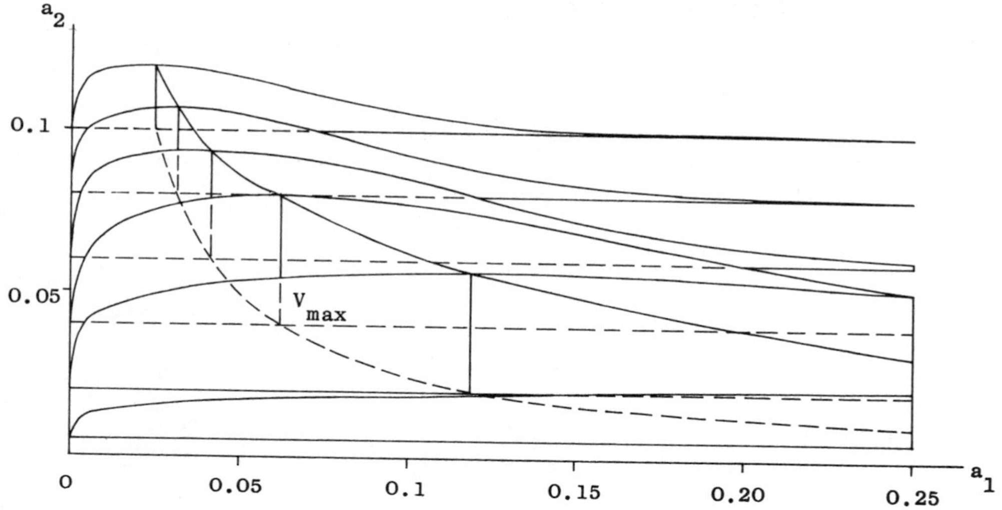

Figure 1

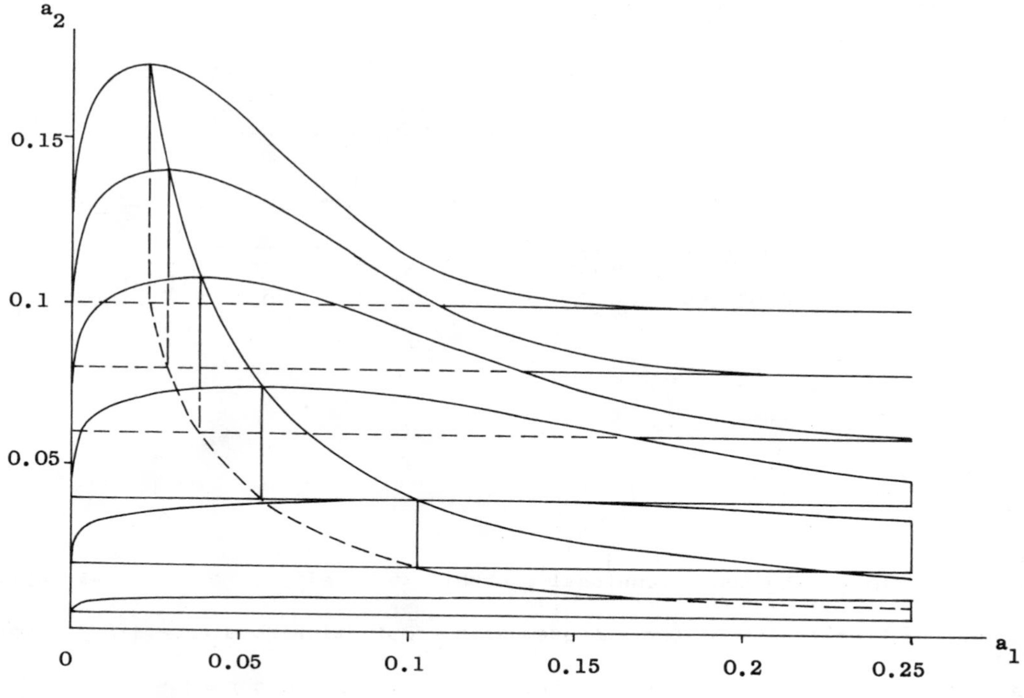

Figure 2

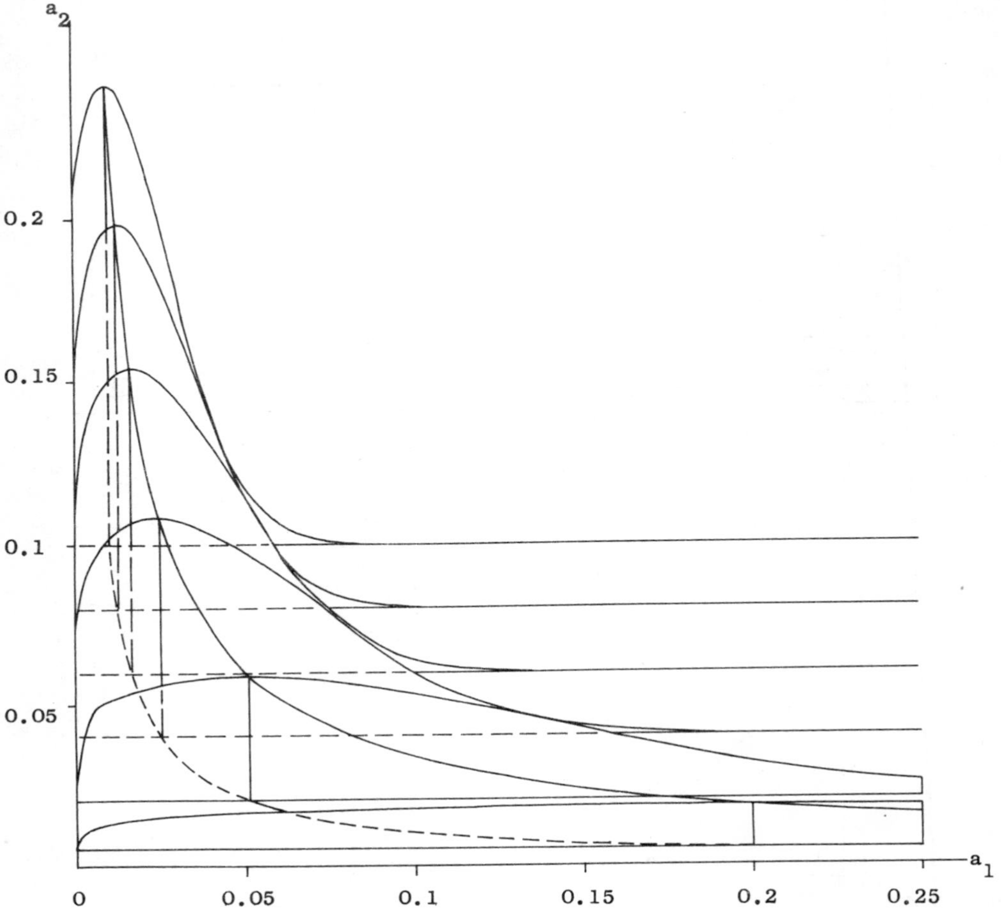

Figure 3

Analogously, the two-dimensional probability density of the partial amplitudes has been evaluated for a system with combination resonance. For comparison, the probability densities in the case of narrow-band parametric excitation have been investigated.

REFERENCES

1 L. Arnold, Stochastische Differentialgleichungen. München/ Wien: R. Oldenbourg Verlag (1973).

2 G.K. Baxter, The nonlinear response of mechanical systems to parametric random excitation. Ph.D Thesis, Syracuse Univ. New York (1971).

3 R.S. Haxton and A.D.S. Barr, The autoparametric vibration absorber. J. Eng. Ind., Tr. Am. Soc. Mech. Eng., Ser. B, 94, 119 (1972).

4 R.A. Ibrahim and J.W. Roberts, Broad band random excitation of a two-degree-of-freedom system with autoparametric coupling. J. Sound Vibr. 44, 335 (1976).

5 G. Schmidt, Parametererregte Schwingungen. Berlin: Dt. Verlag d. Wiss (1975).

6 R.L. Stratonovich, Topics in the theory of random noise. New York: Gordon and Breach (1967).

Professor Dr. G. Schmidt, Akademie der Wissenschaften der DDR ZI für Mathematik und Mechanik, DDR 108 Berlin, Mohrenstrabe 39,

DISCUSSION

The conditions of Lin [1] and Caughey (1963) [2] for solution of the Fokker-Plank equation for two degrees of freedom were mentioned (Ibrahim), in particular it was suggested that the presence of inertial coupling in the present work might cause problems. Professor Schmidt replied that the Fokker-Plank equations could in fact be solved for the two dimensional case by successive approximations but that the analysis was tedious.

It was pointed out (Ariaratnam) that after averaging, the Itô equations are generally coupled in the sense that the amplitude equations $da_i = \ldots.$ contain phase terms θ_i particularly in the present case of parametric excitations. How were these phase terms removed from the Fokker-Plank equation? Professor Schmidt replied that the averaging was done in the F-P equation rather than before (the Itô equations are not averaged) and that the phase terms of the type $2\theta_1 - \theta_2$ can be ignored because of their small size. In a related question (Lin) it was stressed that although the four dimensional solution vector $(a_1, a_2, \theta_1, \theta_2)$ is Markov the two dimensional amplitude vector (a_1, a_2) obtained by omitting phase is generally not Markov and that Fokker-Plank equations cannot therefore strictly be written for amplitude alone. Professor Schmidt replied that the approximation of ignoring phase terms was necessary if one was to proceed at all.

A discussion then ensued on the use of Gaussian closure in truncating the infinite series of moment equations arising in nonlinear problems. Professor Ariaratnam mentioned that he had counter-examples to illustrate errors arising from the use of Gaussian closure and Drs. Ibrahim and J.W. Roberts subsequently provided the written contribution. Later in the discussion it was pointed out (Grossmayer) that the method of cumulants might be used for closing the moments, and mentioned (Ariaratnam) that Bolotin had also used cumulant functions.

Professor Ariaratnam then presented an alternative approach to Schmidt's problem

$$\ddot{x} + 2\varepsilon\ (\beta_1 + 4\beta_2\ x^2)\ \dot{x} + (1 + \varepsilon^{\frac{1}{2}}\ f(t))\ x = 0\ ,$$

which led to predictions differing from those of Schmidt and earlier work of Bolotin. Using the usual transformation $x = a \cos (t+\theta)$, $\dot{x} = -a \sin (t+\theta)$ and the Khasminskii limit theorem one finally obtains an Itô equation

$$da = \varepsilon \left[\frac{350}{16} - (\beta_1 + \beta_2\ a^2)\ a \right] dt + \varepsilon^{\frac{1}{2}} \left(\frac{S_o}{8}\right)^{\frac{1}{2}} a\ dB$$

in which the amplitude equation is decoupled from phase. Here S_o indicates the spectral height of the white noise f(t). After applying a transformation $u = \log a$ this becomes

$$du = \varepsilon \left[\left(\frac{S_o}{8} - \beta_1\right) - \beta_2\ \rho^{2u} \right] dt + \varepsilon^{\frac{1}{2}} \left(\frac{S_o}{8}\right)^{\frac{1}{2}} dB\ .$$

One can obtain a stationary solution of the associated Fokker-Plank equation

$$p(u) = C_o \exp \left\{ \frac{(2S_o - 16\beta_1)}{S_o} u - \frac{8\beta_2}{S_o} \rho^{2u} \right\}$$

where

$$C_o = \left(\frac{8\beta_2}{S_o}\right)^{\left(\frac{S_o - 8\beta_1}{S_o}\right)} \Gamma^{-1} \left(\frac{S_o - 8\beta_1}{S_o}\right)\ .$$

The stationary mean square amplitude $\langle a^2 \rangle$ can then be found as

$$\langle a^2 \rangle = \frac{\beta_1}{2\beta_2} \left(\frac{S_o}{4\beta_1} - 2\right)$$

which is finite only if $\frac{S_o}{4\beta_1} > 2$. This prediction can be compared with those of Schmidt and Bolotin obtained by equivalent linearization (i.e. assuming that nonlinear terms are small and using Gaussian closure; effectively assuming 'Rayleigh' amplitude). The discrepancy was discussed somewhat and the discussion closed with the remark (Barr) that the predictions of Schmidt and Bolotin seemed more plausible physically (see Figure A) than that of Ariaratnam. Further material on this subject was presented in the closing discussion of the symposium (q.v.)

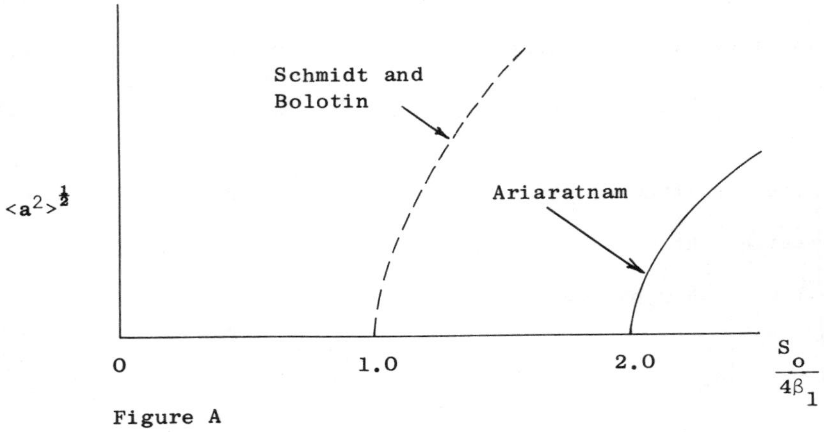

Figure A

WRITTEN CONTRIBUTION TO DISCUSSION

by Dr. R.A. Ibrahim and Dr. J.W. Roberts

In the course of this meeting Professor Ariaratnam made some critical remarks concerning the use of the Gaussian assumption for truncating the infinite hierarchy of moment equations in non-white or nonlinear parametric excitation problems. He referred in particular to our use of this method in a recent study of a two degree of freedom autoparametric system [7]. We would like to thank Professor Ariaratnam for his interest and his valuable comments. We recognise that the method is an approximation with no rigorous basis. The method has been used previously by Bolotin [8] and by Newland [9] for randomly excited systems. In our case we were highly encouraged by the fact that the results obtained showed an excellent qualitative agreement with our experimental observation in its ability to predict the effects of internal resonance. This experimental work is continuing at the University of Edinburgh and we hope that extensive quantitative comparisons will be possible in the near future.

In addition, our attention has been drawn during the symposium to a number of methods which might be applicable to our system as alternatives to the Gaussian truncation scheme. These are:

(1) The averaging method due to Khasminskii, as used by Professor Ariaratnam for linear systems.

(2) The perturbation series technique of Wedig [10] for linear parametric systems.

(3) The averaging approach used by Schmidt, in which the phase variables are eliminated in the averaging process. Some doubt has been cast on the validity of this method by Ariaratnam.

We hope to evaluate the application of these methods to problems of random excitation of systems with nonlinear coupling under internal resonance conditions.

REFERENCES

7 R.A. Ibrahim and J.W. Roberts, Broadband random excitation of a two degree of freedom system with autoparametric coupling. J. Sound and Vibration 44(3), 335-348, February 1976.

8 V.V. Bolotin, Reliability theory and stochastic stability, Chapter 11 in Study No. 6, 'Stability', University of Waterloo, 385-422, (1972).

9 D.E. Newland, Energy sharing in random vibration of nonlinearly coupled modes, J. Inst. of Maths. and Appl. 1(3), 199-207, September 1965.

10 W. Wedig, Stability conditions for oscillations with parametric filtered noise excitation. ZAMM 52(3), 161-166 (1972).

REPLY BY PROFESSOR G. SCHMIDT

In my contribution, the averaging method is used as a first approximation of the Stratonovich method (equations 3 and 4). The remark of Professor Ariaratnam that my results are obtained by equivalent linearization (using Gaussian closure) is incorrect. Professor Ariaratnam only repeats with his 'alternative approach to Schmidt's problem' my analysis for a special case. Therefore he gets the same results. It is not trivial to notice this because of mistakes in Professor Ariaratnam's formulae and because of the fact that he writes down the probability density of $u = \log a$ instead of the probability density (5) of the amplitude in my contribution. (5) yields, after a simple integration, the stationary mean square amplitude

$$\langle a^2 \rangle = \frac{1}{b} \left(\frac{\gamma^2}{\omega_1^2} - 4\beta_1 \right) .$$

Professor Ariaratnam gets, in his notations, the same result. His conclusion of a discrepancy in the mean square amplitude, the base of the doubt about the validity of the method of Ibrahim and Roberts, is incorrect. In this way, I have to thank Professor Ariaratnam for confirming my results.

M SHINOZUKA, H IMAI, Y ENAMI and K TAKEMURA

Identification of aerodynamic characteristics of a suspension bridge based on field data

INTRODUCTION

It is well known that suspension bridges are highly wind-sensitive and therefore careful considerations should be given to their design in order to assure the aerodynamic integrity. Wind forces acting on suspension bridges consist of self-excited forces (including forces due to vortex shedding) and buffeting forces. The first is particularly important since it can lead to a catastrophic failure of the bridge structure as is believed to be the case for the original Tacoma Narrows Bridge. In this respect, it is essential to perform wind tunnel tests using models of proposed bridge deck sections to investigate their aerodynamic stability although usually under uniform flow. Obviously, however, the natural wind has fluctuating (turbulent) velocity components and hence buffeting forces should also be considered in evaluating the aerodynamic response of suspension bridges.

Recently in Japan, a field experiment was performed in natural wind environments using large scale models of suspension bridge decks under the auspices of the Honshu-Shikoku Bridge Authority charged with planning, design and supervision of a number of bridges to be constructed as part of the railroad and highway network linking two of the major islands of Japan; Honshu Island and Shikoku Island. The purpose of this field experiment is to confirm the result of wind tunnel tests performed on smaller scale models and at the same time investigate the effect of turbulence on bridge stability, thus augmenting the data base that can be utilized in design.

In the present paper, first an analytical model is introduced to describe the response behaviour of these bridge models under self-excited and buffeting forces produced in natural wind environments. Then, using the wind velocity and response data obtained from the field experiment, unknown parameters involved in the analytical model are estimated with the aid of system identification techniques and the results are discussed.

EXPERIMENTAL MODEL AND FACILITY

A test facility is constructed on the beach at the tip of the Boso Peninsula located approximately 100 km south of Tokyo where strong seasonal winds are expected in winter and the ground surface is sufficiently smooth providing a general topographical similarity with respect to the construction site of the proposed suspension bridges. A large-scale two-dimensional model of the stiffening truss of a proposed suspension bridge with a centre span of 1,500 m was constructed and tested at this facility. The model has a scale ratio of 1:10.52 in cross-section and is 3.42 m wide and 1.33 m high as shown in Figure 1. Its length of 8 m is considered appropriate for the purpose of two-dimensional model tests to be performed. Other relevant data on the prototype and the model are listed in Table 1.

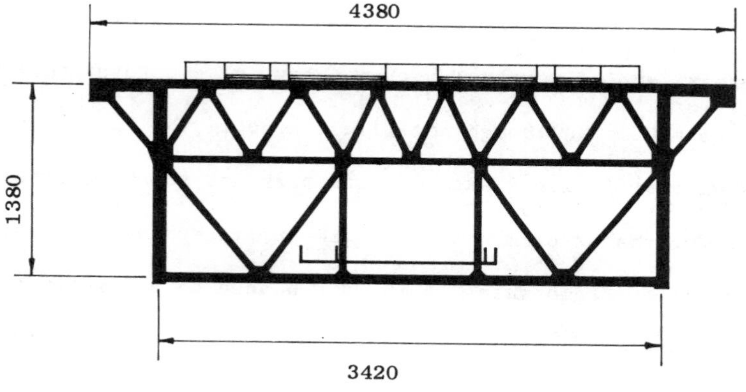

Figure 1 Cross-section of Model (unit: mm)

	PROTOTYPE	MODEL
Scale	1	1/10.52
Length	(Main Span) 1500 m	8.00 m
Width	36 m	3.42 m
Exposed Area	----	3.802 m
Weight	45 t/m/br	3239 kg/model
Polar Moment of Inertia	1242 $t.m.s^2$/br	782 $kg.m.s^2$/model
Frequency — Vertical	0.11 Hz	0.21 Hz
Frequency — Torsional	0.22 Hz	0.43 Hz
Frequency Ratio	2.0	2.0
Logarithmic Damping — Vertical	0.03	0.06
Logarithmic Damping — Torsional	0.03	0.02
Scale of Wind Velocity — Vertical	1	5.4
Scale of Wind Velocity — Torsional	1	5.4

Table 1   Prototype and Model

The model is suspended by a coil spring-balance system as shown in Figure 2. This figure also indicates positions of anemometers, tensionmeters, accelerometers and potentiometers installed for the measurement of wind velocity components and of vertical motion (heaving) and rotational motion (pitching) of the model. Although not illustrated in the figure, the supporting system also includes exciters, dampers and shock absorbers that are also needed for the experimentation. The entire system is housed in a streamlined supporting frame which is in turn placed on a turntable so that the model can be rotated to face the wind direction as indicated in Figure 2.

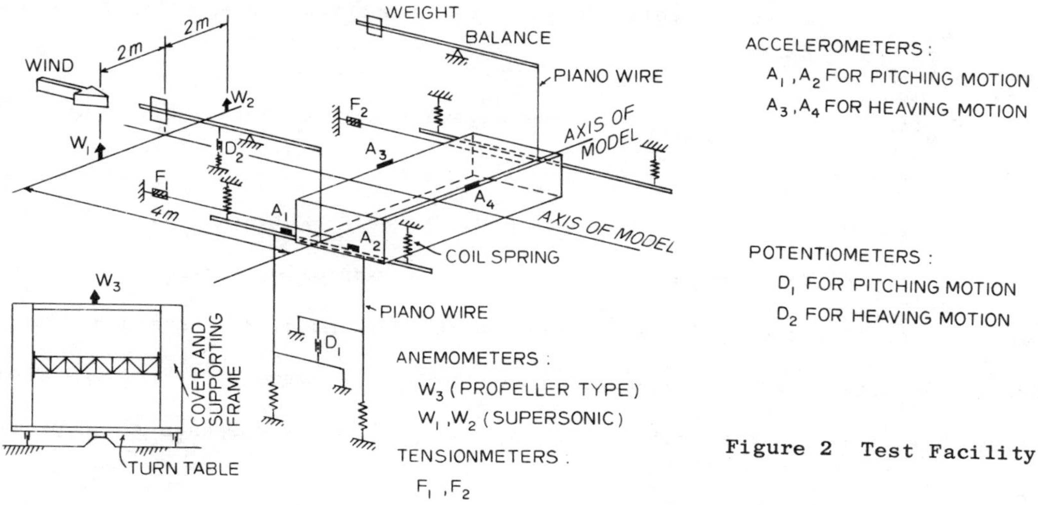

Figure 2 Test Facility

Measurement of longitudinal (along-the-wind) and vertical components of wind velocity to be used in the present analysis are taken at stations W_1 and W_2. These measurements together with the response (pitching and heaving) data are first recorded on an analogue magnetic tape, and then digitized and stored on a digital magnetic tape, to be used with an HP 5451A Fourier Analyzer System at Columbia University.

EQUATIONS OF MOTION

Referring to Figure 3 and dealing only with the pitching motion $\alpha(t)$ and heaving motion $h(t)$ measured from the equilibrium position of the deck model under the mean wind velocities $\overline{U}$ and $\overline{V}$, the governing equations for $\alpha(t)$ and $h(t)$ are given by

$$\begin{aligned} M(\ddot{h} + 2\zeta_h\omega_h\dot{h} + \omega_h^2 h) &= F_{se}(t) + F_b(t) \\ I(\ddot{\alpha} + 2\zeta_\alpha\omega_\alpha\dot{\alpha} + \omega_\alpha^2\alpha) &= Q_{se}(t) + Q_b(t) \end{aligned} \quad (1)$$

where M,I = mass and mass moment of inertia of the deck model respectively.

ζ_h,ζ_α = mechanical damping ratios associated with heaving and pitching motions respectively.

ω_h, ω_α = natural circular frequencies of heaving and pitching motions, respectively.

$F_{se}(t)$, $F_b(t)$ = lift forces due to self-excitation and buffeting effects, respectively.

$Q_{se}(t)$, $Q_b(t)$ = twisting moments due to self-excitation and buffeting effects, respectively. In equation (1), the same approximation as in [1] has been made so that the wind loads acting on the deck model are considered as additively consisting of self-excited and buffeting forces [1]. Figure 3 also illustrates the sign convention applied to wind velocity components and deck motions.

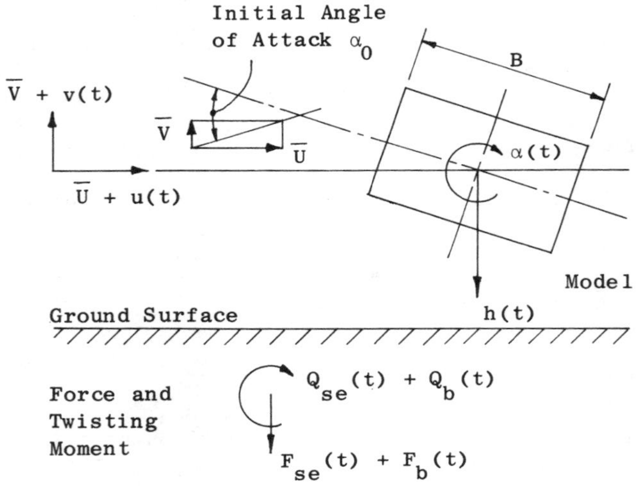

Figure 3 Displacements and Forces

Self-excited forces are represented by using non-dimensional aerodynamic coefficients H_i^* and A_i^* (i=1,2,3,4) as follows

$$F_{se}(t) = \frac{\ell}{2}\rho\bar{U}^2 (2B)\{kH_1^* \frac{\dot{h}}{\bar{U}} + kH_2^* \frac{B\dot{\alpha}}{\bar{U}} + k^2H_3^*\alpha + k^2H_4^* \frac{h}{B}\}$$

$$Q_{se}(t) = \frac{\ell}{2}\rho\bar{U}^2 (2B^2)\{kA_1^* \frac{\dot{h}}{\bar{U}} + kA_2^* \frac{B\dot{\alpha}}{\bar{U}} + k^2A_3^*\alpha + k^2A_4^* \frac{h}{B}\} \quad (2)$$

where ℓ = model length B = model width $\bar{U}$ = mean wind velocity

ρ = air density , k = reduced frequency = $B\omega/\bar{U}$ ($\omega=\omega_h$ for h and $\dot{h}$ terms and $\omega=\omega_\alpha$ for α and $\dot{\alpha}$ terms).

Aerodynamic coefficients H_i^* and A_i^* are usually determined by wind tunnel test under uniform flow. Examples of H_i^* and A_i^* thus obtained for some bridge decks are given in [1], where it can be observed that these coefficients are highly sensitive to the cross-sectional shape of the bridge deck.

Buffeting forces acting on the model are caused by the three dimensional turbulence effects, and assumed to be given by the following equations involving wind velocity components:

$$F_b(t) = -\ell\rho\bar{U}^2 B \frac{u(t)}{\bar{U}} L_{11}^* - \frac{\ell}{2}\rho\bar{U}^2 B \frac{v(t)}{\bar{U}} L_{12}^*$$

$$Q_b(t) = \ell\rho\bar{U}^2 B^2 \frac{u(t)}{\bar{U}} L_{21}^* + \frac{\ell}{2}\rho\bar{U}^2 B^2 \frac{v(t)}{\bar{U}} L_{22}^* \qquad (3)$$

where

u(t) = fluctuating component of longitudinal wind velocity

v(t) = fluctuating component of vertical wind velocity

and L_{11}^*, L_{12}^*, L_{21}^* and L_{22}^* are unknown coefficients which include the effect of the temporal non-uniformity and of the spatial variation of the wind velocity (i.e. aerodynamic admittance and joint acceptance, respectively). Since the length (span) of the deck model is small, however, the spatial variation of the wind velocity along the spanwise direction is neglected, and the average time history between the time histories at stations W_1 and W_2 is used for u(t) and v(t) throughout. Furthermore, since the model width is in general much smaller than the dominant wave lengths of the wind velocity fluctuation, both in the longitudinal and vertical directions, the aerodynamic admittance is neglected. The turbulent force coefficients L_{ij}^* may then be represented by:

$$L_{11}^* = C_L(\alpha), \qquad L_{12}^* = \left.\frac{dC_L}{d\alpha}\right|_\alpha + \frac{A}{B} C_D(\alpha)$$

$$L_{21}^* = C_M(\alpha), \qquad L_{22}^* = \left.\frac{dC_M}{d\alpha}\right|_\alpha \tag{4}$$

where C_L, C_M and C_D denote the lift, moment and drag coefficients and depend on angle of attack α. Expressions in Equation (4) are all evaluated at $\alpha = \alpha_o$, which indicates the angle between the mean wind direction and the model axis at equilibrium (see Figure 3). Also, in Equation 4, A = exposed area of the model per unit length and B = width of the model.

Substituting Equations 2 and 3 into Equation 1, one obtains

$$\ddot{h} + J_{11}\dot{h} + J_{12}\dot{\alpha} + K_{11}h + K_{12}\alpha = L_{11}u + L_{12}v$$

$$\ddot{\alpha} + J_{21}\dot{h} + J_{22}\dot{\alpha} + K_{21}h + K_{22}\alpha = L_{21}u + L_{22}v \tag{5}$$

where

$$J_{11} = 2\zeta_h\omega_h - \frac{\rho B^2\omega_h}{m} H_1^* \qquad J_{12} = -\frac{\rho B^3\omega_\alpha}{m} H_2^*$$

$$J_{21} = -\frac{\rho B^3\omega_h}{I_o} A_1^* \qquad J_{22} = 2\zeta_\alpha\omega_\alpha - \frac{\rho B^4\omega_\alpha}{I_o} A_2^*$$

$$K_{11} = \omega_h^2 - \frac{\rho B^2\omega_h}{m} H_4^* \qquad K_{12} = -\frac{\rho B^3\omega_\alpha^2}{m} H_3^*$$

$$K_{21} = -\frac{\rho B^3\omega_h^2}{I_o} A_4^* \qquad K_{22} = \omega_\alpha^2 - \frac{\rho B^4\omega_\alpha^2}{I_o} A_3^* \tag{6}$$

and

$$L_{11} = -\frac{\rho B\bar{U}}{m} L_{11}^* \qquad L_{21} = \frac{\rho B^2\bar{U}}{I_o} L_{21}^*$$

$$L_{12} = -\frac{\rho B\bar{U}}{2m} L_{12}^* \qquad L_{22} = \frac{\rho B^2\bar{U}}{2I_o} L_{22}^* \tag{7}$$

with m and I_o being respectively the mass and mass moment of inertia of the model per unit length.

Equation (5) can be expressed more concisely in matrix form:

$$\ddot{\xi} + J\dot{\xi} + K\xi = Lf$$

$$\xi = \begin{bmatrix} h \\ \alpha \end{bmatrix}, \qquad f = \begin{bmatrix} u \\ v \end{bmatrix}$$

$$J = \begin{bmatrix} J_{11} & J_{12} \\ J_{21} & J_{22} \end{bmatrix}, \qquad K = \begin{bmatrix} K_{11} & K_{12} \\ K_{21} & K_{22} \end{bmatrix}, \qquad L = \begin{bmatrix} L_{11} & L_{12} \\ L_{21} & L_{22} \end{bmatrix} \tag{8}$$

PARAMETER IDENTIFICATION

The equation of motion described by Equation (8) includes the damping matrix J and the stiffness matrix K both of which depend on the aerodynamic coefficients and the force coefficient matrix L which depends on coefficients L^*_{ij}. Since J, K and L are all regarded as unknown, the problem becomes that of identifying J, K and L by utilizing the observed records of the response ($\alpha(t)$ and $h(t)$) and the excitation ($u(t)$ and $v(t)$). The identification techniques used in this study and the results are described below.

Define the state vector X such that

$$X = \begin{bmatrix} \xi \\ \dot{\xi} \end{bmatrix} = \begin{bmatrix} h & \alpha & \dot{h} & \dot{\alpha} \end{bmatrix}' \tag{9}$$

where a prime indicates matrix transposition. With this definition of X, Equation (8) becomes

$$\dot{X} = \Phi X + \Gamma f \tag{10}$$

where

$$\Phi = \begin{bmatrix} 0 & I \\ -K & -J \end{bmatrix} \qquad \Gamma = \begin{bmatrix} 0 \\ L \end{bmatrix}$$

Since the field data consists of excitation and response values digitized at discrete sampling time instants at Δt (= 0.2 sec) intervals,it is analytically more convenient to modify Equation (10) into the corresponding discrete form:

$$X(i + 1) = AX(i) + Bf(i) \tag{11}$$

where

$$A = e^{\Phi\Delta t}, \qquad B = \left[\int_{0}^{\Delta t} e^{\Phi\tau}\, d\tau \right] \Gamma$$

and $X(i) = X(i\Delta t)$ with the same simplified notation being used for other functions of time throughout.

For the purpose of identification, however, Equation (11) should be used together with the following observation equation:

$$Y(i) = CX(i) + \eta(i) \tag{12}$$

where

$Y(i)$ = observation vector at $t = i\Delta t$

$\eta(i)$ = observation error vector at $t = i\Delta t$

$C = [I, 0]$.

Equation (12) simply indicates that what is observed as the pitching (or heaving) motion consists of the true pitching (or heaving) motion plus the error involved in the measurement of motion.

By introducing X_c, A_c, B_c and R such that

$$X_c = RX, \qquad B_c = RB$$

$$A_c = RAR^{-1} = \begin{bmatrix} 0 & I \\ F_2 & F_1 \end{bmatrix}, \qquad R = \begin{bmatrix} C \\ CA \end{bmatrix} \tag{13}$$

Equations (12) and (13) can be combined to produce

$$X_c(i+1) = A_c X_c(i) + B_c f(i) \tag{14}$$

$$Y(i) = CX_c(i) + \eta(i) \tag{15}$$

By eliminating $X_c(i+1)$ and $X_c(i)$ from Equations (14) and (15), one can arrive at the following autoregressive moving average (ARMA) model for the observation:

$$Y(i) = F_1Y(i-1) + F_2Y(i-2) + G_1f(i-1) + G_2f(i-2) + W(i) \tag{16}$$

where

$$\begin{bmatrix} G_1 \\ G_2 \end{bmatrix} = \begin{bmatrix} I & 0 \\ -F_1 & I \end{bmatrix} B_c \tag{17}$$

and

$$W(i) = \eta(i) - F_1\eta(i-1) - F_2\eta(i-2) \tag{18}$$

With the aid of Equation (16), the identification of matrices Φ and Γ (and hence K, J and L) proceeds as follows: (1) use parameter identification techniques to identify matrices F_1, F_2, G_1 and G_2, (2) construct matrices Φ_c and Γ_c such that

$$\Phi_c = \frac{1}{\Delta t}\ln A_c, \qquad \Gamma_c = \left[\int_0^{\Delta t} e^{\Phi_c\tau}\, d\tau\right]^{-1} B_c \tag{19}$$

where A_c and B_c can be obtained from Equations (13) and (17) using matrices F_1, F_2, G_1 and G_2 just identified and (3) perform coordinate transformations such that

$$\Phi = R^*\Phi_c R^{*^{-1}}, \qquad \Gamma = R^*\Gamma_C = \begin{bmatrix} L_e \\ L \end{bmatrix} \tag{20}$$

where

$$R^* = \begin{bmatrix} C \\ C\Phi_c \end{bmatrix} \tag{21}$$

In theory, the matrix L_e in Equation (20) is supposed to be zero.

Referring to the first step described above, the following three methods are considered.

Ordinary Least Square (OLS) method [2,3]

The ARMA model given in Equation (16) can be rewritten in the form

$$Y(i) = \Theta\Psi(i) + W(i) \quad (i=3,4,\ldots,N) \tag{22}$$

with

$$\Theta = [F_1, F_2, G_1, G_2] \tag{23}$$

$$\Psi(i) = [Y'(i-1)\ Y'(i-2)\ f'(i-1)\ f'(i-2)]' \tag{24}$$

where N is the number of data points. Equation (22) can further be simplified into the form

$$Y_N = \Theta\Psi_N + W_N \tag{25}$$

with

$$Y_N = [Y(3)Y(4)\ \ldots\ Y(N)] \tag{26}$$

$$\Psi_N = [\Psi(3)\Psi(4)\ \ldots\ \Psi(N)] \tag{27}$$

$$W_N = [W(3)W(4)\ \ldots\ W(N)] \tag{28}$$

If the "equation error" is defined as

$$e = Y_N - \overline{\Theta}\Psi_N \tag{29}$$

and if the criterion function for the estimate $\overline{\Theta}$ of Θ is constructed as

$$J = \sum_{i=3}^{N} ||e(i)||^2 = tr[e'e] \tag{30}$$

then it can be shown that the following least square estimate $\overline{\Theta}_{LS}$ of Θ minimizes J;

$$\overline{\Theta}_{LS} = Y_N\Psi'_N\,[\Psi_N\Psi'_N]^{-1} \tag{31}$$

where $tr[e'e]$ indicates the trace of the matrix $[e'e]$ and

$$e(i) = Y(i) - \overline{\Theta}\Psi(i) \tag{32}$$

It can also be shown that the least square estimate $\overline{\Theta}_{LS}$ is not a consistent estimate in the sense that the "estimation error" given by

$$\tilde{\Theta} = \overline{\Theta}_{LS} - \Theta = W_N\Psi'_N\,[\Psi_N\Psi'_N]^{-1} = [\tfrac{1}{N}\,W_N\Psi'_N]\,[\tfrac{1}{N}\,\Psi_N\Psi'_N]^{-1} \tag{33}$$

will not approach zero in probability as N approaches infinity since

$$\underset{N\to\infty}{\text{P lim}} \frac{1}{N} W_N \Psi'_N \neq 0 \quad (34)$$

although $\Psi_N \Psi'_N/N$ can be shown to be positive definite where the limiting process in Equation (34) indicates the limit in probability.

Instrumental Variable (IV) method [4,5,6]

If a matrix Σ_N is found so that

$$\underset{N\to\infty}{\text{P lim}} \frac{1}{N} W_N \Sigma'_N = 0 \quad (35)$$

then, a consistent estimate $\overline{\Theta}_{IV}$ can be obtained as

$$\overline{\Theta}_{IV} = Y_N \Sigma'_N \left[\Psi_N \Sigma'_N\right]^{-1} \quad (36)$$

and Σ_N is called the instrumental variable matrix.

To find Σ_N and $\overline{\Theta}_{IV}$, the following iterative procedure is taken: (i) start with an estimate $\overline{\Theta}$ of Θ (use, for example, $\overline{\Theta}_{LS}$), (ii) evaluate $\overline{Y}(i)$ by

$$\overline{Y}(i) = \overline{\Theta}\overline{\Psi}(i) \quad (i=3,4,...,N) \quad (37)$$

with

$$\overline{\Psi}(i) = \left[\overline{Y}'(i-1)\ \overline{Y}'(i-2)\ f'(i-1)\ f'(i-2)\right]' \quad (38)$$

(iii) construct $\overline{\Sigma}_N$ as follows

$$\overline{\Sigma}_N = \left[\overline{\Psi}(3)\ \overline{\Psi}(4)\ \dots\ \overline{\Psi}(N)\right] \quad (39)$$

(iv) obtain a revised estimate $\overline{\Theta}_{IV}$ from Equation (36) by using $\overline{\Sigma}_N$ just evaluated in place of Σ_N and (v) repeat the steps above until a satisfactory convergence is observed.

Limited Information Maximum Likelihood (LIML) method [7,8,9]

Consider the following criterion function J^* in the form of a determinant

$$J^* = \left| \sum_{i=3}^{N} \left[Y(i) - \overline{Y}(i|i-1)\right]\left[Y(i) - \overline{Y}(i|i-1)\right]' \right| \quad (40)$$

with $\bar{Y}(i|i-1)$ being the conditional expectations:

$$\bar{Y}(i|i-1) = E\{Y(i)|Y(1),\ Y(2),\ \ldots,\ Y(i-1),$$
$$f(1),\ f(2),\ \ldots,\ f(i-1),\ \Theta, V_\eta\} \tag{41}$$

where V_η is the covariance matrix of $\eta(i)$. Then, assuming that the conditional probability of $Y(i)$

$$p\{Y(i)|Y(1),\ Y(2),\ \ldots,\ Y(i-1),\ f(1),\ f(2),\ \ldots,\ f(i-1),\ \Theta, V_\eta\}$$

is Gaussian, it can be shown that for a stable system and for a large value of $N, \bar{Y}(i|i-1)$ are recursively evaluated from

$$\bar{Y}(i|i-1) = F_1\bar{Y}(i-1|i-2) + F_2\bar{Y}(i-2|i-3) + G_1f(i-1) + G_2f(i-2) \tag{42}$$

and that the estimate $\bar{\Theta}_{ML}$ of Θ which minimizes J^* in Equation (40) is the maximum likelihood estimate with respect to the likelihood function

$$L_N(\Theta, V_\eta) = p\{Y(N),\ Y(N-1),\ \ldots,\ Y(3)|Y(2),$$
$$Y(1),\ f(N-1),\ f(N-2),\ \ldots,\ f(1),\ \Theta, V_\eta\} \tag{43}$$

To find the estimate $\bar{\Theta}_{ML}$ that minimizes J^*, however, an iteration procedure has to be implemented. In each iteration, corresponding to a perturbation imposed on the current estimate of Θ, Equation (42) is used to generate revised values of $\bar{Y}(i|i-1)(i=3,4,\ldots N)$ which are in turn used in Equation (40) to evaluate the revised value of J^*. The procedure is terminated only after no change in J^* is found upon perturbing the latest estimate of Θ.

This iterative procedure is found to be an extremely time consuming task requiring such numerical techniques as the gradient method involving sixteen unknown components of the matrices F_1, F_2, G_1 and G_2. In the present study, therefore, the limited information maximum likelihood (LIML) method is used for the parameter estimation. As applied to the problem at hand, the LIML method modifies the estimation procedure in the following way:

The estimate $\overline{\Theta}_{LIML}$ of Θ is called the limited information likelihood estimate if it minimizes

$$J^*_m = \sum_{i=3}^{N} \{Y_m(i) - \overline{Y}_m(i|i-1)\}^2 \quad (m=1,2) \tag{44}$$

with

$$\overline{Y}_m(i|i-1) = F_{1,mm}\overline{Y}_m(i-1|i-2) + F_{2,mm}\overline{Y}_m(i-2|i-3)$$
$$+ F_{1,mn}Y_n(i-1) + F_{2,mn}Y_n(i-2)$$
$$+ \sum_{n=1}^{2} G_{1,mn}f_n(i-1) + \sum_{n=1}^{2} G_{2,mn}f_n(i-2) \tag{45}$$

where m, n=1 or 2($m \neq n$), $Y_1(i) = h(i) + \eta_1(i)$, $Y_2(i) = \alpha(i) + \eta_2(i)$ and $F_{1,mn}$ = the m-n component of matrix F_1, etc. The use of Equations (44) and (45) implies in essence the artificial uncoupling of heaving and pitching motions for the purpose of easier parameter estimation by reducing the number of unknown parameters to eight in each of the two iterative procedures (for Y_1 and Y_2).

An analytical simulation study performed at Columbia University indicates that the LIML method appears to provide the best estimation capability under a wide range of intensity of contamination associated with the excitation (u and v) as well as the response (h and α). Therefore, the LIML method has been used for parameter estimation in the present study.

EXAMPLES AND DISCUSSION

Seventeen sets of excitation and response data are considered for identification purposes. Of these seventeen, eight sets are for the case where only the pitching motion is permitted while the heaving motion is constrained; thus the motion has one degree of freedom. The remaining sets are for the cases where the motion has two degrees of freedom with both pitching and heaving motions being permitted. Figure 4 indicates recorded

Experiment ID Number	$\overline{U}$ (m/s)			$\overline{V}$ (m/s)			σ_u(m/s)			σ_v(m/s)		
	W_1	W_2	Av.	W_1	W_2	Av.	W_1	W_2	Av.	W_1	W_2	Av.
9501a	6.72	6.60	6.66	.001	-.208	-.104	1.42	1.48	1.26	.886	.873	.562

<u>Notes</u>

σ_u = standard deviation (longitudinal)

σ_v = standard deviation (vertical)

W_1 = measurement at station W_1

W_2 = measurement at station W_2

Av. = Average

Number of data points = 1,201 (240s)

Table 2 Wind Velocity Data (Experiment No. 9501a)

ARMA Model	J	$F_{1,22}$	$F_{2,22}$	$G_{1,21}$	$G_{1,22}$	$G_{2,21}$	$G_{2,22}$	
	.012828	1.70783	-.99542	.00009	.00045	-.00019	-.00047	
Differential Equation	K_{22}	J_{22}	L_{21}	$L_{e_{21}}$	L_{22}	$L_{e_{22}}$	A_2^*	A_3^*
	7.38638 (ω_α^*=2.718)	.02295 (ζ_α^*=0.00422)	-.00253	.00076	-.00039	.00243	-.01052	-.09536

Table 3 Estimated Parameters (LIML Method)

excitation and response data for Experiment No. 9501a in which the heaving motion is constrained. The first record shows the average value of the fluctuating components of horizontal wind velocity measured at stations W_1 and W_2, the second is the similar average of the fluctuating components of vertical wind velocity and the third is the pitching (response) motion $\alpha(t)$, all in the interval of t = 0 to 240 sec. Other pertinent data on the wind velocities are listed in Table 2. Table 3 indicates the result of identification of the ARMA model parameters $F_{1,22}$, $F_{2,22}$, etc., by means of the LIML method. The parameters of the equation of motion K_{22}, J_{22}, etc., are recovered from the ARMA model parameters and are listed also in Table 3 together with aerodynamic coefficients A_2^* and A_3^* as well as with ω_α^* and ζ_α^* values. Note that $F_{1,21}$, $F_{2,21}$, J_{21}, and K_{21} do not appear since this case involves only the pitching motion. Figure 5 compares the simulated responses with the observation (the top). The second record indicates the simulated pitching motion when the ARMA model with the LIML estimate of Θ is solved under the excitation u(t) and v(t), while the third illustrates the simulated motion when the differential equation with the LIML estimates of J,K, and L is solved under the same excitation. The simulations of the ARMA model and the differential equation may differ significantly depending upon how closely the estimation of the matrix L_e is confined in the neighbourhood of zero, although in Figure 5 the difference does not appear to be significant except for a faster decay of the response obtained by the differential equations. The last record shows the simulation of pitching motion when turbulences, u(t) and v(t) are not considered. In this case, the simulation by either method produces an identical result since the matrix L_e plays no part in either the ARMA model or in the differential equation. It is interesting to observe that in this particular example, the turbulence appears to help sustain the pitching motion. Figures 6 and 7 respectively plot the estimated values of A_2^* and $\omega_\alpha^*/\omega_\alpha$ based on all seventeen sets of experimental data (of both one - and two - degrees of freedom), where $\omega_\alpha^* = (K_{22})^{1/2}$ = natural frequency in torsional mode with the effect of A_3^* considered (see Equation (6)). Figure 6 indicates a better than reasonable agreement between the results obtained from wind tunnel tests and from the current field experiment, while Figure 7 confirms the common belief that the stiffness of the structural system is not significantly sensitive to the self-excitation.

ACKNOWLEDGEMENT

The authors are deeply grateful to the US National Science Foundation (Grant No. NSF-ENG-75-13761) and the Honshu-Shikoku Bridge Authority of Japan for their support of this work.

REFERENCES

1 R.H. Scanlan — Analytical models of suspension bridge response in 'Reliability Approach in Structural Engineering' Eds. A.M. Freudenthal et al. pp 299-314. (1975).

2 R.C.K. Lee — Optimal estimation, identification and control. Research Monograph 28, MIT Press. (1964).

3 R. Isermann et al — Comparison on six on-line identification and parameter estimation methods. Automatica 10, pp 81-103. (1974).

4 I.H. Rowe — A bootstrap method for the statistical estimation of model parameters. Int. J. Contr. 10, N.5, pp 721-738. (1970).

5 K.Y. Wong and E. Polak Identification of linear discrete time systems using the instrumental variable method. IEEE Trans on Auto. Contr. AC-12 No.6, pp 707-718. (1967).

6 P.C. Young An instrumental variable method for real time identification of a noisy process. Automatica 6, pp 271-287. (1970).

7 R.L. Kashyap Maximum likelihood identification of stochastic linear systems. IEEE Trans. on Auto. Contr. AC-15, No. 1, pp. 25-34. (1970).

8 K.J. Astrom and T. Bohlin Numerical identification of linear dynamical systems from normal operating records. IFAC Symp. Theory of Self-Adaptive Systems, Teddington, England. (1975).

9 R.L. Kashyap and R.E. Nasburg Parameter estimation in multivariate stochastic difference equation. IEEE Trans. AC-19, pp 784-797. (1974).

Professor M. Shinozuka, Department of Civil Engineering and Engineering Mechanics, Columbia University in the City of New York, New York, NY 10027, USA.

Professor H. Imai, Kyoto University, Kyoto, Japan.

Mr. Y. Enami, The Honshu-Shikoku Bridge Authority, Tokyo, Japan.

Mr. K. Takemura, Steel Structure Division, Kawasaki Heavy Industries Ltd., Kakogawa, Japan.

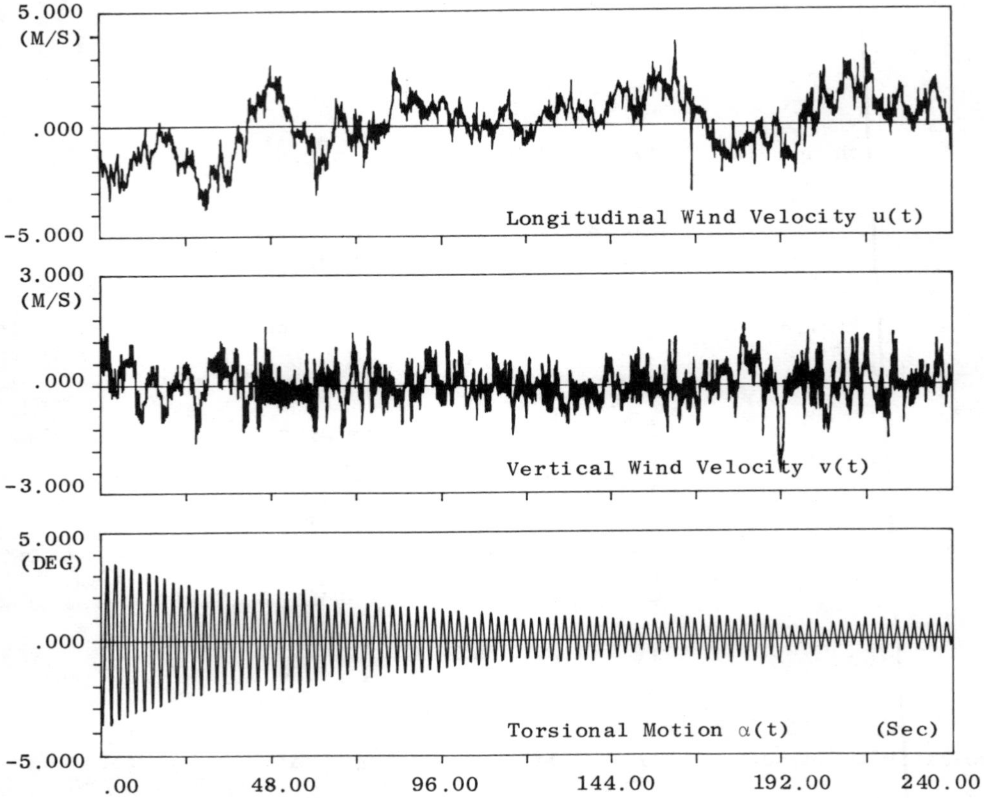

Figure 4 Wind Velocity and Torsional Motion Records

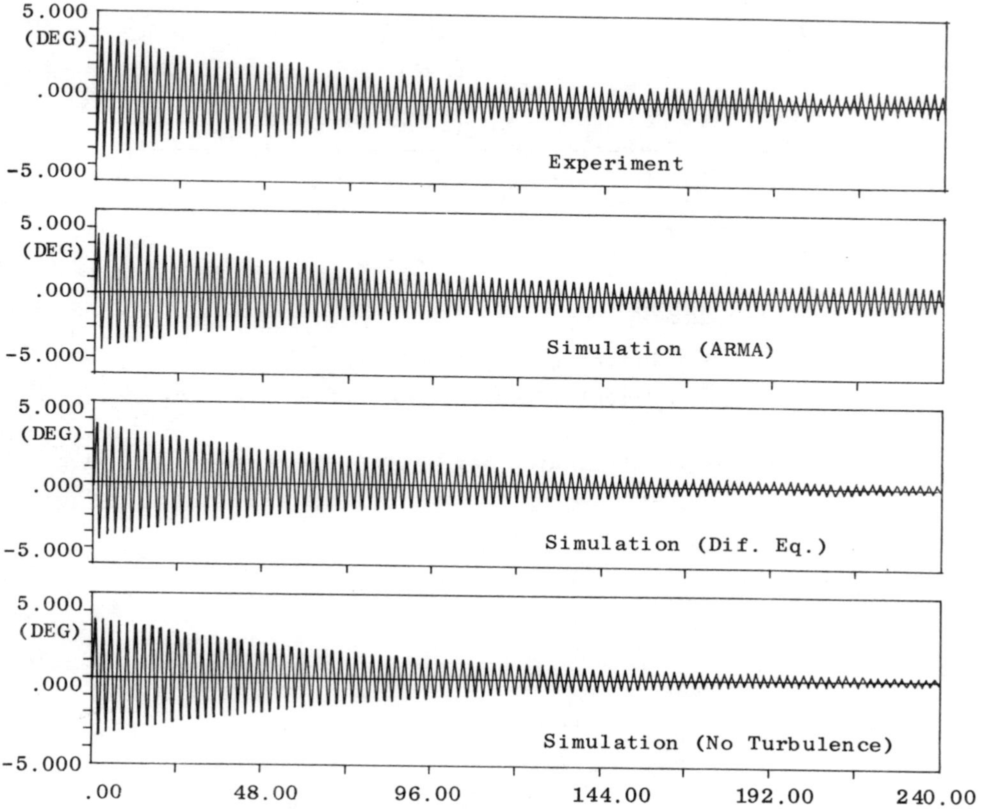

Figure 5 Simulated Torsional Motions

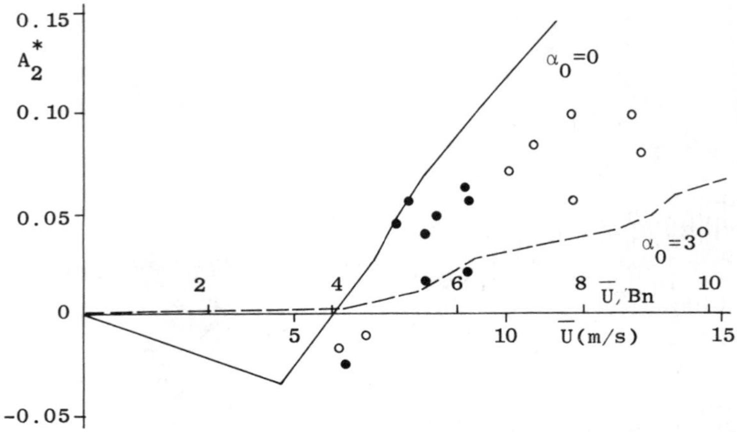

——— Wind Tunnel Test
– – –

○ Pitching Motion only

● Pitching and Heaving Motion

Figure 6 Estimated Aerodynamic Coefficient A_2^*

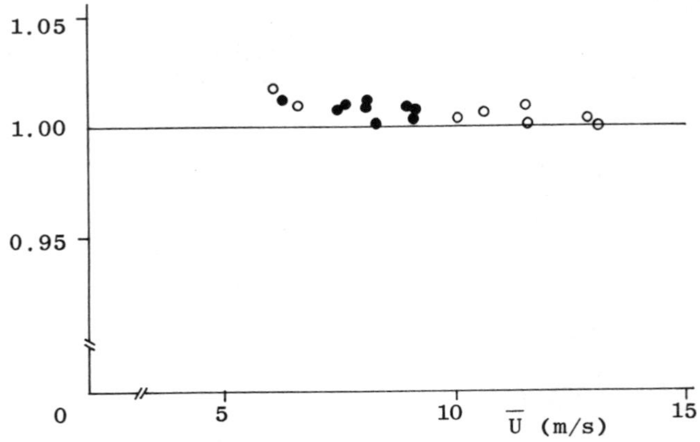

○ Pitching Motion only

● Pitching and Heaving Motion

Figure 7 Estimated Torsional Frequency

DISCUSSION

When asked (Bendat) if measures of cross spectra and coherency had been made the author replied that computations had been done but the results had not been interpreted.

In reply to queries (Crandall, Lin) concerning the correlation of u, v, the author emphasized that these were time histories that were obtained from measurements and the effect of correlation was automatically taken into account in the process of identification. But it was noted that some work had been done (not in the text) on assessing the performance of the three methods of identification when wide band noise was added artificially to the inputs to the structure.

When questioned (Akaike) about the importance of taking the spatial distribution of wind into account the author confirmed that measurements from the model indicated some spatial variation but that this had not been accounted for, so that the assumption of uniform flow across the structure implied taking an average. The assumption of uniformity of flow over the entire span of the prototype (1.5 km) was questionable but likely to be a conservative estimate, though it was thought (Murzewski) that such an assumption might be too conservative.

In reply to a question concerning the identification of damping (Crandall) the author pointed out that the aerodynamic damping was being identified but that structural damping was being measured in advance. The author also confirmed (Kozin) that the results from experiment and estimated model (Figure 5) always exhibited the close similarity of behaviour.

The author also noted (Lin) that the variation of estimated parameters with mean wind speed were consistent with the results obtained in wind tunnel tests by Scanlan .

Reference was made (Weissgerber) to the Tacoma Narrows collapse and the work of Bolotin suggesting parametric excitation as the cause. The author replied that there were substantial differences in the cross section of the present bridge (relatively stable to start with) compared to the Tacoma Bridge, and the method presented seemed a reasonable approach to the identification problem. It was noted (Lin) that the results might be used for later flutter analysis or indeed (Crandall) the sectional characteristics might be used to build up a model of the complete structure which would be made up of many such sections, thereby permitting an analysis of the entire bridge.

R F DRENICK

On a class of non-robust problems in stochastic dynamics

INTRODUCTION

This paper deals with certain problems in the reliability or safety of mechanical systems which will fail if the magnitude of the response exceeds a certain limit. These problems are often treated by probability theory. It is then assumed that the system responses are simple functions from a random process, usually a Gaussian one, and an attempt is made at calculating or at least estimating the exceedance probability of that limit.

The first point of this paper is that this procedure is not robust, and that it can easily lead to very misleading results. The supporting evidence is presented in the second section. The conclusion that is reached there should not be overly surprising. Failures are, or had better be, rare events in most instances. They are therefore events whose probabilities are strongly dependent on the shapes of the tails of the underlying distributions which are usually the least known portions, and those least accessible to statistical estimation. The failure probabilities that are calculated from them are subject to large errors.

The third section raises the question of what one should do in such problems. The answer that is reached there is, by what seems to be a fairly generally valid argument, that one should perform a combination of probabilistic and worst-case analyses. The probabilistic portion should more particularly utilize all information that is available regarding the statistics of the underlying random process and that possesses the desired level of assurance. The worst-case analysis is then used to obtain bounds on the failure probability that are consistent with that information.

The fourth and fifth sections present an example taken from earthquake engineering, which, according to rather recent work, appears to be producing practical results. This is the assessment of the seismic resistance of structures. The ground motions during earthquakes probably form a very good example of a random process whose probabilistic structure is poorly known, especially on the tails of its distributions. It is therefore natural to apply the general ideas presented in the third section. This is done for elastic structures in the fourth section and for inelastic ones in the fifth section.

NON-ROBUSTNESS PROBLEMS

The kind of problem to be discussed in this paper is one in which a failure of a system is brought about by an excessively severe response. It will be convenient to assume that the 'severity' of a response variable y is measured by the norm $\|y\|$. The norm that seems most useful in practice is

$$\|y\| = \sup_t |y(t)| \tag{1}$$

where t ranges over the interval T of interest. One can then define system failure as an event of the form $\{ \|y\| > L\}$ where L is some failure limit.

In the stochastic treatment of the problem one seeks to determine the probability

$$P\{ \|y\| > L\} = P\{\sup_t |y(t)| > L\} \tag{2}$$

of the event of failure. In order to do so one assumes that the response variable y forms a random process with a completely known probability measure. (For simplicity, the symbol y will be used in what follow to denote the random process as well as an individual sample function; the wording of the text will, it is hoped, avoid misunderstandings due to this imprecise notation). It is in fact usually assumed that the probability measure is Gaussian, and the probability in (2) is then calculated, or at least estimated, on that assumption.

The point to be made in this section is that the value of the failure probability (2) is very sensitive to the assumption of Gaussianity ('ill-conditioned', in the language of numerical analysis): small changes away from it can produce very large changes in the value of the failure probability. The evidence to be presented indicates more particularly that the failure probability is most sensitive to those characteristics of the underlying random process y that are least likely to be well known, namely the behaviour for large $|y(t)|$.

With a few exceptions, closed expressions for the distribution of $\|y\|$ in (1) are known only when the random process y is stationary and Gaussian. One that was derived fairly recently by Pickands [1] is typical of most others. It is of the familiar double exponential form

$$P\{\ \|y\| > L\} = 1 - \exp\{-\exp\left[-\frac{L}{\sigma}(2\log 2n)^{\frac{1}{2}} + \eta\right]\} \tag{3a}$$

in which n and η are constants. The first is more specifically a coefficient in the Maclaurin series for the autocovariance $R_y(\tau)$ of y, i.e.

$$R_y(\tau) = R_y(0)\ \left[1 - n\left|\frac{\tau}{T}\right| + 0\ (\frac{\tau}{T})\right] \tag{3b}$$

and the second is

$$\eta = 2\log 2n + \tfrac{1}{2}(\log\pi - \log\log 2n)\ . \tag{3c}$$

Formulae (3) are asymptotically valid for large L and large n, in the sense that terms of order $O(L/\sigma)^{-1}$ and $O(\log n)^{-1}$ are neglected. It is of interest that the expression (3a) is essentially the same as for the exceedance probability of n independent Gaussian variables, each with the same density as y(t). (The only discrepancy is in the factor of $\log\pi$ in (3c)).

The derivation of (3) rests very heavily on the Gaussian nature of the process y. Accordingly no similar expressions are known for non-Gaussian processes, to the writer's knowledge. However, it may be at least plausible to expect an equivalence to exist, between non-Gaussian processes and a suitable number n of independent non-Gaussian variables, which is of the kind that has just been described for Gaussian ones. There exists no mathematical proof of the equivalence, but it is difficult to think of any reasons why it should fail to hold, at least if all conditions are satisfied under which (3) is valid and if the departure from Gaussianity is small.

If one can accept this equivalence, one can proceed further. This will be done here, at any rate. To be more specific, a non-Gaussian random process will be considered whose one-dimensional probability density p(y) has the following properties.

(1) Between two limits ($\pm y_o$), p(y) is Gaussian:

$$p(y) = \frac{a_1}{\sigma} \exp\left(-\frac{y^2}{2\sigma^2}\right), \quad (|y| < y_o). \tag{4a}$$

(2) Beyond these two limits, p(y) is 'of the exponential type' (in the terminology of Gumbel [2, p.120]):

$$p(y) = \frac{a_2}{s} \left|\frac{y}{s}\right|^m \exp\left(-\frac{1}{r}\left|\frac{y}{s}\right|^r\right), \quad (|y| > y_o,\ r > 1) . \tag{4b}$$

(3) At $|y| = y_o$, p(y) is continuous.

(4) n is large.

(5) L is large.

The last two assumptions are fairly traditional in the theory of extremes [3, p.374]. The first three are made here in order to be able to postulate a random process which is Gaussian in a region in which observational data are available (namely for y(t) values which are not very large) but which may depart from Gaussianity where such data are scarce and where such a departure would be difficult to ascertain statistically. The case of no departure is included: one merely sets

$$r=2, \quad m=0, \quad a_1=a_2 = (2\pi)^{-\frac{1}{2}}, \quad s=\sigma \ . \tag{5}$$

Suppose now also, as suggested above, that the exceedance probability $P\{ \|y\| > L\}$ in (2) for the random process y is the same as of n independent variables $y_1, y_2, \dots y_n$, each with the density (4).

Under these assumptions, one can derive a formula for the exceedance probability $P\{ \|y\| > L\}$ which is analogous to (3). The derivation is laborious but straightforward. It is simplified if one can make a sixth assumption, namely $L \geq y_0 > s$, which is not unreasonable and which will be made here. One then finds

$$P\{ \|y\| > L\} = 1 - \exp\{-\exp\left[- \frac{L}{s}(r \log 2n)^{\frac{r-1}{r}} + \eta'\right]\} \tag{6a}$$

with

$$\eta' = \log a_2 + r \log 2n + \frac{m-r+1}{r} (\log r + \log \log 2n). \tag{6b}$$

This is again valid asymptotically for large (L/s) and large n, but in the sense that terms of orders $O(L/s)^{-r}$ and $O(\log n)^{-1}$ are negligibly small.

Expression (6a) for the exceedance probability is of roughly the same double exponential form as its counterpart (3a). Since all parameters of the underlying density p(y) enter into the second exponent, and some even exponentially so, the probability is very sensitive to even small changes in them.

The changes that are of interest here are those in the parameters a_1, a_2, s, m and r, away from the values (5) which they take if the random process y is Gaussian. Their effect on the failure probability $P\{ \|y\| > L\}$ can be evaluated by a conventional perturbation calculation. If p_g is used to denote the value of this probability when y is Gaussian, and δp the change induced by small departure δs, δr, and δm from Gaussianity one finds

$$\frac{\delta p}{p_g} = \left[M_1 \frac{\delta s}{\sigma} + M_2\ \delta m + M_3\ \delta r\right] \log p_g \tag{7}$$

where

$$M_1 = \left(\frac{y_o}{\sigma}\right)^2 - 1 - \frac{L}{\sigma}\,(2 \log 2n)^{\frac{1}{2}}$$

$$M_2 = \log \frac{y_o}{\sigma} + \tfrac{1}{2}\,(\log 2 + \log \log 2n) \tag{8}$$

$$M_3 = \tfrac{1}{4}\left(\frac{y_o}{\sigma}\right)^2 \left(1 - 2 \log \frac{y_o}{\sigma}\right) + \log 2n$$

$$+ \tfrac{1}{4}\,\frac{L}{\sigma}\,(2 \log 2n)^{\frac{1}{2}}(1 - \log 2 - 4 \log \log 2n)$$

$$- \tfrac{1}{4}\,(1 + \log 2 + 4 \log \log 2n).$$

The expressions are valid if, as before, terms of orders $O(L/\sigma)^{-1}$ and $O(\log n)^{-1}$ are considered negligible relative to 1, and if the same is true of terms of order $O(y_o/\sigma)^{-2}$.

A mere inspection of (7) and (8) shows that even small changes in the one-dimensional density of the process y are prone to produce large changes δp in the failure probability. Numerical work confirms this. Suppose, for example, that a system had been designed on the assumption that y is Gaussian, and for a failure probability of $p_g = .05$. This would mean that L/σ would have been set at $L/\sigma = 1.64$. In order to simplify the formulae (8), suppose further that $y_o = L$ (i.e. that the departure from Gaussianity is most pronounced beyond the failure limit L), and that n = 20 (i.e. that the random process is equivalent to 20 independent random variables). In that case one finds

$$\frac{\delta p}{p_g} = 15.3\,\frac{\delta s}{\sigma} - 22.4\ \delta r - 4.50\ \delta m.$$

This shows that merely a change in m alone from 0 to 1, produces a change in the failure probability by a factor of 4.5.

Such a change would be extremely difficult to detect statistically, on the level of confidence which one would often wish to attach to an estimate of the failure probability. The usual statistical tests in particular which aim at the estimation of certain mean values of the density p(y) of y, are known to yield no information regarding the behaviour for large values of y [4].

The evidence presented here therefore indicates that a reliable estimation of the failure probability will often be very difficult, basically of course because it depends on the behaviour of the underlying random process for large values of its sample functions. That, however, is the region that is the least accessible to robust statistical tests.

PROBABILISTIC AND WORST-CASE ANALYSIS COMBINED

The discussion in the preceding section has, it is hoped, made a reasonably persuasive case for the non-robust nature of the probability of a system failure which is induced by the magnitude of its response. Unless the stochastic characteristics of the latter are very well known precise pronouncements regarding the former will often be impossible. Under the circumstances, one may have to settle for weaker statements regarding this probability, especially upper bounds, and seek to make these as robust as possible. In order to do so, one may have to follow a line of reasoning which seems to be quite generally valid and which, the writer believes, will frequently be inevitable. It leads to a cross between probability theory and worst-case analysis.

In this procedure, one would first of all utilize any information which is known on the desired level of statistical confidence and which bears on the probabilistic structure of the random process y under study. It is possible in principle that this information characterizes the random process completely. This is unlikely however, for in that case it would have to specify the probability measure on the sigma algebra Σ of all (measurable) sets of sample functions of the process. More often, the reliable information will be incomplete, in the sense that it specifies with the desired assurance the probability measure only on the sets of a family Σ' within Σ. (Σ' will in fact either be a coarser subsigma algebra of Σ, or else will have to be embedded in one).

What matters here is that, so far as any statements regarding the random process y are concerned, they cannot be made on the sets in Σ but only on those in the coarser Σ'. They will be correspondingly weaker statements, and the best thing to do is to make Σ' as fine as it can be made, consistently with the available information.

The next question is how to arrive at those weaker statements. One can, and may even be forced to, proceed as follows.

Suppose that S is a set in Σ', and more particularly one that consists of, or at least contains, all sample functions that are of interest in a particular problem. The question that is considered in this paper is the probability that some among those sample functions induce failure. In other words, it is desired to know the probability of the intersection

$$\{y: \|y\| > L\} \cap \{y \in S\} , \tag{9}$$

using the somewhat imprecise notation introduced above. This probability clearly obeys the inequality

$$0 \leq P\{ \|y\| > L | \; y \in S\} \, P(S) \leq P(S) . \tag{10}$$

The upper bound is attained if almost all sample functions $y \in S$ of the process exceed the failure limit L, i.e., if the intersection (9) is essentially equal to S; the lower bound applies if almost none do, i.e., if the intersection is essentially empty.

One can now use these two bounds towards statements such as 'the failure probability of a system will not exceed P(S) when $y \in S$', or 'the system will not fail under this condition'. Moreover, these statements will carry the same degree of assurance as the information that led to the definition of the set S in the first place. They may however, be rather extreme. The first one in particular may be extremely conservative, in fact, even pessimistic in many cases: the upper bound P(S), as just mentioned, is attained only if essentially all sample functions in S produce failure. The second one will, for similar reasons, be attained only rarely.

The point to be made here is that, pessimistic or not, it often is impossible to do much better. There will, of course, be the temptation of reducing the conservatism of these statements or, which is saying the same thing, estimating the magnitude of the factor $P\{ \|y\| > L \mid y \in S\}$ in (10). This is actually usually done. The tacit argument in such cases is that it is better to avoid excessive conservatism than to avoid unreliable information. Consequently, various assumptions are made which are thought to be reasonable and which allow a calculation, or at least an estimation, of $P\{ \|y\| > L|y \in S\}$. The moral of the discussion of the preceding section, however, is that this is risky business: it will often be better to make only those statements that can be made on a level of confidence that is consistent with the one attached to the data, and to let the resulting conservatism fall where it may. These are then statements of the kind that have been suggested above. They amount to setting

$$P\{\ \|y\|\ > L\ |\ y \varepsilon S\} = 0 \text{ or } 1 \tag{11}$$

depending on whether the intersection (9) is, or is not empty.

The problem then becomes one of first making the family Σ' as fine as possible, i.e. of utilizing all information that is considered to be reliable enough to be used. In this way, the upper bound P(S) in (10) will be tightened as much as possible. Also, the achievement of the lower bound, namely zero, will be made more likely. Secondly, the intersection (9) must be studied: if it is found empty, the lower bound applies; if not, the upper.

This procedure is in effect a combination of probability theory with worst-case analysis: probability theory is used in setting the measures P(S) of the sets $S \varepsilon \Sigma'$; the worst-case analysis complements it, via (11), by allowing no probabilities other than 0 and 1.

In practice, the probabilistic part seems to be more difficult than the second. It is often doubtful just what information is available, and also reliable enough to be used in the determination of the sets S. A further complication often is that the information which is available does not pertain directly to the response process y of a system, but to the excitation process x. The transformation from one to the other can be difficult. By contrast, the question of whether the value of 0 applies in (11) or 1, seems relatively easy to settle in practice.

The example to be treated in the next two sections will, it is hoped, bear out these comments.

THE CRITICAL EXCITATIONS AND RESPONSES OF LINEAR SYSTEMS

An example which illustrates the general remarks just made will be discussed in this section and the next. It arose from a problem in earthquake engineering.

The ground motions during earthquakes form an almost ideal example of a random process whose precise statistics are very imperfectly known and unlikely to be well-known in the near future. It has been customary in recent years to make the assumption that the ground motions form a Gaussian random process. However, very little evidence in this direction has ever been presented and what evidence exists, apparently does not support the assumption [5]. This uncertainty, of course, is transmitted to the response. As a consequence, and as explained in the second section, any statement regarding structural failure or survival is liable to be in serious error.

Based on the above remarks, one should next inquire what information concerning the statistics of ground motion during earthquakes is well enough established, to be used towards the prediction of structural failure. One can perhaps say that the distribution of ground motion intensities is based on a sample of sufficient size to qualify in this respect. Such information has been accumulated over many years, as pointed out by Housner [6, p.97-99]. There are admittedly many possible definitions for the term 'intensity'. In this paper it will be convenient to define it as the L_2-norm $\|x\|$ of the ground acceleration x, i.e. by

$$\|x\|^2 = \int_{-\infty}^{\infty} x^2(t)\, dt.$$

(Other norms, in particular the maximum ground acceleration, could be used equally well and might even seem more natural here.) One can then perhaps assume that the distribution of $\|x\|$ can be equally well documented no matter which definition is adopted and in fact that it is of the roughly exponential form that has been pointed out by Housner [6, ibid.].

Suppose now that the distribution of $\|x\|$ is actually all that is reliably known regarding the stochastic nature of the ground motion.

Suppose further that the response of a variable, such as the base shear or base moment is of interest in an elastic structure. This response is then related to the ground acceleration by the Duhamel integral

$$y(t) = \int_{-\infty}^{\infty} h(t - \tau)\, x(\tau)\, d\tau \tag{12}$$

in which h is the impulse response of the variable. The sets S introduced in the preceding section are of a special form in this example, namely

$$S = \{y : y(t) = \int_{-\infty}^{\infty} h(t-\tau)\, x(\tau)\, d\tau ; \quad \|x\| \leq M\} . \tag{13}$$

Each consists of all responses that are generated by ground motions with intensities $\|x\| \leq M$. The probability measure of each S is

$$P(S) = P\{ \|x\| \leq M\} \tag{14}$$

which, as was just assumed, is all that is known regarding the statistics of the ground motions.

The question now, according to (11) is whether or not there are any responses in S whose peaks $\|y\|$ exceed the failure limit L. The answer can be given quite easily, by a straightforward use of the Schwarz inequality [7]. One finds that a response in S with the highest peak $\|y\|$ is

$$y^*(t) = \frac{M}{N} \int_{-\infty}^{\infty} h(t-\tau)\, h(-\tau)\, d\tau \quad ,$$

and that all others differ from y* only by time shifts or by a change in sign. Here, N has been used for

$$N^2 = \int_{-\infty}^{\infty} h^2(t)\, dt < \infty .$$

The peak of y*, namely

$$y^*(0) = MN \tag{15}$$

occurs at t=0. It is generated by the excitation

$$x^*(t) = \frac{M}{N}\, h(-t)$$

which is, except for the constant factor, (M/N), the time-reversed impulse response of the variable under consideration. The pair x^*, y^* have been called the 'critical excitation' and the 'critical response' of the structural variable; relative to the set S. Hence, the title of this section.

It should be added that the set S, as defined in (13) developed to be too large in many cases in practice: the critical response peaks (15) were often unrealistically large. However, the fact that the term 'unrealistic' can be used here at all implies that some information is available regarding ground motions, other than merely the distribution (14) of their intensities, as has been assumed here. For, if that were really all that is known it would not be possible to disqualify some of the response peaks as being excessively large. There has been some speculation of what this additional information might be. Shinozuka [8] has suggested the envelope of the Fourier amplitude spectrum as one possible item, and Iyengar [9] the envelope of the time history of the ground acceleration as another. Either suggestion amounts to a restriction of the sets (13) or, equivalently, a refinement of the kind that has been advocated in the third section for Σ', as being helpful towards the reduction of excessive conservatism. The writer and his colleagues have experimented with yet another restriction, which seems to be successful in that the residual conservatism is quite well consistent with good structural design practice [10].

All of the restrictions mentioned here, however, suffer from the same defect, namely, that it is very difficult to say just what the probabilities P(S) of the resulting sets S are. Beyond that, the approach has been criticized on several counts, for instance, the fact that each structural variable has its own critical excitation and response and hence must in principle be analyzed individually, or the implicit assumption that all uncertainties in the response statistics are imputed to the ground motion and none to the structure. Work is under way which will, it is hoped, meet these and other objections.

THE CRITICAL EXCITATIONS AND RESPONSES OF NONLINEAR SYSTEMS

A recent generalization [11] of the result mentioned in the preceding section from linear to nonlinear systems may be of sufficient interest to be reported on here briefly.

What has been shown more specifically is this. The critical excitation x^* and response y^* of a nonlinear system obey two sets of simultaneous equations. One is, of course, the set which defined the system under consideration. The second set is obtained from the first by

(a) linearizing it about x^* and y^*

(b) replacing x with $k\delta$ where δ is the unit impulse function and where k is so determined that $\|x\| = M$

and

(c) reversing time, i.e., replacing t with $(-t)$.

For example, if the system under consideration is given by a single differential equation of the form

$$g\left(\frac{d^n y}{dt^n}, \frac{d^{n-1} y}{dt^{n-1}}, \ldots y\right) = x \tag{16}$$

the critical excitation and response obey two differential equations, namely

$$g\left(\frac{d^n y^*}{dt^n}, \frac{d^{n-1} y^*}{dt^{n-1}}, \ldots y^*\right) = x^* \tag{17}$$

which merely expresses the fact that y^* is the response to x^*, and

$$a_n(t)\,\frac{d^n x^*}{dt^n} + a_{n-1}(t)\,\frac{d^{n-1}x^*}{dt^{n-1}} + \ldots + a_0(t)=k\delta \tag{18}$$

where

$$a_k(t) = (-1)^k \left.\frac{\partial^k g}{\partial y^{(k)}}\right|_{y=y^*} , \quad (y^{(k)} = \frac{d^k y}{dt^k})$$

which is necessary for the criticality of x^* and y^*.

Another way of stating this result which brings out the parallel with linear systems is the following. The critical excitation x^* is again, except for the constant factor k, a time-reversed impulse response. However, by contrast to linear systems, it is not the impulse response of the given system but of a linearized version of it. The linearization must more particularly be around the critical excitation/response pair.

The result holds not only for systems that are, or can be, defined by a single differential equation, such as (16). On the contrary, substantially more general excitation/response relationships are admissible than nonlinear differential equations. In particular history-dependent failure mechanisms, such as material fatigue, are subsumed under it.

The result is derived in roughly the following way. The system is first assumed to be specified by its Volterra series [12], rather than by its differential equations. This is done partly for sake of greater generality and partly to preserve the analogy to the Duhamel integral in (12). The result then follows very quickly by a variational argument. Some attention must be paid to the fact that Volterra series frequently have small radii of convergence, and to the transition from those series to other system representations, such as (16).

The solution that is obtained in this way is valid under fairly general conditions. It has, however, certain drawbacks as well. Among those are, to begin with, all those mentioned in the preceding section in connection with linear systems. In addition, the solution for nonlinear ones need not be unique. There may, in other words, be more than one excitation/response pair that satisfies equations (17) and (18), or others like these. Finally, and perhaps most importantly, these equations unfortunately cannot be solved simultaneously; the obstacle develops to be time-reversal in the second equation, as one recognizes quite easily. The solution can often be carried out by successive approximation, however. On the basis of some limited computational experience, there is in fact hope that the approximation will converge quite rapidly in many problems of practical interest.

ACKNOWLEDGEMENT

The work on which this paper is based was supported by the National Science Foundation, Washington, D.C., USA under Grant No. GK14550 and AEN72-00219 A 01. This support is gratefully acknowledged.

REFERENCES

1 J. Pickands, Asymptotic properties of the maxima of a stationary Gaussian process. Trans. Amer. Math. Soc. 145, 75 (1969).

2 E.J. Gumbel, Statistics of Extremes. Columbia U Press (1958).

3 H. Cramér, Mathematical Methods of Statistics. Princeton U Press (1946).

4 R.R. Bahadur and L.J. Savage, The non-existence of certain statistical procedures in nonparametric problems. The Annals of Math. Stat. 27, 1115 (1956).

5 T. Kobori et al, Statistical properties of earthquake accelerograms and equivalent earthquake excitations. Bull Disaster Prevention Inst. U Kyoto, (1965).

6 G. Housner, Design Spectrum. Earthquake Engineering (R.L. Wiegel ed) (1970).

7 R.F. Drenick, Model-free design of aseismic structures. Jour. Eng. Mech. Div. Amer. Soc. Civil Eng. 96, 483 (1970).

8 M. Shinozuka, Maximum structural response to seismic excitations. Jour. Eng. Mech. Div. Amer. Soc. Civil Eng. 96, 729 (1970).

9 R.N. Iyengar, Matched inputs. Center of Applied Stochastics Purdue U. Rep. 47, Ser. J (1970).

10 P-C Wang and R.F. Drenick, The critical excitation and response of structures. Proc. Sixth World Conf. Earthquake Eng. (to appear) (1976).

11 R.F. Drenick, The critical excitation of nonlinear systems (submitted for publication).

12 V. Volterra, Theory of Functionals and of Integral and Integro-Differential Equations. Dover Publications (1959).

Professor R.F. Drenick, Polytechnic Institute of New York, 333 Jay Street, Brooklyn, New York 11201.

DISCUSSION

Answering a first question (Grossmayer) on the quantity M of ground motion intensity (fourth and fifth sections), Professor Drenick explained that there exists a great deal of data for this quantity from earthquake observations. Using this information the random excitation x had been defined under the assumption of a roughly exponential distribution for M^2 as usual in the literature.

In a comment Dr. Mark pointed out that the equivalence between the results of the worst-case analysis in the paper and results from matched filter theory of radar-sonar signals. There one knows that among all filters with the same integral squared values of their impulse responses the one which maximizes the peak response to all signals with the same energy is equal to the time reversed signal.

Concerning a proposal of Dr. Mark to constrain the energy of the earthquake inputs only within the duration of the input comparable to the duration of the impulse response of the structure one is interested in instead of constraining the total energy, Professor Drenick said this was actually done in computational practice.

In a further discussion (with Murzewski) Professor Drenick pointed out that the asymptotic distribution of extrema as well as the probability of exceedence are sensitive to the probability distribution of the input of the system. There is the broad family of distributions of the input which leads to the same double exponential distribution of extrema (comment Kozin), but the latter contains the parameters which are the ones that cause sensitivity. Further on Professor Drenick stated clearly that until now he had shown the sensitivity of the distribution of extrema with respect to the parameters m, r and σ, but not yet with respect to n. Here Dr. Mark supposed that there might be no severe sensitivity with respect to n pointing out the analogy that the probability distribution of the maximum response of a linear oscillator with white noise excitation over a time interval large in comparison with the decay time of the oscillator is rather insensitive with respect to the observation time.

In a final comment Professor Vanmarcke suggested to use more than only the quantity M from existing earthquake data. He briefly reported upon the work at MIT where there are many efforts to get better knowledge about the major parameters of ground motion, i.e. ground motion duration, dominant ground frequencies, and ground motion intensity. Thus one should be able to give more precise answers to the question of seismic risks. It had turned out that by means of Gaussian models of ground motion and classical reliability theory probabilities of exceeding certain boundaries might be determined very well. Finally Professor Vanmarcke pointed out that it was possible to check the validity of the Gaussian assumption with the ground motion comparing theoretical results from random vibration solutions of linear systems based on the assumption of a Gaussian input and results of numerical simulations based on many actual ground motion records scaled all to the same intensity level. Good agreement had been found, but naturally no statements could be given about the tails of the distribution of ground motion since no data exist.

J B ROBERTS

Probability of first passage failure for lightly damped oscillators

INTRODUCTION

When studying the response of a mechanical system to random excitation, the estimation of the probability of system failure is usually a matter of principal concern. One of the simplest modes of failure is the 'first-passage' type, where it is assumed that failure occurs at the instant the response amplitude first reaches a critical level; recently considerable efforts have been made to develop analytical methods of estimating the probability of first-passage failure (e.g. see references in [1] and [2]).

In this paper the important case of a lightly damped, single degree of freedom oscillator excited by white noise is discussed. Here the response is a two-dimensional Markov process and the powerful, general theory of Markov processes is applicable. However, no exact analytical solution to the problem of calculating the distribution of the time to first-passage failure is yet in sight and recourse must be made to approximate methods. Most of the approaches which have so far been advocated depend on the assumption of a Gaussian response and thus apply only to the case of a linear oscillator.

Here it will be shown that one simple, approximate method, based on the energy envelope of the response, is particularly suitable when the oscillator is lightly damped. Further, it has the great advantage that it is applicable in cases where the oscillator is non-linear. The method is introduced by reviewing some results for the linear oscillator. It is then shown how the theory can be extended to the non-linear case and some numerical results are given for particular cases. The accuracy of the theoretical results is assessed by a comparison with digital simulation estimates.

THE LINEAR OSCILLATOR

A linear oscillator is considered with light viscous damping, subjected to normal, stationary, white noise excitation n(t) with zero mean. The equation of motion is

$$\ddot{X}(t) + 2\zeta\omega_o \dot{X}(t) + \omega_o^2 X(t) = n(t) \tag{1}$$

where ζ is the damping ratio, ω_o is the undamped natural frequency and X(t) is the displacement response process.

The energy envelope, V(t), of the oscillator is defined as

$$V(t) = \frac{\dot{X}^2(t)}{2} + \frac{\omega_o^2 X^2(t)}{2} \tag{2}$$

The Continuous Time Envelope

It has been known for many years that V(t) is a close approximation to a one-dimensional Markov process when $\zeta << 1$ (e.g. see [3]). This fact enabled Helmstrom [4] and subsequently others [5,6,7] to calculate the first-passage probability P_f that the continuous time process V(t) exceeds some level, h, at least once in the interval O-T. The result is of the form

$$P_f = 1 - \sum_{k=1}^{\infty} A_k \exp(-\zeta\omega_o \gamma_k T) \tag{3}$$

where γ_k are the roots of the equation

$${}_1F_1\left(-\frac{\gamma_k}{2} ; 1 ; \frac{\eta^2}{2}\right) = 0 \tag{4}$$

Here ${}_1F_1$ is the confluent hypergeometric equation and η is the non-dimensional barrier height

$$\eta = b/\sigma \quad (h = \omega_o^2 b^2/2) \tag{5}$$

where σ is the standard deviation of the stationary response. The A_k coefficients can be related to hypergeometric functions [4].

Usually the case where T is large is of most concern in practice: the first term in the summation in equation (3) then becomes dominant and one has

$$P_f = 1 - \exp(-\alpha T) \tag{6}$$

where α (= $\zeta\omega_o \gamma_1$) is the 'limiting decay rate'. The variation of α with η is discussed in Reference [8]. There it is demonstrated that the theoretical result agrees very closely with simulation estimates, obtained by generating realisations of V(t) for values of ζ of 0.01 and 0.08. This agreement strongly supports the adoption of a one-dimensional Markov process, as a model of the energy envelope V(t), when the damping is light.

The Discrete Time Envelope

In Helmstrom's analysis, failure occurs when V(t) first reaches a fixed level; thus, in the phase plane, the region of safe operation is enclosed by a circular 'barrier'. The barrier is analytically convenient but in practice one is usually interested in either; (a) a single-sided barrier, such that failure occurs when X(t) first exceeds a fixed, constant amplitude level, b (which may be positive or negative) or (b) a double-sided barrier, such that failure occurs when $|X(t)|$ first exceeds a critical level, b.

If the response of the oscillator is represented by the components V(t) and $\theta(t)$, where $\theta(t) = \tan^{-1}[\dot{X}(t) / \omega_o X(t)]$ then, in the case of a single-sided barrier, failure occurs when $V(t) = h \sec^2\theta(t)$, where $h = \omega_o^2 b^2/2$. Thus the barrier as 'seen' by V(t) will appear as shown in Figure 1(a). For a lightly damped oscillator the times at which $\theta(t) = 0$ (i.e. the barrier reaches its lowest value, h) will have only a small dispersion about the equi-spaced times t_n, where $t_n = 2n\pi/\omega_o$; further, large macroscopic changes in V(t) are unlikely in the intervals between these times. Thus, for light damping, a reasonable approximation is to replace the h $\sec^2\theta(t)$ barrier by the simpler barrier shown in Figure 1(b) [8]. This barrier allows failure to occur at the discrete equi-spaced times t_n; this is equivalent to replacing the continuous time envelope (CE process) by a discrete time envelope (DE process), for the purposes of first-passage analysis.

For the DE process one can write

$$q\ (v,\ t_{n+1}) = \int_0^h q(v_0,\ t_n)\ p\ (v|v_0;\Delta t)\ dv_0 \tag{7}$$

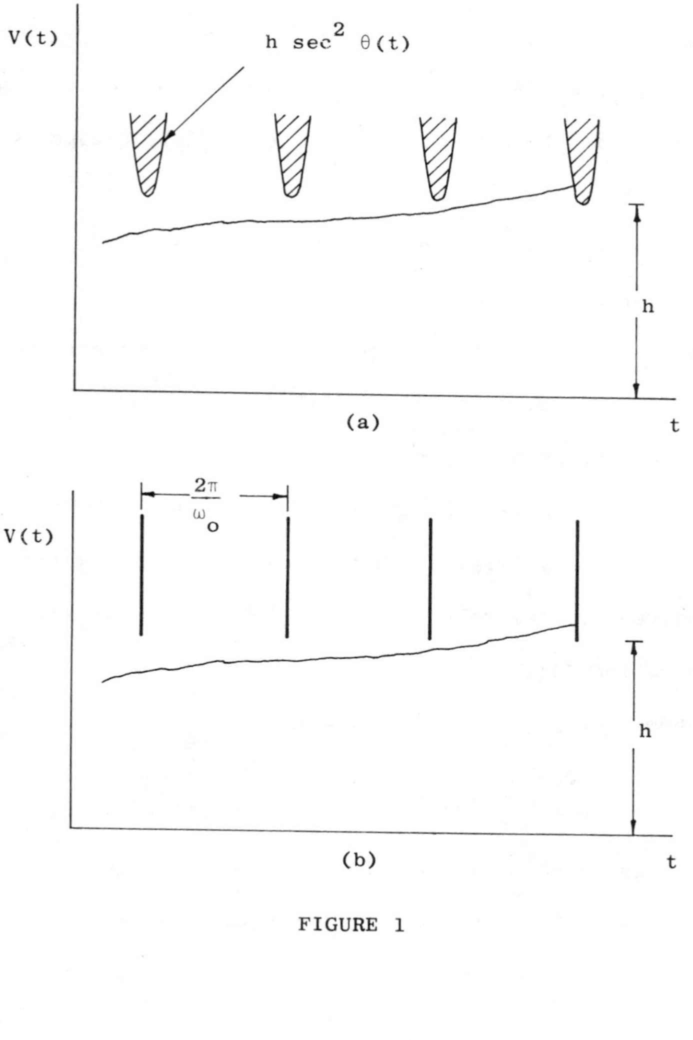

FIGURE 1

where $q(v, t_n)\, dv$ is the probability that $V(t)$ reaches the differential element centred at v at time t_n, without crossing the barrier and $p(v|v_o;\Delta t)$ is the transition density for $V(t)$, with $\Delta t = 2\pi/\omega_o$. $p(v|v_o;\Delta t)$ can be expressed analytically (e.g. see [8]). The probability of failure is then simply

$$P_f = 1 - \int_o^h q\ (v,\ t_n)\ dv \tag{8}$$

In the case of a double-sided barrier, similar arguments apply, but here $\Delta t = \pi/\omega_o$ and $t_n = n\pi/\omega_o$.

It is noted that some work reported by Mark [9] is based on an integral equation similar to equation (8).

As shown in Reference [8], two properties of the DE process may be deduced from the fact that, in the expression for $p\ (v|v_o;\Delta t)$, ζ appears only as a product with t. These are:

(1) P_f depends only on η, $\mu\zeta$ and ζN where $\mu = 1$ for a single-sided barrier and $\mu = \frac{1}{2}$ for a double-sided barrier ($N = \omega_o\ T/2\pi$).

(2) P_f values for the DE process approach the corresponding values for the CE process as $\zeta \to 0$.

Equation (7) can be solved by a simple marching technique, if η is not too large [8]. For large η a better approach is to expand $p\ (v|v_o;\Delta t)$ and $q(v,t_n)$ in terms of Laguerre polynomials. This leads to a simple matrix equation, which can be solved iteratively [8]. When T is large the solution is of the same form as equation (6).

It can be shown that, for the DE process,

$$\alpha \to \frac{\omega_o}{2\mu\pi} \exp\ (-\ \frac{\eta^2}{2}) = \nu\ \text{(say)} \tag{9}$$

as $\eta \to \infty$ [8]. ν is equal to average frequency of barrier crossings from below. This is an exact asymptotic result for $X(t)$ (e.g. see [1]).

Comparison with Simulation Results

Figures 2 and 3 show the variation of α/ν with η, for $\mu\zeta = 0.04$ and 0.02, respectively, as predicted by the CE and DE methods. Also shown are some simulation estimates for X(t), obtained from References [2] and [10], together with results for X(t) calculated by the method of non-approaching points (NAP) [2]. The DE results are seen to be in excellent agreement with the simulation results at low values of η. At high values of η, where the 'in and exclusion' series is rapidly convergent and the NAP estimates are consequently accurate, there is close agreement between the DE results and the NAP estimates. It can be concluded that the DE method gives an accurate prediction of the variation of α with η, over the range shown. In contrast, the CE results are a poor guide, except when ζ is very small and η is small.

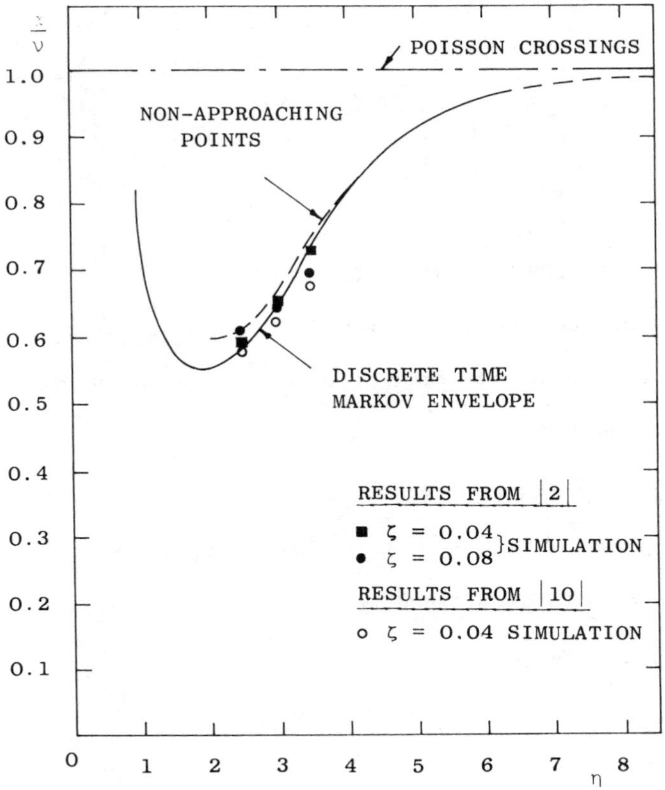

FIGURE 2

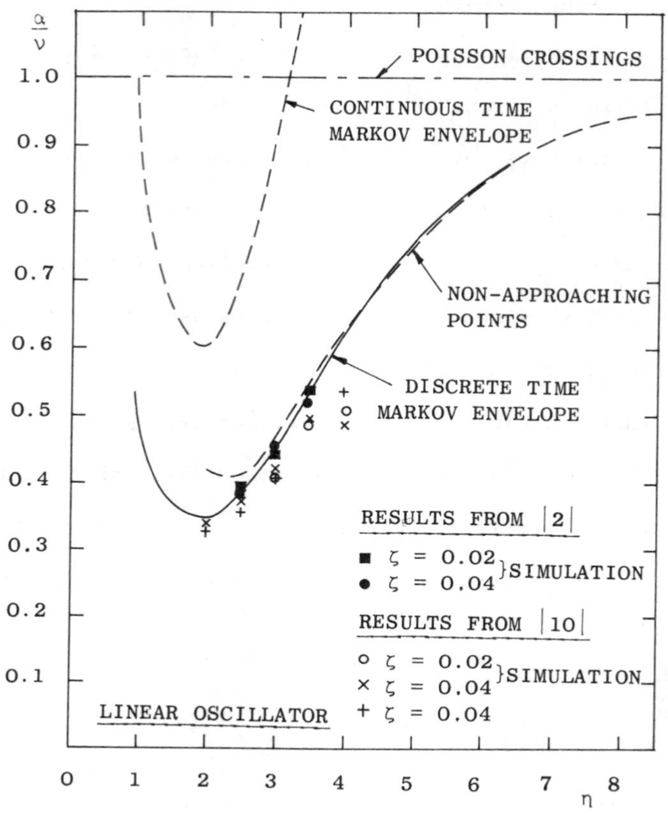

FIGURE 3

CONTINUOUS TIME ENVELOPE FOR NON-LINEAR OSCILLATORS

The discussion will now be widened to include oscillators with light non-linear damping and non-linear restoring forces. The equation of motion will be assumed to be of the form

$$\ddot{X}(t) + \beta F\left[X(t), \dot{X}(t)\right] + G\left[X(t)\right] = n(t) \qquad (10)$$

where β is a small parameter and $F[\;]$ is an odd function with respect to $\dot{X}(t)$.

The energy envelope V(t) is defined as

$$V(t) = \dot{X}^2(t)/2 + U\left[X(t)\right] \qquad (11)$$

where

$$U(x) = \int_0^x G(\xi)\, d\xi \qquad (12)$$

In the linear case equation (11) reduces to equation (2).

The Fokker-Planck Equation

The transition density for the joint process X(t), $\dot{X}(t)$ is governed by a two-dimensional Fokker-Planck equation. By transforming this equation from displacement, velocity variables to displacement, energy variables, omitting one term which is negligible when β is small, and then integrating with respect to the displacement variable, it is possible to derive the following one-dimensional Fokker-Planck equation for V(t) [11]

$$\frac{\partial p}{\partial t} = \frac{\partial}{\partial v}\left[\{ \beta B(v) - \frac{I}{2} \} p \right] + \frac{I}{2}\frac{\partial^2}{\partial v^2}\left[C(v)\ p \right] \tag{13}$$

where $p\ (v|v_0;\ t)$ is the transition density for V(t) and

$$B(v) = \frac{1}{\sqrt{2}\ A(v)} \int_R F\ \{x,\ \sqrt{2(v-U(x))}\}\ dx \tag{14}$$

$$C(v) = \frac{1}{A(v)} \int_R \sqrt{v - U(x)}\ dx \tag{15}$$

$$A(v) = \frac{1}{2} \int_R \frac{dx}{\sqrt{v - U(x)}} \tag{16}$$

The integration range, R, here is such that $U(x) < v$. I, in equation (13) is the 'strength' of the white noise, n(t) - i.e.,

$$E\ \{ n(t)\ n(t + \tau) \} = I\ \delta(\tau) \tag{17}$$

Equation (13) has been obtained previously by Stratonovitch [12], using a different approach. However, his derivation is unsatisfactory since it involves the incorrect assumption that the joint process X(t), V(t) is governed by a two-dimensional Fokker-Planck equation.

It can be shown [11] that, in the linear case, equation (13) is identical to the Fokker-Planck equation derived by Helmstrom [4]. This equation is thus a suitable basis for a generalisation of the linear results discussed earlier.

Analytical Solutions

From equation (13) it follows that the probability $W(v_o;t)$ that first passage failure will not occur in the interval 0-t, for trajectories starting at v_o, is governed by the equation (11)

$$\frac{\partial W}{\partial t} = - \left[\beta\, B\, (v_o) - \frac{I}{2} \right] \frac{\partial W}{\partial v_o} + \frac{I}{2}\, C(v_o)\, \frac{\partial^2 W}{\partial v_o^2} \qquad (18)$$

with the boundary condition W (h;t) = 0. Here h = U(b).

Equation (18) can be solved analytically only in special cases. In Reference [11] it is shown that for an oscillator with linear damping ($F(X,\dot{X}) = \dot{X}$) and a spring force of the power law form

$$G\,(X) = k\, |\, X\, |^{\nu}\, \text{sign}\,(X) \qquad (19)$$

where k is a constant, an analytical solution can be found in terms of hypergeometric functions. The result is of the form

$$P_f = 1 - \sum_{k=1}^{\infty} B_k \exp\,(-\,\beta\,\lambda_k\, T) \qquad (20)$$

where λ_k are the roots of the equation

$$_1F_1\,(-\frac{\lambda_k}{\alpha}\,;\,\frac{1}{\alpha}\,;\,\frac{\mu}{\gamma}) = 0 \qquad (21)$$

Here

$$\alpha = \frac{2(\nu+1)}{(\nu+3)}\,, \qquad \gamma = \frac{1}{\nu+1} \left\{ \frac{\Gamma(\frac{1}{\nu+1})}{\Gamma(\frac{3}{\nu+1})} \right\}^{\frac{\nu+1}{2}} \qquad (22)$$

and

$$\mu = \frac{\eta^{\nu+1}}{(\nu+1)} \qquad (23)$$

The B_k coefficients are related to hypergeometric functions [11]. In the linear case, $\alpha = \gamma = 1$ and equations (20) and (21) coincide with equations (3) and (4), respectively.

Moment Equations

Although equation (18) cannot, in general, be solved analytically, and may be difficult to solve numerically, the moments of the first passage time

$$M_n (v_o) = E \{ T^n (v_o) \} \tag{24}$$

where $T(v_o)$ is the time to reach the boundary, for trajectories starting at v_o, can be readily obtained from the following equations [11]:

$$- nM_{n-1} = - \left[\beta B(v_o) - \frac{I}{2} \right] \frac{dM_n}{dv_o} + \frac{I}{2} C(v_o) \frac{d^2M_n}{dv_o{}^2} \tag{25}$$

where $M_o (v_o) = 1$. In the linear case this result reduces to the equations given by Ariaratnam and Pi [13].

Equations (25) can be used to generate M_n recursively; i.e. first solving for M_1, then M_2, and so on. They are readily solved numerically.

Effect of Non-linear Stiffness

To illustrate the effect of non-linearities in the spring stiffness on the first passage time for the continuous energy envelope, V(t), it is convenient to consider an oscillator of the Duffing type. The equation of motion is

$$\ddot{X}(t) + \beta \dot{X} (t) + k X(t) (1 + \frac{X^2(t)}{L^2}) = n(t) \tag{26}$$

where k and L are spring constants.

From equation (25) one finds that, if the mean time to first passage failure, $M_1(0)$, is measured in units of βt, then it depends only on the parameters η' and ε_s, where

$$\eta' = b/\sigma' , \qquad \varepsilon_s = \sigma'^2/L^2 \tag{27}$$

and σ' is the stationary standard deviation of the oscillator response, with $\varepsilon_s = 0$. Figure 4 shows the computed variation of $M_1(0)$ with η', for $\varepsilon_s = 0$, 0.05 and 0.2. As one would expect, even a small value of ε_s has a pronounced effect on the first passage time when η' is large. Also shown in Figure 4 are simulation estimates of $M_1(0)$ obtained by generating realisations of V(t),

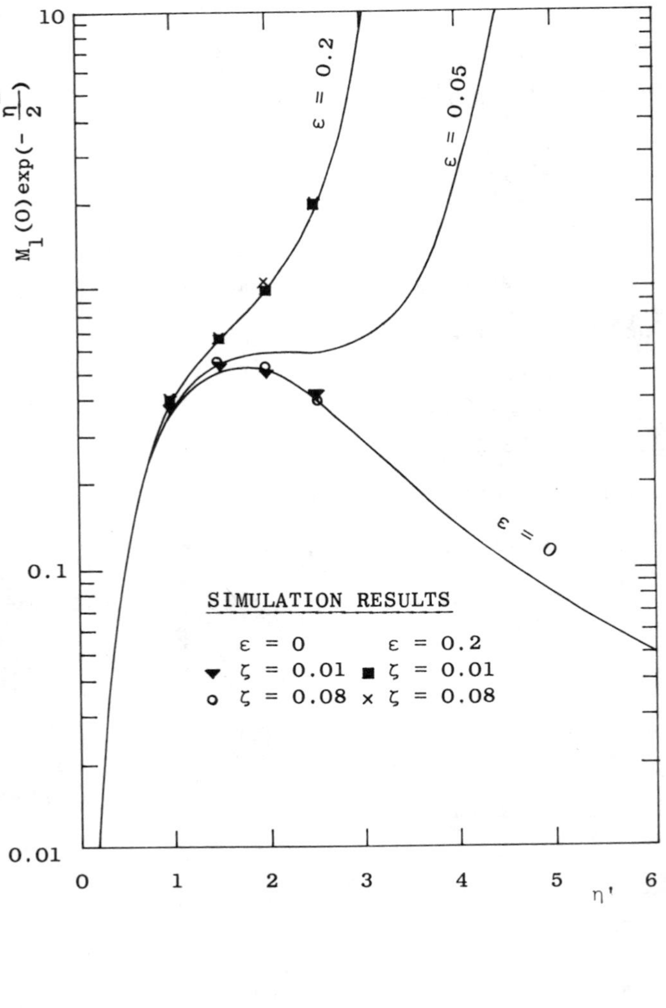

FIGURE 4

for $\zeta = 0.01$ and 0.08 ($\zeta = \beta/(2\sqrt{k})$). Good agreement exists between the theoretical results and the simulation estimates, for both $\varepsilon_s = 0$ and $\varepsilon_s = 0.2$.

Further comparisons are given in Reference [11], for both $M_1(0)$ and $M_2(0)$.

Effect of Non-linear Damping

To illustrate the effect of non-linearities in the damping term, an oscillator with the following equation of motion is considered:

$$\ddot{X}(t) + \beta\dot{X}(t)\left[1 + \frac{|\dot{X}(t)|}{V}\right] + k\,X(t) = n(t) \tag{28}$$

Here V is a damping parameter, with velocity dimensions.

In this case $M_1(0)$, with time in units of βt again, depends only on η' and ε_D, where

$$\varepsilon_D = \sigma'\,\omega_0/V \tag{29}$$

Figure 5 shows the variation of $M_1(0)$ with η' for various values of ε_D, from 0 to 0.5. Again, simulation estimates of $M_1(0)$, obtained by generating realisations of V(t), show good agreement with the theoretical results, for $\varepsilon_D = 0.5$ and $\zeta = 0.01$ and 0.08.

DISCRETE TIME ENVELOPE FOR NON-LINEAR OSCILLATORS

Although results relating to the first-passage probability of the continuous envelope V(t) can be obtained fairly readily, in the non-linear case, as shown in the preceding section, they are of limited value. As in the linear case, simulation results indicate that P_f values of X(t), for both single and double-sided barriers, deviate considerably from the corresponding P_f values of V(t), unless the barrier height is low and β is very small [11].

In view of the excellent agreement between the discrete time envelope theory and simulation results for single and double-sided barriers it seems

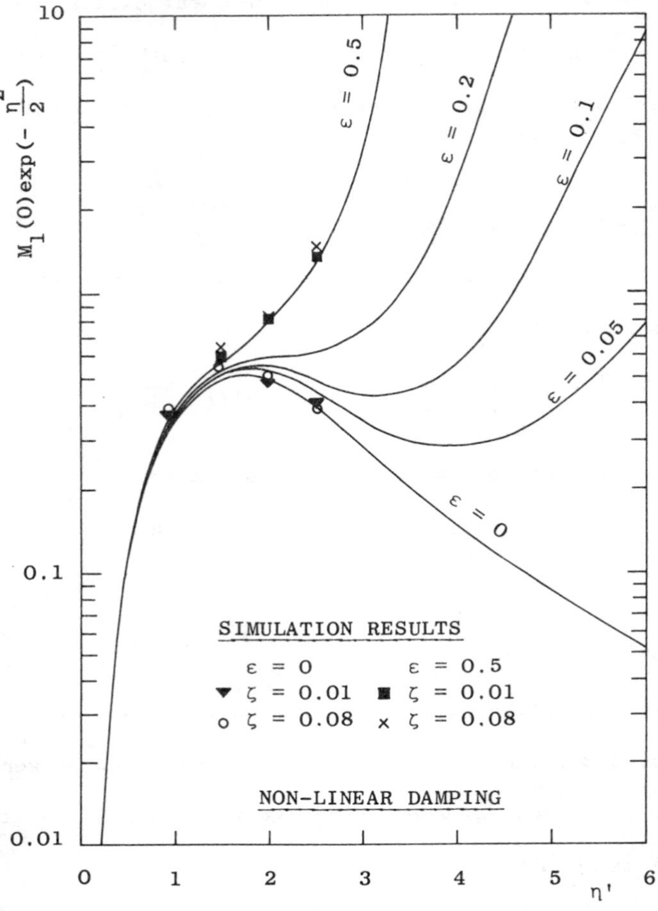

FIGURE 5

worthwhile to explore the possibility of extending this method to the non-linear case. Equation (7) is again applicable, with h = U(b), but two difficulties now arise:

(1) An expression for the transition density function $p(v|v_o;t)$ is required. In principle this can be found by solving equation (13), with the boundary condition

$$p(v|v_o;t) = \delta(v-v_o) \text{ as } t \to o \tag{30}$$

(2) A suitable value of Δt must be chosen. This choice is not immediately obvious in the case of non-linear stiffness, since the period of a cycle is amplitude dependent.

The Transition Density Function

It can be shown (e.g. see [12] that, in general, the transition density function can be expressed as

$$p(v|v_o;t) = \sum_{m=0}^{\infty} \frac{V_m(v)\ V_m(v_o)}{V_o(v_o)} e^{-\lambda_m t} \tag{31}$$

where $V_m(v)$ and λ_m are, respectively, eigenfunctions and eigenvalues associated with the Fokker-Planck equation. $V_o(v)$ is just the stationary solution: i.e.,

$$V_o(v) = p_s(v) = \lim_{t \to \infty} p(v|v_o;t) \tag{32}$$

and $\lambda_o = 0$. The range of problems for which eigenfunctions and eigenvalues can be found analytically appears to be very limited (e.g., see [14]). However, an analytical solution can be found in the case of an oscillator with linear damping but a non-linear spring of the power law form given by equation (19). On replacing v by the variable

$$z = \frac{v}{\gamma k \sigma^{\nu+1}} \tag{33}$$

the result can be written as

$$p(z|z_o;t) = \exp(-z)\ z^{\rho} \sum_{m=0}^{\infty} \frac{\Gamma(m+1)}{\Gamma(m+1+\rho)} L_m^{\rho}(z)\ L_m^{\rho}(z_o) e^{-m\alpha\beta t} \tag{34}$$

where $L_m^{\rho}(z)$ are generalised Laguerre polynomials and

$$\rho = \frac{1}{\alpha} - 1 \tag{35}$$

Equation (34) can be cast into a closed form expression by using the Hille-Hardy formula (e.g. see [15]. This gives

$$p(z|z_o;t) = \frac{1}{(1-q)} \left(\frac{z}{z_o q}\right)^{\rho/2} \exp\left\{ - \frac{(z+qz_o)}{(1-q}\right\} I_\rho \left\{ \frac{2\sqrt{zz_o q}}{1-q} \right\} \tag{36}$$

where

$$q = \exp\{ - \alpha\beta t \} \tag{37}$$

In the linear case, $\alpha = \gamma = 1$, $\rho = 0$ and equation (36) reduces to the expression given by Helmstrom [4]. It is interesting to note that in the case $\nu = 0$, the so-called 'bang-bang' oscillator, $\rho = \frac{1}{2}$ and $p(z|z_o;t)$ can be expressed in terms of elementary functions, as follows:

$$p(z|z_o;t) = \frac{1}{\sqrt{\pi z_o q(1-q)}} \exp\left\{ - \frac{(z+qz_o)}{(1-q)} \right\} \sinh\left\{ \frac{2\sqrt{zz_o q}}{1-q} \right\} \tag{38}$$

Choice of Time Interval

In the case of an oscillator with light non-linear damping, but linear stiffness, the period of the cycles in the response process will be virtually independent of amplitude, as in the linear case. A natural choice for Δt is then $2\pi\mu/\omega_o$.

For oscillators with non-linear stiffness the choice of Δt requires more careful consideration. It is noted that, for one time step, realisations of $V(t)$ are only likely to exceed the barrier at time t_{n+1} if they are already near the boundary at time t_n. Thus, when applying equation (7), the change in P_f from t_n to t_{n+1} is mainly derived from that part of the integration range where v_o is near to h, (see equation (8)). This suggests that a suitable choice of Δt is μ times the undamped natural period of the oscillator, when it has an amplitude of b, where $h = U(b)$. On integrating equation (10), with $\beta = 0$ and $n(t) = 0$, one finds that

$$\Delta t = 4\mu \int_0^b \frac{dx}{\sqrt{2\left[U(b)-U(x)\right]}} = 2\sqrt{2}\, A(h)\mu \tag{39}$$

where the A( ) function is defined by equation (16).

Asymptotic Behaviour

It is interesting to examine the effect of this choice of Δt on the limiting decay rate α, in the asymptotic case where b becomes large.

Assuming a separable solution

$$q(v,t_n) = \emptyset(v)\ f\ (t_n) \tag{40}$$

to equation (7) is possible, one finds that

$$\emptyset(v) = \lambda \int_0^h \emptyset(v_o)\ p\ (v|v_o;\ \Delta t)\ dv_o \tag{41}$$

where

$$\lambda = f(t_n)/f(t_{n+1}) \tag{42}$$

Thus, the general solution can be written in the form

$$q(v,t_n) = \sum_{k=1}^{\infty} \emptyset_k(v)\ e^{-\gamma_k t_n} \tag{43}$$

where

$$\gamma_k = \frac{1}{\Delta t} \log\ (\lambda_k) \tag{44}$$

λ_k is the kth eigenvalue of equation (41) and $\emptyset_k(v)$ is the corresponding eigenfunction.

The limiting decay rate α is simply γ_1 and, as $b \to \infty$, $\emptyset_1(v)$ will approach $p_s(v)$. Setting $\emptyset_1(v) = p_s(v)$ in equation (41), using equation (31) for $p(v|v_o;\ \Delta t)$ and the fact that [12]

$$\int_{0}^{\infty} V_m (v)dv = 0 \text{ for } m > 0 \tag{45}$$

one finds that

$$\lambda_1 \to \left[\int_{0}^{h} p_s(v)\, dv \right]^{-1} \qquad \text{as } b \to \infty \tag{46}$$

When b is large λ_1 will be very close to unity and hence, from equation (44),

$$\alpha = \gamma_1 \to \frac{1}{\Delta t} \int_{h}^{\infty} p_s(v)dv \qquad \text{as } b \to \infty \tag{47}$$

The stationary envelope distribution $p_s(v)$ can be found from the Fokker-Planck equation (13), by setting $\partial p/\partial t = 0$. The result is [11]

$$p_s(v) = K\, A(v) \exp \{ - \frac{2\beta}{I} H(v) \} \tag{48}$$

where

$$H(v) = \int_{0}^{v} \frac{B(\xi)}{C(\xi)} d\xi \tag{49}$$

and K is a constant found by normalisation. Thus

$$K = \left[\int_{0}^{\infty} A(v) \exp \{ - \frac{2\beta}{I} H(v) \} dv \right]^{-1} \tag{50}$$

In the case of linear damping $H(v) = v$ and equation (48) coincides with the exact solution, obtained from the two-dimensional Fokker-Planck equation. In the case of non-linear damping, equation (48) is a close approximation, as a comparison with simulation results testifies [16].

Restricting attention now to the case of linear damping, but a general non-linear stiffness, one finds, on substituting equation (48) into equation (47), using equation (39), and integrating by parts, that

$$\alpha \rightarrow \frac{KI}{4\sqrt{2\beta\mu}} \exp\left\{ - \frac{2\beta h}{I} \right\} + \frac{KI}{4\sqrt{2\beta A(h)\mu}} \int_h^\infty \exp\left\{ - \frac{2\beta v}{I} \right\} A'(v)dv \tag{51}$$

Also, in this case, the expression for K can be recast into the simpler form

$$K = \frac{2}{\sqrt{\pi}} \sqrt{\frac{2\beta}{I}} \left[\int_{-\infty}^{\infty} \exp\left\{ - \frac{2\beta}{I} U(x) \right\} dx \right]^{-1} \tag{52}$$

The first term on the right hand side of equation (51) can be identified with ν, the expected number of crossings of X(t) of a barrier of height b, from inside to outside, per unit time [17]. The second term is zero in the linear case, and then equation (51) reduces to equation (9). In the non-linear case it appears that the second term will be negligible as b becomes large, in most cases. For example, if A(v) behaves like v^a, when v is large, then the second term will behave like $\exp\{ - 2\beta h/I \}/h$ when b is large. Thus one can expect that

$$\alpha \rightarrow \nu, \quad \text{as } b \rightarrow \infty \tag{53}$$

$\alpha = \nu$ is the result obtained if the barrier crossings are Poisson distributed and it is well known that this is the correct asymptote in the linear case, where X(t) is Gaussian. In the non-linear case, although a rigorous proof is not available, it seems reasonable to suppose that the Poisson assumption is again valid in the limit as $b \rightarrow \infty$, and hence that equation (53) is the correct asymptotic result. It can be concluded, therefore, that the DE method proposed here yields the correct asymptotic value for α, as $b \rightarrow \infty$.

Evaluation of α

The simplest way of computing α is to march in time, using equation (7) to redistribute $q(v,t_n)$ at each time t_n. $q(v,t_n)$ becomes proportional to the first eigenfunction, $\emptyset_1(v)$, as n becomes large and if

$$Q(n) = \int_{0}^{h} q(v,t_n)\, dv \tag{54}$$

then the ratio

$$R(n) = Q(n)/Q(n+1) \to \lambda_1 \text{ as } n \to \infty \tag{55}$$

λ_1 can thus be estimated by marching until R(n) reaches its limiting value.

Numerical difficulties arise with this simple method when h is large, since λ_1 is then very close to unity and γ_1 becomes sensitive to errors in λ_1. A much better approach is to use a generalised version of the matrix method described in Reference [8]. This involves expanding $q(v,t_n)$ in terms of the eigenfunctions $V_n(v)$ occurring in equation (31). When h is large only a few terms in the eigenfunction expansions of $p(v|v_o;t)$ and $q(v,t_n)$ are required.

The procedure is now illustrated in the case of an oscillator with linear damping, and a non-linear spring of power-law form. Here $p(z|z_o:t)$ is given by equation (34) where z is proportional to v (see equation (33)). On substituting this result into equation (7), together with the expansion

$$q(z,t_n) = \exp\,(-\,z)\; z^{\rho} \sum_{k=0}^{\infty} C_k^{(n)}\; L_k^{\rho}\,(z) \tag{56}$$

it is found that

$$C_k^{(n+1)} = \exp\,\{-\,k\alpha\beta\Delta t\}\; \frac{\Gamma(k+1)}{\Gamma(k+1+\rho)} \sum_{j=0}^{\infty} C_j^{(n)}\; A_{jk} \tag{57}$$

where

$$A_{jk} = \int_{0}^{\mu} L_j^{\rho}\,(z)\; e^{-z}\; z^{\rho}\; L_k^{\rho}\,(z)\; dz \tag{58}$$

and $\mu = h/\gamma k \sigma^{\nu+1}$.

Starting with a set of coefficients $C_k^{(1)}$, one can use equation (57) to generate successively the coefficients $C_k^{(n)}$ for n = 2,3,.... As n becomes large $C_k^{(n)} \rightarrow C_k$, where C_k are the coefficients corresponding to the first eigenfunction, and

$$\frac{C_o^{(n)}}{C_o^{(n+1)}} \rightarrow \lambda_1 \quad \text{as} \quad n \rightarrow \infty \tag{59}$$

This provides a method of calculating λ_1, and hence α.

The coefficients A_{jk} in equation (58) can be expressed as

$$A_{jk} = \frac{\Gamma(k+1+\rho)}{\Gamma(k+1)} \delta_{jk} - B_{jk}$$

where δ_{jk} is the Kronecker delta and

$$B_{jk} = \int_{\mu}^{\infty} L_j^{\rho} (z) \; e^{-z} \; z^{\rho} \; L_k^{\rho} (z) \; dz \tag{60}$$

B_{jk} can be generated recursively, using the equations

$$kB_{ok} = (k-1) \; B_{o,k-1} - e^{-\mu} \; \mu^{\rho+1} \; L_{k-1}^{\rho} \; (\mu) \tag{61}$$

and

$$B_{jk} = \{(k-j-1)/k\} B_{j,k-1} + \{(j+\rho)/k\} \; B_{j-1,k-1} - (\mu^{\rho+1}/k) e^{-\mu} \; L_{k-1}^{\rho} \; (\mu) \; L_j^{\rho} \; (\mu) \tag{62}$$

Comparison with Simulation Results

In the case of an oscillator with a linear damper, $F(X,\dot{X}) = \dot{X}$, and a non-linear spring force, $G(X)$, as given by equation (19), the non-dimensional limiting decay rate α/ν depends only on the parameters $\mu\chi$ and $\eta(= b/\sigma)$, where

$$\chi = \frac{\beta}{2\sqrt{k\sigma^{\nu-1}}} \tag{63}$$

is a non-dimensional damping coefficient (in the linear case, $\chi = \zeta$).

Figures 6 and 7 show the variation of α/ν with η, for $\mu\chi = 0.02$ and 0.04, respectively, in the case of the bang-bang oscillator (ν=o). Here the solid line is the result obtained by the DE method, using the time step given by equation (39). It is seen to be in excellent agreement with simulation estimates of α/ν, obtained by generating realisations of X(t), for both single and double sided barriers. The DE result approaches a value of unity when η is large, as predicted earlier. For comparison purposes, the corresponding results obtained from the CE method are also shown in Figures 6 and 7. These were obtained by finding the first root of equation (21), as a function of η. It is observed that, as in the linear case, the CE results only approach the DE results when the damping is extremely light and η is small.

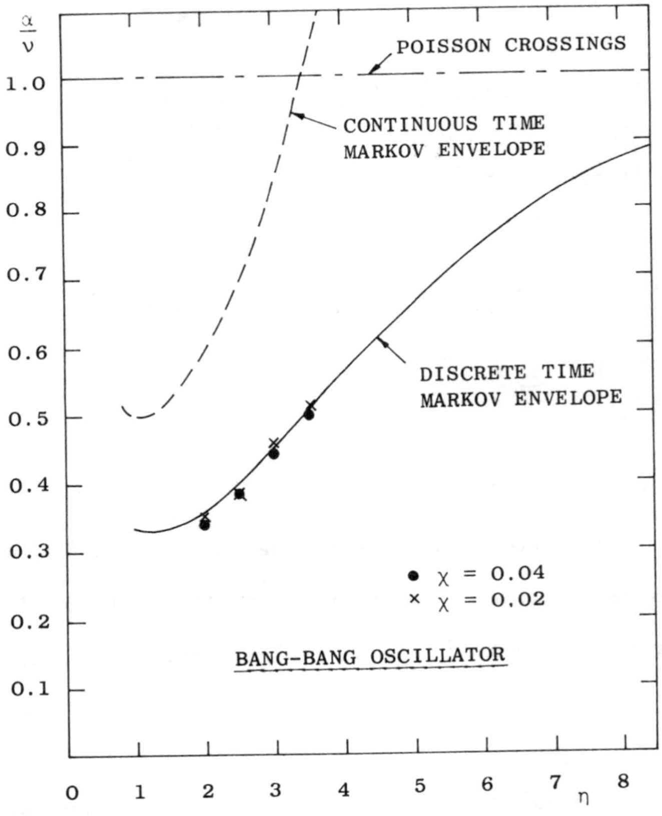

FIGURE 6

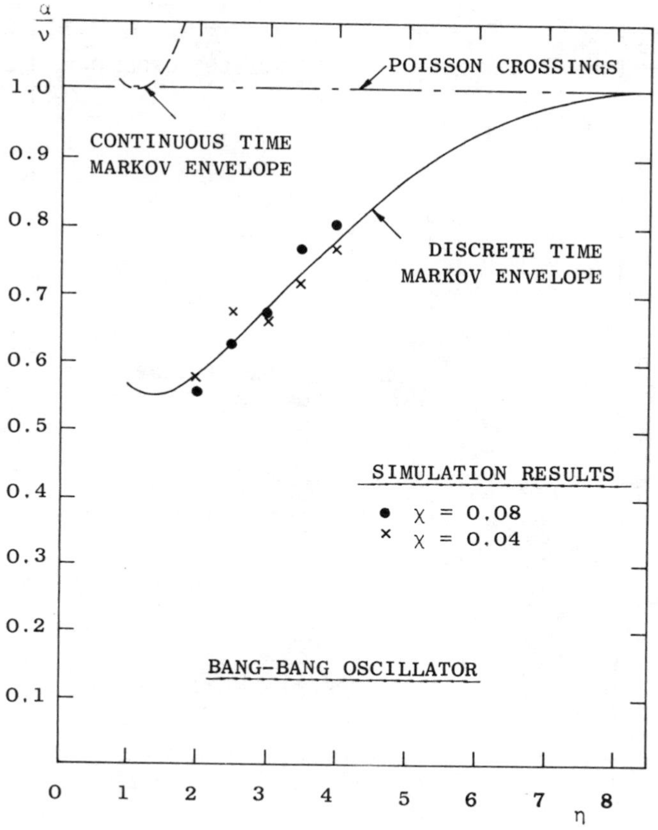

FIGURE 7

Finally, Figure 8 shows the variation of α/ν with η, for $\mu\chi = 0.02$, in the case of an oscillator with a cubic-law spring (ν=3). Again the DE result is in excellent agreement with simulation estimates, for both single and double-sided barriers. The corresponding CE result is also shown here, for comparison purposes.

CONCLUSIONS

The main conclusions are summarised as follows:

(1) The energy envelope of a lightly damped oscillator can be modelled satisfactorily as a one-dimensional Markov process.

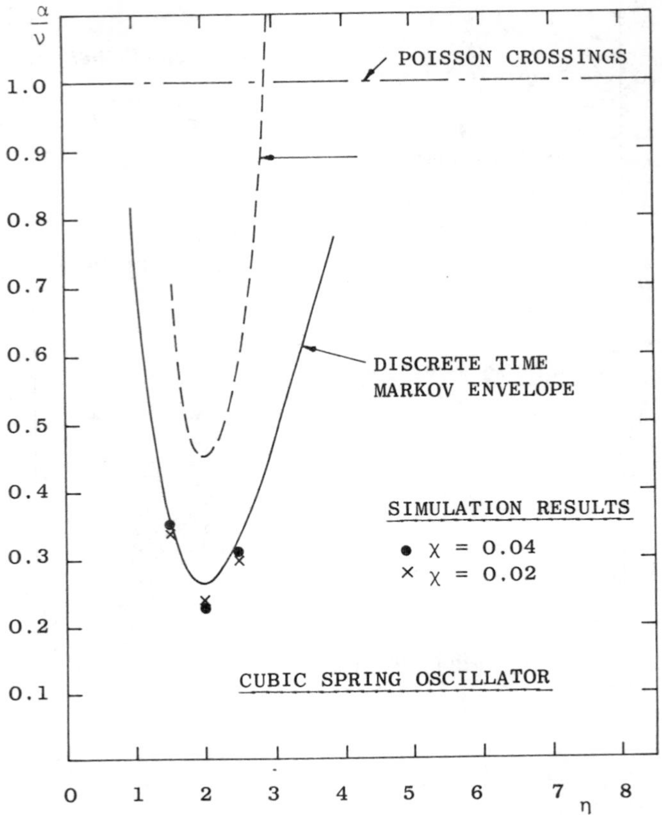

FIGURE 8

(2) The discrete time Markov envelope (DE) method gives accurate estimates of the probability of first passage failure, P_f, for both single and double sided barriers, in typical cases.

(3) Results obtained by the continuous time Markov envelope (CE) method deviate considerably from the corresponding DE results, unless η is small and $\mu\zeta$ is extremely small.

REFERENCES

1 S.H. Crandall, First-crossing probabilities of the linear oscillator. J. Sound Vib. 12, 285 (1970).

2 J.B. Roberts, Probability of first passage failure for stationary random vibration. AIAA J 12, 1636 (1974).

3 J.N. Pierce, A Markoff envelope process. IRE Trans. Inf. Theory IT-4, 163 (1958).

4 C.W. Helmstrom, Note on a Markoff envelope process. IRE Trans. Inf. Theory IT-5, 139 (1959).

5 E. Rosenblueth and J. Bustamente, Distribution of structural response to earthquakes. Trans. ASCE J. Eng. Mech. Div. 88, 75 (1962).

6 A.H. Gray, First-passage time in a random vibrational system. Trans. ASME J. App. Mech. 33, 187 (1966).

7 W.C. Lennox and D.A. Fraser, On the first-passage distribution for the envelope of a non-stationary narrow-band stochastic process. Trans. ASME J. App. Mech. 41, 793 (1974).

8 J.B. Roberts, First passage time for the envelope of a randomly excited linear oscillator. J. Sound Vib. 46, 1, (1976).

9 W.D. Mark, On false-alarm probabilities of filtered noise. Proc. IEEE 54, 316 (1966).

10 S.H. Crandall, K.L. Chandiramini and R.G. Cook, Some first passage problems in random vibration. Trans. ASME J. App. Mech. 33, 532 (1966).

11 J.B. Roberts, First passage probability for non-linear oscillators. J. Eng. Mech. Div. ASCE 102, 851 (1976).

12 R.L. Stratonovitch, Topics in the theory of random noise, Vol. 1, Gordon and Breach, New York (1963).

13 S.T. Ariaratnam and H.N. Pi, On the first passage time for envelope crossing for a linear oscillator. Int. J. Control 18, 89 (1973).

14 E. Wong and J.B. Thomas, On polynomial expansions of second order distributions. SIAM J. App. Math. 10, 507, (1962).

15 A. Erdelyi (ed), Higher transcendental functions, Vol. 2, 189, McGraw-Hill, New York (1953).

16 J.B. Roberts, Stationary response of oscillators with non-linear damping to random excitation. J. Sound Vib. 50, 145 (1977).

17 S.H. Crandall, The envelope of random vibration of a lightly damped non-linear oscillator. Zagadnienia drgań nieliniowych (non-linear vibration problems), 5, 120 (1964).

Dr. J.B. Roberts, Applied Sciences Laboratory, University of Sussex, Falmer, Brighton BN1 9QT.

DISCUSSION

When questioned (Lin) about the transformation of the problem involving a two-dimensional Markov vector $(X(t), \dot{X}(t))$ to a one dimensional problem the author pointed out that it was not being claimed that the joint distribution of (V(t), X(t)) was exactly Markovian, but that an approximation had to be made to get down to the one-dimensional equation. This approximation was difficult to justify by a rigorous theoretical argument but seemed plausible on the basis of simulation results.

It was noted (Vanmarke) that the comparison of simulation results and theoretical predictions in the nonlinear case amounted to a comparison of the mean value of the first passage times and the choice of the first passage probability in the nonlinear case was queried. The author replied that in the nonlinear case a histogram of first passage times had been computed to enable the limiting decay rate to be estimated. This procedure was more accurate than merely estimating the decay rate from the mean time, for which one might be in error by about 10%.

It was suggested (Kozin) that for the case of no damping the first few moments could be computed exactly (Reference [18] p.232) and it might be interesting to compare these with the simulations for small β.

It was commented (Mark) that the approach to the linear oscillator problem was essentially similar to that of Reference [9]. The author replied that it had been difficult to compare the methods but that on the basis of the graphs presented in [9] there seemed to be close agreement.

On being asked (Grossmayer) if there was any possibility of extending the method to include hysteretic damping the author expressed doubts and it was added (Vanmarke) that systems with such damping exhibit a drift of the mean value which is altogether a different phenomenon.

ADDITIONAL COMMENT BY THE AUTHOR

In the case of an oscillator with a linear restoring force, but non-linear damping, it can be shown (Reference [16]) that the method of stochastic averaging, due to Stratonovitch (Reference [12]) yields a one-dimensional equation for the energy envelope which is identical to that given in the present paper (Equation (13)). This agreement provides further theoretical support to the proposed method, since the stochastic averaging method can be justified rigorously, as shown by Khasminskii [19] and Papanicolaou and Kohler [20].

Fuller References

18 K. Ito and H.P. McKean, Diffusion Processes and their Sample Paths. Springer-Verlag, Berlin (1965).

19 R.Z. Khasminskii, A limit theorem for solutions of differential equations with a random right-hand side. Theory Prob. Applications, Vol. 11, 390 (1966).

20 G.C. Papanicolaou and W. Kohler, Asymptotic theory of mixing stochastic ordinary differential equations. Communications on Pure and Appl. Maths 27, 641 (1974).

R GROSSMAYER

On the application of various crossing probabilities in the structural aseismic reliability problem

INTRODUCTION

Earthquake Engineering is a modern and wide field for application of random vibration theory. Structural aseismic design presents many problems like the uncertainty of the excitation, the task of finding a proper structural model, the interaction between the structure and its foundation or surrounding structures. Although more and more data from measured records are becoming available the next earthquake expected to occur on a certain site, cannot be foreseen. Seismic risk maps enable one to choose an expected earthquake magnitude depending on a given probability of occurrence. Other important characteristics of earthquake records are their duration (usually about twenty seconds), their frequency content (the main frequencies lie usually in the range from 1 to 10 Hz) and their nonstationarity (after a short build-up period of about two seconds there is a nearly stationary strong-motion period taking about five to ten seconds followed by a period of decaying motion) and, last but not least, their random nature. In the following, a stochastic model for the nonstationary excitation process will therefore be used.

Various types of structures should be built strong enough to resist earthquakes, e.g., nuclear power plants, tall buildings or radio transmitters. In many cases a lumped-parameter model is chosen. The fundamental frequencies of the structures are about 0.4 Hz for a power plant structure, or 4 Hz for a tall building with forty floors. Experimental investigations show an equivalent viscous damping coefficient between 1 and 7% of critical damping. Therefore, the modal analysis will be used in the following.

The structure will be called reliable, if the structural response (e.g. a deflection or a stress) at some critical point remains within prescribed safety bounds during the lifetime of the structure. The reliability problem is intimately related to the first-crossing problem. If the structural response leaves the safety region at least once, structural collapse will be assumed to occur. The safety bounds remain constant within the time of operation. In the following we will restrict ourselves to the two-sided barrier problem, where the safety region is given by $x \leq \pm \lambda$.

Because of its importance the first-passage problem has attracted many workers in the past and various approaches and approximations have been proposed.

In the case of a Markov process X(t) a partial differential equation can be set up for the reliability U(t) [1]. In the stationary case the so-called Pontryagin-equation can be solved either exactly in only very simple cases [2] or approximately with the help of the Galerkin's procedure. In the non-stationary case this equation has time-dependent coefficients and no solutions are known. An equivalent representation is given by an integral equation, that has been solved numerically for a special case [3].

A widely used general expression for the first passage density $\theta(t)$ has the form [4]

$$\theta(t) = A \exp(-\alpha(t)) \tag{1}$$

where A is the initial failure probability. This expression becomes exact, if the crossing of the barriers λ are statistically independent. In this case and for a zero mean process $\alpha(t)$ is determined from the mean rate of upward-crossings of λ, n_λ^+, and downward-crossings of $-\lambda$, n_λ^-

$$\alpha(t) = \int_0^t (n_\lambda^+(\tau) + n_\lambda^-(\tau))\, d\tau = \int_0^t 2n_\lambda^+(\tau)\, d\tau \tag{2}$$

In the limit $\lambda \to \infty$ the barrier crossings become independent for both stationary [5] and nonstationary processes [6]. The main advantage of (2) lies in its simplicity, but it leads to either conservative or non-conservative results, depending on the bandwidth of the system, thus being unsuited for design purposes.

It should be noted, that in spite of the assumption of structural collapse after the first exceedance of the tolerable barrier level the structural response X(t) is assumed to remain unchanged. Thus, various concepts of crossing statistics yield meaningful results for the first-passage probability.

An infinite number of upper and lower bounds is obtained from the 'inclusion-exclusion-series' [7].

$$\theta(t) = n_\lambda(t) - \int_0^t n_\lambda(t,s)\,ds + \int_0^t\int_u^t n_\lambda(t,s,u)\,ds\,du - + - + \dots \qquad (3)$$

The major difficulty lies in the computation of the multiple simultaneous mean crossing rates of the process X(t) at time instants t,s,u... Computations have been performed for the stationary [8] and the transient response [9] of a single-degree-of-freedom oscillator, taking into account the first three terms of the series. For large time intervals (0,t) the series does not converge satisfactorily. This difficulty can be removed by a representation like equation (1), where a series for α is obtained [8]. In the stationary case it is possible, by inserting various assumptions, to compute the higher terms in this series from the first and second only [6]. Some authors suggest introducing a point process approximation, where the points indicate either a barrier-crossing [10] or a local extreme value [11], [12]. This approximation is particularly useful for narrow-band processes in the stationary case and in the nonstationary case only, if the mid-frequency does not change significantly with time.

The first term in equation (3) coincides with the upper bound given by Shinozuka [13].

$$\theta(t) \leq n_\lambda(t) \tag{4}$$

Shinozuka also derived a lower bound for the first-passage probability of a Gaussian process,

$$H(t) = 1 - \phi(b^*) \tag{5}$$

where ϕ is the Gaussian error function

$$\phi(t) = \sqrt{\frac{2}{\pi}} \int_0^t \exp(-\frac{x^2}{2})\, dx \tag{6}$$

and b^* is the barrier level λ normalised to the maximum of the rms-function σ_1 of X(t)

$$b^* = \frac{\lambda}{\sigma_1^*} = \max_t b \tag{7}$$

These bounds are very well suited for nonstationary processes because of the rapid convergence of the upper bound

$$H(t) \leq \int_0^t n_\lambda\ (\tau)\ d\tau \tag{8}$$

with [1]

$$n_\lambda = n_o^+ \left[\exp(-\frac{b^2}{2(1-\rho_{12}^2)}) + v_1\sqrt{\frac{\pi}{2}}\ (1+\phi(v_1))\exp(-\frac{b^2}{2})\right] \tag{9}$$

where the time-dependency of the rms-functions $\sigma_1(t)$, $\sigma_2(t)$ of the process X(t) and its derivative $\dot{X}(t)$ and of their joint correlation coefficient $\rho_{12}(t)$ was suppressed. The abbreviations

$$n_o^+ = \frac{\sigma_2}{2\pi\sigma_1}\sqrt{1 - \rho_{12}^2}$$

$$v_1 = \frac{b\rho_{12}}{\sqrt{1-\rho_{12}^2}} \tag{10}$$

were used. Relatively little statistical information is needed for the computation of these bounds. They give satisfactory results in many cases; but particularly in the case of narrowband-processes X(t) the bounds are not close enough. Therefore, new crossing rates must be derived for the nonstationary process X(t).

ENVELOPE CROSSING RATES

Improved upper and lower bounds must take into account that X-crossings of the barrier level λ occur clumpwise. Therefore a nonstationary envelope process is introduced.

Following Rice [14]

$$X(t) = A(t) \cos(\omega_m t + \Omega(t)) \tag{11a}$$

and taking the mid-frequency ω_m equal to two orthogonal processes

$$U(t) = A(t) \cos \Omega(t)$$

$$V(t) = A(t) \sin \Omega(t) \tag{11b}$$

are obtained. For a Gaussian process X(t), which will be assumed exclusively in the following, the phase Ω is uniformly distributed in $(0,2\pi)$ and the envelope A(t) as well as U(t) and V(t) are Rayleigh-distributed. The cross-correlation from U and V must be determined from the spectral representation of X(t).

Priestley found that many nonstationary processes possess an evolutionary power spectrum [15]

$$X(t) = \frac{1}{2\pi} \int_{-\infty}^{\infty} A(t,\omega) e^{i\omega t} \, dZ(\omega) \tag{12}$$

The time-dependent spectral density is given by

$$S_t(\omega) = | A(t,\omega) |^2 E \{dZ^2(\omega)\} \tag{13}$$

with statistically independent increments $dZ(\omega)$ as in the stationary case. Using the spectral representations

$$U(t) = \frac{1}{\pi} \int_0^\infty A(t,\omega) e^{i\omega t} \, dZ(\omega)$$

$$V(t) = - \frac{1}{\pi} \int_0^\infty i \, \mathrm{sgn}(\omega) \, A(t,\omega) e^{i\omega t} \, dZ(\omega)$$

and denoting the complex conjugate with a bar, crosscorrelation

$$E \{ U(t)V(t+\tau) \} = r(\tau)\sigma_1(t)\sigma_1(t+\tau)$$

$$= \frac{1}{\pi} \int_0^\infty \bar{A}(t,\omega)A(t+\tau,\omega) \sin \omega\tau \, S(\omega) d\omega \tag{14}$$

and autocorrelation

$$E \{ U(t)U(t+\tau) \} = k(\tau)\sigma_1(t)\sigma_1(t+\tau)$$

$$= \frac{1}{\pi} \int_0^\infty \bar{A}(t,\omega)A(t+\tau,\omega) \cos \omega\tau \, S(\omega) \, d\omega \tag{15}$$

are obtained. $S(\omega)$ is the spectral density of the corresponding stationary process. It is observed, that the definitions of the envelope process (equations (11)) coincide with a generalisation of the definition of Cramér and Leadbetter for the nonstationary case [12].

Following Rice [14], the mean rate of envelope upward crossings is found from

$$m_\lambda(t) = \int_0^\infty \dot{a} p(\lambda,\dot{a};t,t) d\dot{a} \tag{16}$$

The probability density $p(a,\dot{a})$ is obtained from the four-dimensional Gaussian probability density for the variables $u(t_1)$, $v(t_1)$, $u(t_2)$, $v(t_2)$ after a transformation of variables

$$\begin{aligned} u(t_i) &= a_i \cos \phi_i \\ & \qquad i = 1,2 \\ v(t_i) &= a_i \sin \phi_i \end{aligned} \tag{17}$$

and integration over the phase angles. In the limit

$$p(a,\dot{a};t,t) = \lim_{\tau \to 0} \tau \, p(a_1,a_2;t,t+\tau) \tag{18}$$

the following density is obtained

$$p(a,\dot{a}) = \frac{a}{\sigma_1^3 \sqrt{2\pi\Delta}} \exp\left(-\frac{a^2}{2\sigma_1^2}\right) \cdot \exp\left(-\frac{(\dot{a}-ak_1)^2}{2\sigma_1^2 \Delta}\right) \tag{19}$$

The parameter Δ represents the bandwidth of the evolutionary spectrum

$$\Delta = \omega_2 - \omega_1^2$$

$$\omega_2 = \frac{1}{\pi\sigma_1^2} \int_0^\infty \left(\frac{\partial A}{\partial t}\frac{\partial \bar{A}}{\partial t} + \omega^2 \bar{A} A\right) S(\omega)\, d\omega \tag{20}$$

$$\omega_1 = \frac{1}{\pi\sigma_1^2} \int_0^\infty \left(\bar{A}\frac{\partial A}{\partial t} + \omega A\bar{A}\right) S(\omega)\, d\omega$$

where A was written instead of $A(t,\omega)$. Numerical results indicate that the bandwidth reaches very large values within the build-up period of the structural response $X(t)$ and reaches approximately stationary values afterwards (Figure 4). Equation (19) splits up into a Rayleigh-density for the envelope and a Gaussian density for its derivative in the stationary case, where

$$k_1 = \rho_{12} \frac{\sigma_2}{\sigma_1} = \frac{E\{X\dot{X}\}}{\sigma_1^2} \tag{21}$$

vanishes.

Inserting equation (19) into (16) leads to

$$m_\lambda = b\sqrt{\Delta}\,\exp(-\frac{b^2}{2}) \times \left[\frac{1}{\sqrt{2\pi}}\exp(-\frac{v_2^2}{2}) + \frac{v_2}{2}(1+\phi(v_2))\right] \tag{22}$$

with the abbreviations equation (6) and

$$v_2 = \frac{bk_1}{\sqrt{\Delta}}$$

Using this concept, improved upper and lower bounds are obtained. The lower bound for the first passage probability

$$H(t) > \exp(-\frac{b^{*2}}{2}) \tag{23}$$

can be derived from the inequality [1]

$$P\left[\sup |x(\tau)| > \lambda\right] > P\left[a(t) > \lambda\right] > P\left[\,|x(t)| > \lambda\right]$$

The improved upper bound 2 takes into account the mean clumpsize of X(t) proposed by Lyon [16]

$$H(t) < \int_0^t m_\lambda(\tau)\,d\tau \tag{24}$$

It leads to an improvement for low and moderate barrier levels λ. In the case of very high λ-values envelope crossings might occur, that are not followed by crossings of the process itself. Vanmarcke introduced this effect and, for stationary processes, found the mean rate of the so-called qualified envelope crossing rates [4] as

$$q_\lambda = \frac{2n_\lambda^+}{cs_v} \tag{25}$$

from the mean clumpsize

$$cs_v = \frac{n_\lambda}{n_o}\,e^{b^2/2}\left[1 - \exp(-\frac{m_\lambda}{n_\lambda}\;e^{b^2/2})\right]^{-1} \tag{26}$$

It can be shown, that cs_v is always greater than one. These formulae can also be used in the nonstationary case, if the mean rate of zero crossings n_o does not change significantly with time. This can be assumed after the build-up period (Figure 4). Within this period cs_v has to be computed by numerical integration [1].

An improved upper bound 3 is given by

$$H(t) < \int_0^t q_\lambda^+(\tau)\, d\tau \tag{27}$$

INITIAL ENVELOPE CROSSING RATE

Until now the behaviour of the processes X(t) or A(t) was studied at the end of the considered time interval (O,t) only. Taking into account the behaviour of A(t) in at least one more time instant from (O,t) must lead to an improvement

$$\theta(t) < P\left[(a(t-\tau) < \lambda) \cap (a(t) < \lambda) \cap (\sup_{t \le \tau' = t+dt} a(\tau') > \lambda)\right] \quad 0 \le \tau \le t \tag{28}$$

These so-called initial (within the interval $(t-\tau,t)$) envelope crossing rates r_λ are always less or equal to the envelope crossing rates m_λ. The best improvement is obtained from the smallest values of $r_\lambda(t,\tau)$ with respect to τ. Therefore, τ was chosen as [1]

$$\tau = \frac{2\pi}{\sqrt{\omega_2(t)}} \tag{29}$$

with ω_2 the instantaneous mid-frequency of the evolutionary spectrum (Equation 20).

Similarly to equation (16) r_λ is found from

$$r_\lambda(t,\tau) = \int_0^\lambda da_1 \int_0^\infty \dot{a}_2 p(a_1,\lambda,\ \dot{a}_2; t-\tau, t, t)\, d\dot{a}_2 \tag{30}$$

The derivation of the density $p(a_1,a_2,\dot{a}_2)$ starts with the six-dimensional Gaussian probability distribution $p(u_1,v_1,u_2,u_3,v_3)$ for the processes U and V at the three different time instants t_1,t_2,t_3. After a transformation of variables like equation (17) the integration over the variables ϕ_1 and $\xi = \phi_1-\phi_3$ can be performed. In the limit $t_3 \to t_2$ an indefinite expression is obtained and, therefore, the integrand is developed into a series with respect to t_3-t_2. Thus, the following integral is found in the limit $t_3-t_2 \to 0$

$$r_\lambda(t,\tau) = \frac{1}{2\pi} \int_0^{b_1} \frac{da_1}{\sigma_1} \int_0^{2\pi} \frac{a_1 b_2 \sqrt{\psi}}{\sigma_1 \sqrt{2\pi}(1-\rho_1^2)^{3/2}}$$

$$\exp\left[-\frac{a_1^2}{2\sigma_1^2(1-\rho_1^2)} - \frac{b_2^2}{2(1-\rho_1^2)} + \frac{a_1 b_1}{\sigma_1(1-\rho_1^2)}(k_1\cos\eta + r_1\sin\eta) - \frac{v_3^2(\eta)}{2}\right] .$$

$$\{1 + v_3\sqrt{\frac{\pi}{2}}\,(1+\phi(v_3))e^{v_3^2/2}\}\, d\eta$$

The following abbreviations have been used herein:

$$\eta = \phi_2 - \phi_1, \quad b_1 = \frac{\lambda}{\sigma_1}, \quad b_2 = \frac{\lambda}{\sigma_2}$$

$$\gamma(\eta) = \frac{a_2}{\sigma_2}(2k_1' - \rho_2') + \frac{2a_1}{\sigma_1}(\delta_1\cos\eta + \delta_3\sin\eta)$$

$$v_3 = \frac{\gamma(\eta)}{2\sqrt{\psi(1-\rho_1^2)}}$$

$$\psi = (1-\rho_1^2)(s''-2k_1'') - k_2'^2 - r_2'^2 - k_1'^2 - r_1'^2 + 2r_3(r_2'k_1' - r_1'k_2') + 2k_3(k_2'k_1' + r_2'r_1')$$

$$\rho_3' = 2k_2k_3' + 2r_2r_3'$$

$$\delta_1 = k_1k_3' + r_1r_3' + k_2'$$

$$\delta_3 = r_3 k_1' - k_3 r_1' + r_2'$$

and the symbol

$$' = \frac{\partial}{\partial t_3} (\ ) \Big|_{t_3=t_2}$$

has been used. The derivatives of the normalised correlation functions

$$\begin{aligned} \sigma_i \sigma_j k_\ell &= E\{X(t_i)X(t_j)\} \\ & \qquad\qquad\qquad\qquad i,j,\ell = 1,2,3 \\ \sigma_i \sigma_j r_\ell &= E\{U(t_i)V(t_j)\} \end{aligned} \tag{32}$$

$$\rho_i^2 = k_i^2 + r_i^2$$

$$\sigma_3^2 = \sigma_2^2 s$$

can easily be computed, when the spectral representations (equations (14) and (15)) are used. The evaluation of r_λ has to be done numerically. Afterwards an improved upper bound 4

$$H(t) < \int_0^t r_\lambda(w, w-\tau)\ dw \tag{33}$$

can be computed.

NUMERICAL EXAMPLES

The earthquake excitation process was taken from reference [17].

$$\ddot{u}(t) = \int_0^t \psi\ (\tau) h_B(t-\tau) V(\tau) d\tau \tag{34}$$

with $\psi(t) = \exp(-\alpha t) - \exp(-\beta t)$, a deterministic envelope function $h_B(t)$ the impulse response for a damped linear oscillator with frequency ω_B and fraction of critical damping δ, and $V(t)$ a stationary Gaussian white noise with zero mean and intensity D. One sample is plotted in Figure 1.

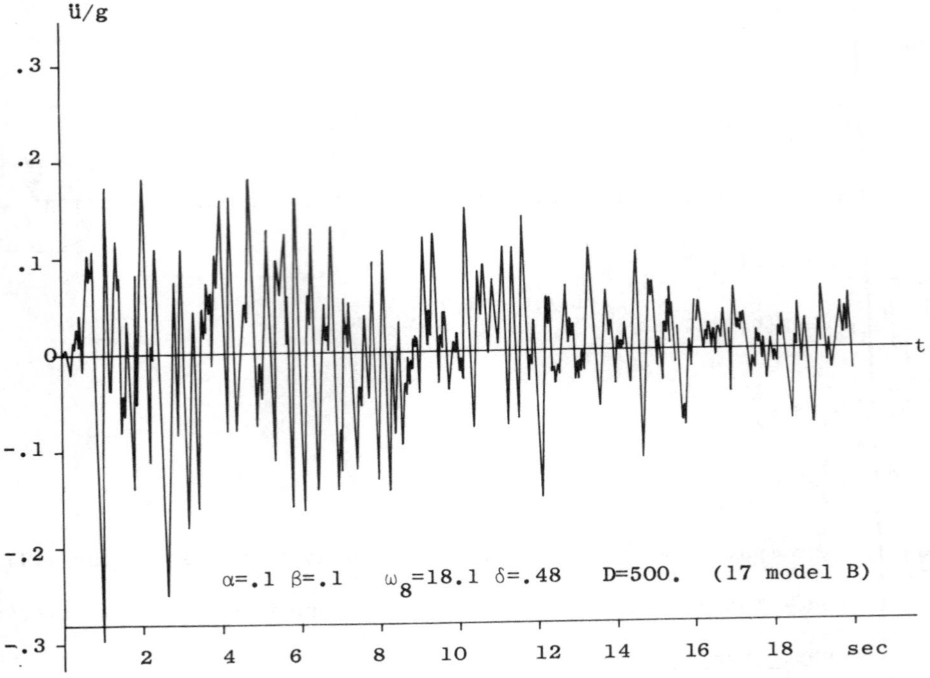

Figure 1 Simulated ground acceleration $\ddot{u}(t)$

The structure was modelled by a single or multi-degree-of-freedom system.

$$\ddot{x}_i + 2\mu_i\dot{x} + \nu_i^2 x = b_i\ddot{u} \qquad i = 1,..N \tag{35}$$

and b_i is the mode participation coefficient. The noise intensity D was determined from the condition

$$\sigma_1^* = 1$$

The nonstationary character of the structural response becomes obvious in Figure 2.

The modulating function $A(t,\omega)$ is found from

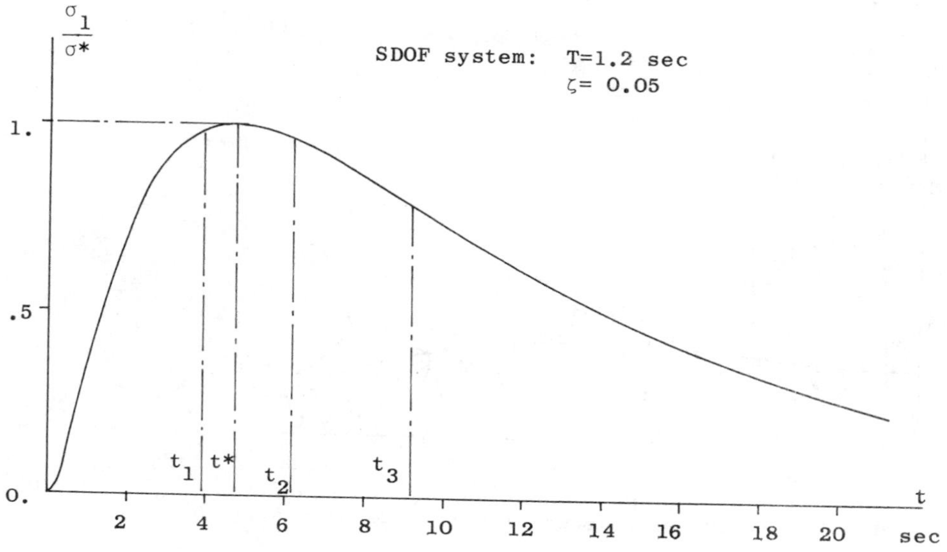

Figure 2 RMS function of structural response X(t)

$$A(t,\omega) = \int_0^t \psi(t-\tau)h_G(\tau)e^{-i\omega\tau}\, d\tau \tag{36}$$

where h_G is the combined impulse response of h_B and the structural modes

$$h_G(t) = \sum_{j=1}^{N} \int_0^t h_j(t-\tau)h_B(\tau)d\tau$$

Thus $A(t,\omega)$ can be easily computed [1].

Figure 3 shows the evolutionary spectral density $S_t(\omega)$ for a shear beam at different time instants. The spectrum is nearly white at 0.5 sec. The resonance peaks at the eigenfrequencies become more and more pronounced with increasing time.

The time-behaviour of the frequencies ω_1, ω_2 and Δ (equation 20) is plotted in Figure 4 for a multi-degree-of-freedom system.

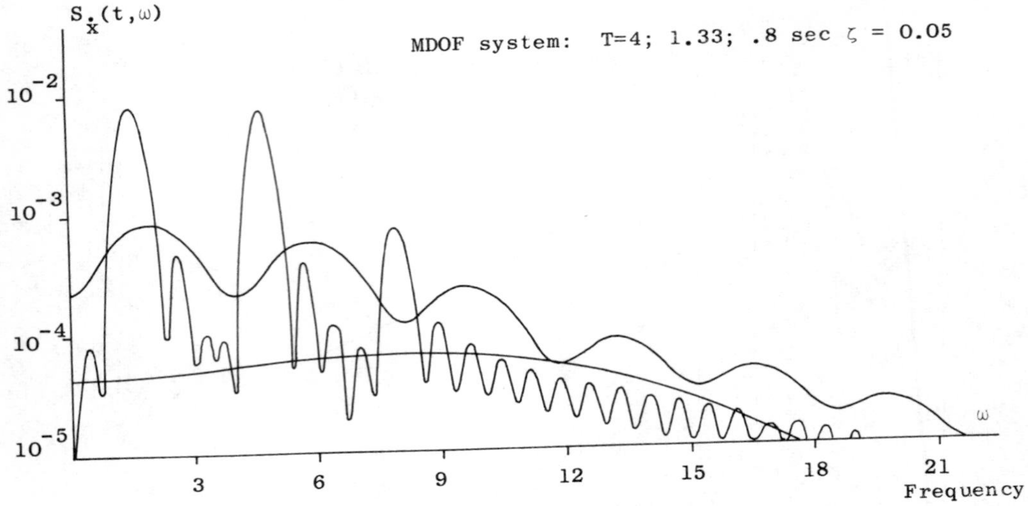

Figure 3 Evolutionary power spectral density $S_{\dot{x}}(t,\omega)$

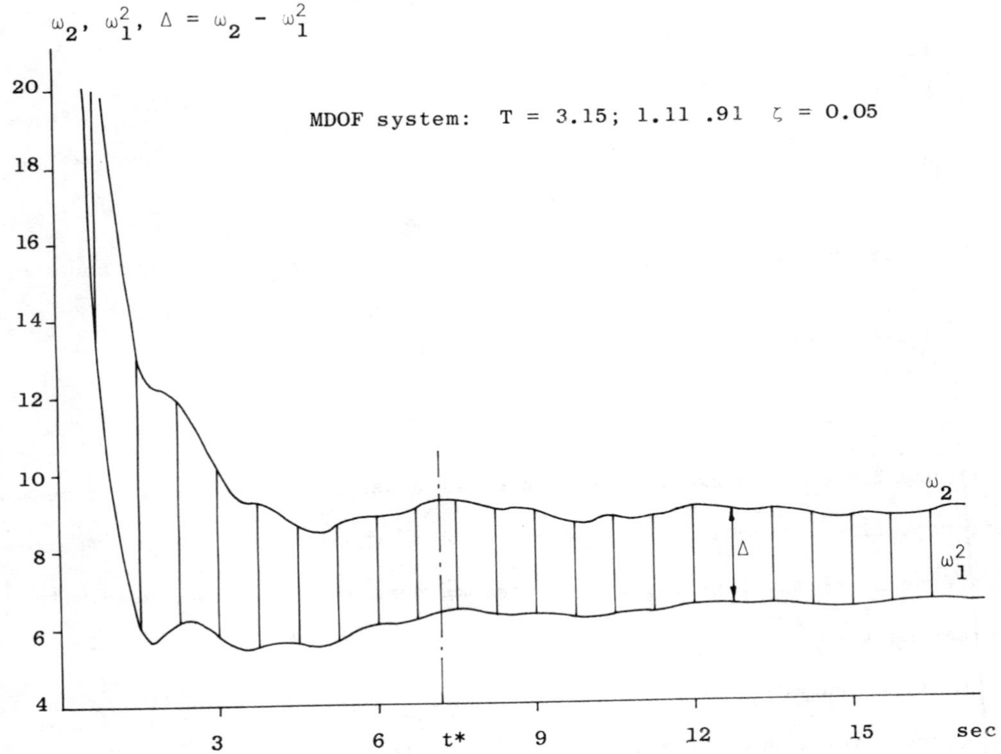

Figure 4 Spectral bandwidth Δ (eqn. 20)

Comparison Between Different Crossing Probabilities

Figure 5 shows the results for a single-degree-of-freedom system at three different time instants. The graphs represent the crossing probabilities n_λ, r_λ and q_λ, all of them normalised to the mean rate of envelope crossings m_λ. The first time instant is chosen before t^*, when the rms-function σ_1 reaches its maximum. The ratio

$$c = \frac{r_\lambda}{m_\lambda} \tag{37}$$

is smaller than 1 only for very low barrier levels $b = \lambda/\sigma^* = \lambda$. This could be expected intuitively, because nearly all samples will remain below λ at t_1, when the rms-function is significantly smaller than one cycle later, at t_2. Hence the improvement from the upper bound 4 over the upper bound 3 will be negligible in the build-up period.

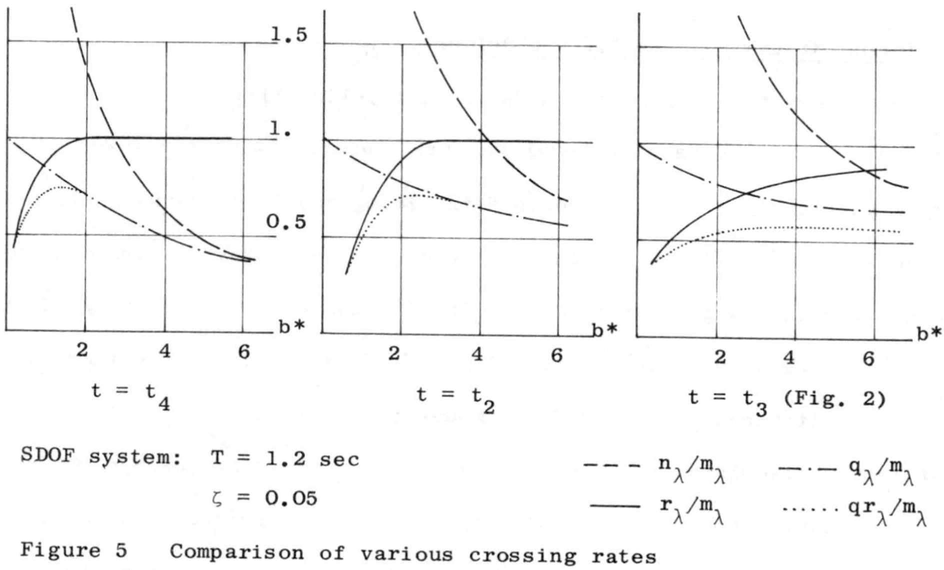

Figure 5 Comparison of various crossing rates

The second time instant is chosen shortly after t^+, when $\sigma_1 = \sigma^*$, and the third time instant lies in the 'dying out' period. More and more envelope samples remain below λ now, if they cross λ from below one cycle later. But it should be kept in mind, that the envelope crossing rate is much smaller in the third instant than in the first.

The qualified envelope crossing rates m_λ are always smaller than the X-process crossing rates n_λ. The upper bound 4 will give better results than the upper bound 1, therefore. For very high barrier levels q_λ tends to n_λ and the crossings become independent.

Assuming statistical independence between the events of initial crossings and qualified crossings a combined crossing rate

$$qr_\lambda(t) = q_\lambda(t)r_\lambda(t) \tag{38}$$

gives an improvement over n_λ for all barrier levels λ.

Discussion of the Results for the Reliability

Figure 6 shows the different bounds for the reliability $U(t \to \infty) = 1 - H(t\to\infty)$ plotted against the barrier level λ. The improvement of the lower bound is very significant. The upper bound 2 involving envelope crossings lies beneath the upper bound 1 within the whole λ-range plotted. Further improvements for low barrier levels are obtained from the initial envelope crossings (upper bound 4). The upper bound 3 involving qualified envelope crossings gives better results than the bound 4 for moderate and large barrier levels. The best result is obtained by a combination of both leading to upper bound 5. The results are checked numerically by a simulation procedure on a digital computer. The improved bounds come close to the simulation results.

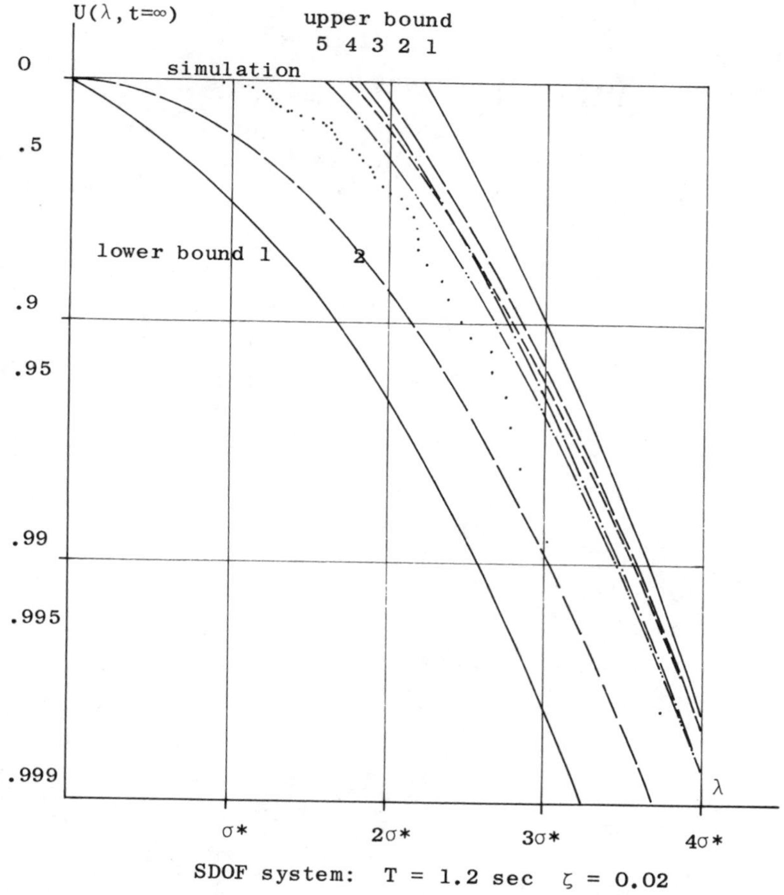

Figure 6 Upper and lower bounds for the reliability

Figure 7 presents the results of a lumped parameter system with three degrees of freedom. The relations between the different bounds are the same as before. But, in contrast to Figure 6, the λ-range of improved bounds involving envelope statistics is smaller. This results from the greater bandwidth Δ of a multi-degree-of-freedom system.

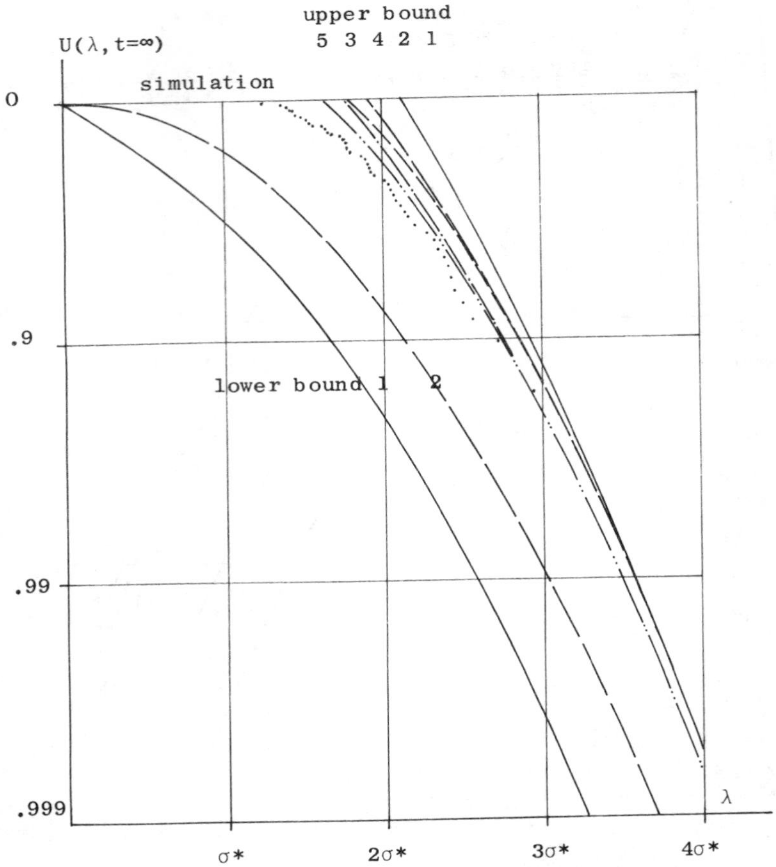

MDOF system: T = 1.11; 1.; .91 sec; ζ = 0.05

Figure 7 Upper and lower bounds for the reliability

A study of systems with various parameters [1] reveals that the best upper bounds are obtained for low barrier levels (0-2.5σ^*) involving initial crossing rates, for moderate λ-values (2.5σ^* - 6σ^*) from the qualified envelope crossing rates.

For large threshold levels the assumption of independent barrier crossings leads to the exact results.

FINAL REMARKS

Bearing in mind the necessity of conservative structural aseismic design improved upper and lower bounds have been derived for linear multi-degree-of-freedom systems. Numerical simulation investigations indicate, that the results remain approximately unchanged for non-linear systems with low eigenfrequencies (e.g., ideal elastic-plastic), if the elastic limit is only rarely exceeded (Figure 8).

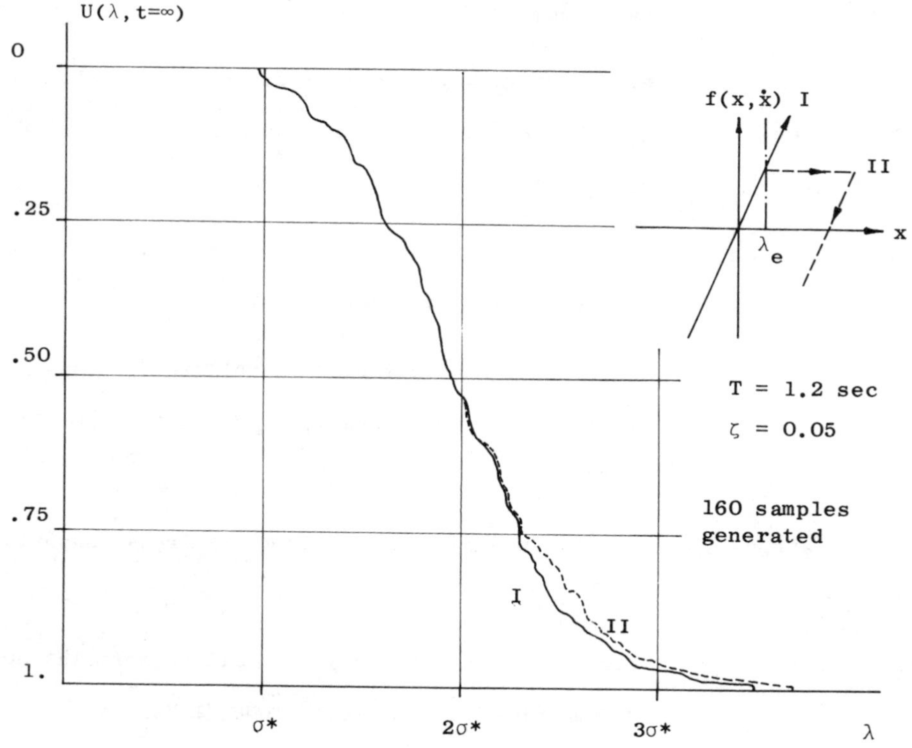

Figure 8 Simulation results for the reliability of an elastic and an elastic-plastic oscillator

The reliability of a structure under earthquake excitation can now be assessed and structural design can be based on these results.

REFERENCES

1 R. Grossmayer, Doctoral thesis, Technical University Vienna Stochastische Zuverlässigkeitsanalyse erdbebenerregter Strukturen. (1975).

2 A.H. Gray, Jr., First passage time problem in a random vibrational system. J. Appl. Mech, 187 (1966).

3 H. Parkus and J.L. Zeman, Some stochastic problems of thermoviscoelasticity. Iutam Symp. on Thermoinelasticity, Glasgow, 532 (1968).

4 E.H. Vanmarcke, On the distribution of the first-passage time for normal stationary random processes. J. Appl. Mech, 215 (1975).

5 H. Cramer and M.R. Leadbetter, Stationary and related stochastic processes. Wiley, New York (1967).

6 J.B. Roberts, Probability of first-passage failure for non-stationary random vibration. J. Appl. Mech, 716 (1975).

7 M.S. Bartlett, Stochastic processes. University Press, Cambridge (1960).

8 J.B. Roberts, Probability of first passage failure for stationary random vibration. AIAA J, 1636 (1974).

9 J.B. Roberts, An approach to the first passage problem in random vibration. J.S. Vibr. 8, 301 (1968).

10 Y.K. Lin, First excursion failure of randomly excited structures. AIAA J 8, 720 and 1888 (1970).

11 J.N. Yang and M. Shinozuka, On the first excursion probability in stationary narrow-band random vibration. J. Appl. Mech, 1017 (1971).

12 J.N. Yang, First excursion probability in nonstationary random vibration. J.S. Vibr. 27, 165 (1973).

13 M. Shinozuka, Probability of structural failure under random loading. J. Eng. Mech. Div., ASCE 90, EM 5, 147 (1964).

14 S.O. Rice, Mathematical analysis of random noise. Selected papers on noise and stochastic processes. New York-Dover (1955).

15 M.B. Priestley, Power spectral analysis of non-stationary random processes. J.S. Vibr. 6, 86 (1967).

16 R.H. Lyon, On the vibration statistics of a randomly excited hard-spring oscillator. J. Ac. Soc. 33, 1395 (1961).

17 R. Levy, F. Kozin and R.B.B. Moorman, Random processes for earthquake simulation. J. Eng. Mech. Div., ASCE 97, EM 2, 495.

NOTATION AND SYMBOLS

$A(t)$ envelope process of the structural response

$b=\lambda/\sigma$ normalised tolerable barrier level

cs	mean clumpsize
ζ	fraction of critical damping
D	intensity of white noise
Δ	bandwidth of the evolutionary spectrum
$\theta(t,\lambda)$	first-passage probability density at t for a barrier level λ
$H(t,\lambda)$	first-passage probability at t for a barrier level λ
$h_B(t)$	impulse response function of the ground filter
k_1	normalised autocorrelation
λ	tolerable barrier level
$\mu = \zeta\nu$	damping coefficient
m_λ	mean rate of envelope upward crossings
MDOF	multi-degree-of-freedom
n_λ^+	mean rate of upward crossings of the process X(t)
ν	undamped natural frequency of the structure
$\psi(t)$	deterministic time multiplier
SDOF	single-degree-of-freedom
q_λ	mean rate of qualified envelope upward crossings
qr_λ	mean rate of qualified initial envelope crossings
r_λ	mean rate of initial envelope crossings
r_1	normalised crosscorrelation
σ_{12}	correlation coefficient of X(t) and $\dot{X}(t)$
σ_1	rms function of X(t)
σ^*	maximum value of σ_1
σ_2	rms function of $\dot{X}(t)$
T	natural period of structure

Dr. R. Grossmayer, II Institut für Mechanik, Karlsplatz 13, A-1040 Wien, Vienna.

DISCUSSION

Interest was expressed (Sobczyk) in the use of Priestley's evolutionary spectral density and in relation to the modulating function $A(t,\omega)$ the author confirmed that one could verify that the process was indeed 'oscillatory' in the sense of Priestley by noting that the Fourier transform of the modulating function (with respect to t and for fixed ω) had an absolute maximum at zero frequency. The author noted that the frequency-time spectral function was useful in computing reliability and had shown that an envelope process could also be introduced without using the Priestley concept.

When questioned (Clarkson) about the choice of uniform modulating function $\psi(\tau)$ the author pointed out that the exponential character of this function might now be altered in favour of the results of some recent work by Kozin in which curve fitting by splines results in a decay which is more rapid than the exponential form [18].

Following queries concerning the use of the evolutionary spectral density and the determination of the modulating function a short list of references is appended (Ed.) [19], [20], [21].

It was noted (Shinozuka) that the results of Reference [13] in the paper should be used for short lengths of data and not for the records considered here which are relatively long in duration and so give bounds which are far apart. Professor Grossmayer replied that various numerical examples [22] indicate, that the distance between the upper and lower bound 1, proposed by Shinozuka, becomes smaller for relatively flexible structures with high damping values. In these cases the improvement of the upper bounds 2 to 5 over the upper bound 1 is not significant. On the contrary, relatively rigid structures, that are lightly damped, and where the original bounds 1 are relatively far apart, lead to a significant improvement indeed. Thus we obtain approximately the same distance between the best upper and lower bound for all structures of interest and the method is therefore generally applicable.

When asked (Crandall) about the assumption that independent crossing rates should be non conservative the author commented that simulation studies for wide band processes with low barrier levels confirmed this. It was noted (Vanmarke) that in a Doctoral Thesis [23] it had been demonstrated through simulation and analysis that for wide band processes and low barrier levels the ratio of crossing rates is above one, the explanation being that when excursions occur above a level they are nothing like pointwise excursions but are long excursions relative to the time below the level, so that the mean time above is greater than the mean time below if the barrier were low enough.

Finally, asked (Crandall) if he would favour simulation or calculation of bounds when confronted with a specific problem the author replied that for linear systems it would be more useful to compute the bounds (upper bound 3 probably being good enough) since then the results obtained would not be as specific as those that would be obtained from simulation using a particular ground motion. For nonlinear systems however simulation might sometimes be the only approach.

REFERENCES

18 F. Kozin, An approach to Characterizing, Modelling and Analyzing Earthquake excitation Records. ASM Course on Random Excitation of Structures by Earthquakes and Atmospheric Turbulence (1976).

19 J.K. Hammond, Evolutionary spectra in random vibrations. J. Roy Stat. Soc. B 35 (2), (1973).

20 Yoshinori Fujimori and Y.K. Lin, Analysis of airplane response to non-stationary turbulence including wing bending flexibility. AIAA J. Vol. 11, No. 3, 334 (1973).

21 W.-Y.T. Chan and H. Tong, A simulation study of the estimation of evolutionary spectral functions. Appl. Statist 24 No. 3, 333 (1975).

22 R. Grossmayer, Aseismic Reliability and First- Passage Failure. CISM Course on Random Excitation of Structures by Earthquakes and Atmospheric Turbulence (1976).

23 O. Ditlevsen, Extremes and First Passage Times with Applications in Civil Engineering, Doct. Thesis, Technical University of Denmark (1971).

J S BENDAT

Procedures for frequency decomposition of multiple input/output relationships

INTRODUCTION

In recent published papers, 'Solutions for the Multiple Input/Output Problem', Reference [1], and 'Partial Coherence in Multivariate Random Processes', Reference [2], new analytical ideas are developed for determining how an arbitrary measured stationary random or transient output record can be due to optimum linear operations from a set of arbitrary measured input records. See Figure 1. These input records may be partially, not fully, coherent with each other and with the output for the models to be well defined. Results

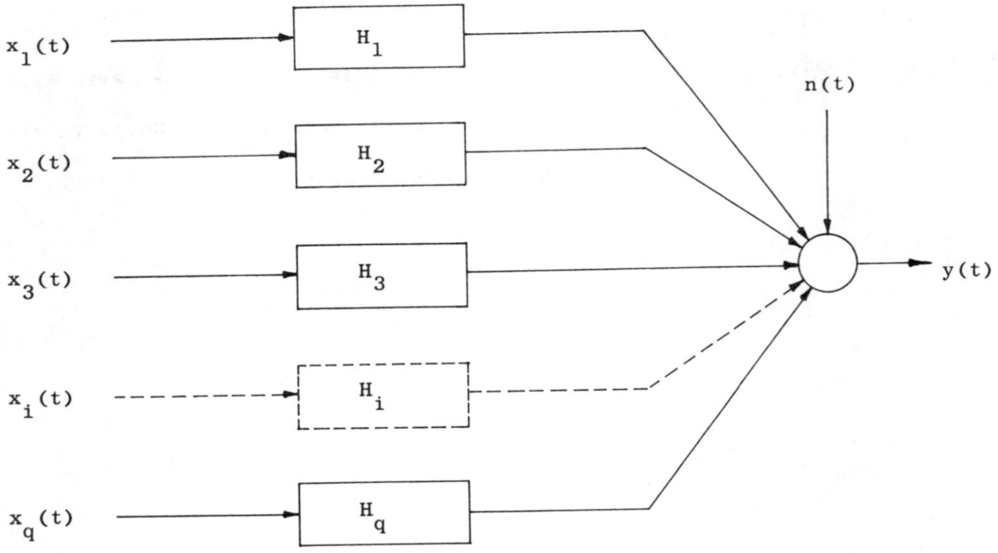

FIGURE 1. Multiple Input/Output Model for Arbitrary Inputs

Results derived in these two papers represent a frequency decomposition of the multiple input/output data in that they show to what extent the spectral content in any record can be linearly related to the other records as a function of frequency.

Computational procedures to perform this analysis can be established using formulas from these papers together with appropriate techniques in Reference [3]. The purpose of this present paper is to outline some specific new procedures to follow to help carry out this analysis.

MULTIPLE INPUT/OUTPUT MODELS FOR ARBITRARY INPUTS

As shown in Figure 1, the input records are assumed to pass through physically realizable constant parameter linear systems described by frequency response functions $\{H_i\}$, $i = 1,2,\ldots,q$. The output record $y(t)$ is assumed to be the sum of the individual outputs due to passage of the individual inputs $\{x_i(t)\}$, plus an unknown independent noise record $n(t)$ which accounts for all unknown non-linear operations as well as extraneous noise effects.

Note in Figure 1 that $q!$ different configurations are possible depending upon which record is chosen as $x_1(t)$, which is then selected as $x_2(t)$, and so on. The analysis to follow is based on choosing a particular ordering of the inputs and sticking with this order. Similar results apply to any other desired ordering. Special attention will be given in this paper to the case of three inputs since the formulas can be listed for this case without difficulty and it is representative of the general case.

Figure 2 gives the result for the total output spectral density function $S_{yy} = S_{yy}(f)$ in terms of other quantities for the three input case, where the dependence upon frequency f has been omitted in all these terms for simplicity in notation. This will also be done in later equations of this paper.

Three Input Case

$$S_{yy} = |H_1|^2 S_{11} + H_1H_2^*S_{21} + H_1H_3^*S_{31}$$
$$+ H_2H_1^*S_{12} + |H_2|^2 S_{22} + H_2H_3^*S_{32}$$
$$+ H_3H_1^*S_{13} + H_3H_2^*S_{23} + |H_3|^2 S_{33} + S_{nn}$$

FIGURE 2 Output Terms in Original Model

Note that S_{yy} can be either a power or an energy spectral density function depending upon whether the data is either stationary random data or transient data. The output noise spectral density function S_{nn} represents the difference between S_{yy} and results predicted from x_1, x_2 and x_3 by passage through any linear systems, H_1, H_2 and H_3. Because of the cross-terms between inputs, it is not clear here how much of the output is due to any particular point.

Optimum linear systems are defined by least-square prediction techniques as those systems which produce minimum mean square system error. This will occur if S_{nn} is minimized as a function of H_i for all $i = 1,2,\ldots,q$, leading, in general, to a set of complicated equations with many interacting input terms. However, for mutually uncoherent inputs, these equations simplify greatly since each optimum linear system can be determined from its own particular input independently of the other inputs.

For the three input case of arbitrary inputs, the three optimum frequency response functions will satisfy the equations listed in Figure 3. The terms shown in the numerators and denominators are conditioned (residual) spectral density functions for which computational formulas are developed in the references and also in this paper. For the single input case, the result is

$H_1 = (S_{1y}/S_{11})$. For the case of two arbitrary inputs, results are $H_1 = (S_{1y.2}/S_{11.2})$ and $H_2 = (S_{2y.1}/S_{22.1})$.

Three Input Case

$$H_1 = \frac{S_{1y.23}}{S_{11.23}}$$

$$H_2 = \frac{S_{2y.13}}{S_{22.13}}$$

$$H_3 = \frac{S_{3y.12}}{S_{33.12}}$$

FIGURE 3

Optimum Frequency Response Functions
For Arbitrary Inputs

In Figure 3, the particular conditioned quantities in H_3 are defined as follows with similar definitions for H_1 and H_2.

$S_{33.12}$ = spectral (power or energy) density function of $x_3(t)$ when the linear effects of both $x_1(t)$ and $x_2(t)$ are removed from $x_3(t)$ by optimum least-squares prediction techniques.

$S_{3y.12}$ = cross-spectral density function between $x_3(t)$ and $y(t)$ when the linear effects of both $x_1(t)$ and $x_2(t)$ are removed from both $x_3(t)$ and $y(t)$ by optimum least-squares prediction techniques.

MULTIPLE INPUT/OUTPUT MODELS FOR CONDITIONED INPUTS

Figure 4 shows a three input/output model for conditioned inputs $x_1(t)$, $x_{2.1}(t)$ and $x_{3.12}(t)$ which are obtained from the original inputs $x_1(t)$, $x_2(t)$ and $x_3(t)$ shown in Figure 1. These conditioned inputs are defined in the following ordered way:

(1) The first input $x_1(t)$ is left alone.

(2) The second input $x_2(t)$ is replaced by $x_{2.1}(t)$ obtained by removing the linear effects of $x_1(t)$ from $x_2(t)$ by optimum least-squares prediction techniques.

(3) The third input $x_3(t)$ is replaced by $x_{3.12}(t)$ obtained by removing the linear effects of both $x_1(t)$ and $x_2(t)$ from $x_3(t)$ by optimum least-squares prediction techniques.

This procedure can be extended to any number of inputs.

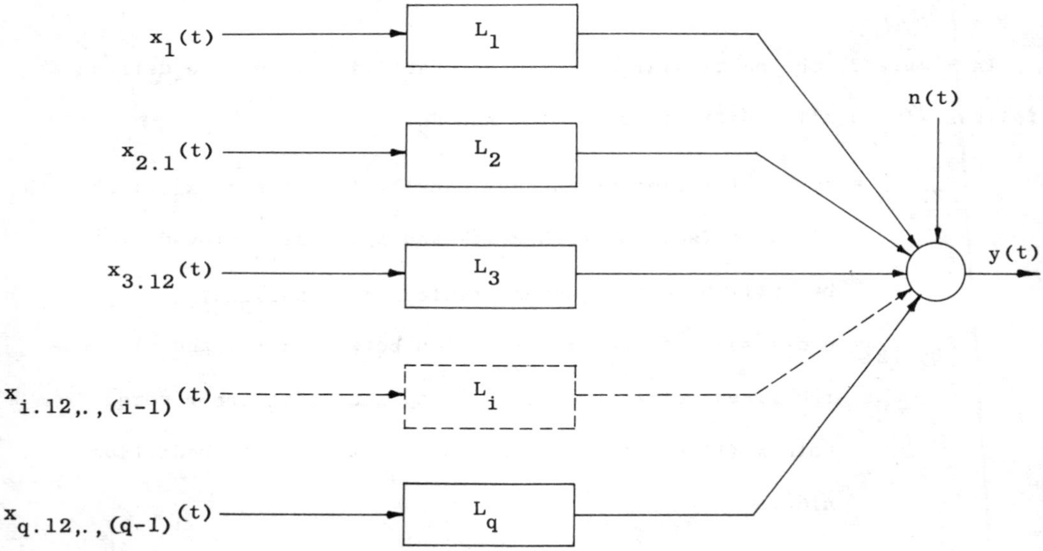

FIGURE 4. Multiple Input/Output Model for Conditioned Inputs

The systems in Figure 4 are denoted by $\{L_i\}$ instead of by $\{H_i\}$ as in Figure 1 to distinguish these two distinct types of models. The terms n(t) and y(t) are exactly the same in both models. Relationships between these systems are described in Reference [1].

The set of conditioned inputs in Figure 4 will be mutually uncoherent. Optimum frequency response functions for this case will satisfy the equations listed in Figure 5. Note that L_1 and L_2 are simpler than H_1 and H_2 in Figure 3 while L_3 is the same as H_3. For the single input case, the result is L_1 alone as shown. For the case of two conditioned inputs, the results would be L_1 and L_2 as shown. If there had originally been four inputs to consider, then L_1, L_2 and L_3 would be unchanged and L_4 would be given by $L_4 = (S_{4y.123}/S_{44.123})$ where L_4 acts on the conditioned input $x_{4.123}$.

<u>Three Input Case</u>

$$L_1 = \frac{S_{1y}}{S_{11}}$$

$$L_2 = \frac{S_{2y.1}}{S_{22.1}}$$

$$L_3 = \frac{S_{3y.12}}{S_{33.12}}$$

FIGURE 5

Optimum Frequency Response Functions

For Conditioned Inputs

In words, the conditioned record $x_{2.1}(t)$ is described as $x_2(t)$ conditioned on $x_1(t)$. Similarly, the conditioned record $x_{3.12}(t)$ is described as $x_3(t)$ conditioned on both $x_1(t)$ and $x_2(t)$. An output record y(t) can also be conditioned, if desired, on $x_1(t)$ alone, on both $x_1(t)$ and $x_2(t)$, and so on to as many combined input records as exist.

CONDITIONED RECORDS FROM ARBITRARY RECORDS

Procedures for obtaining conditioned records from original arbitrary records are shown in Figure 6. These results demonstrate precisely how to change any set of given records into an ordered set of conditioned records which will have optimum linear least-squares properties. In particular, the input records of Figure 1 become the input records of Figure 4 by these operations.

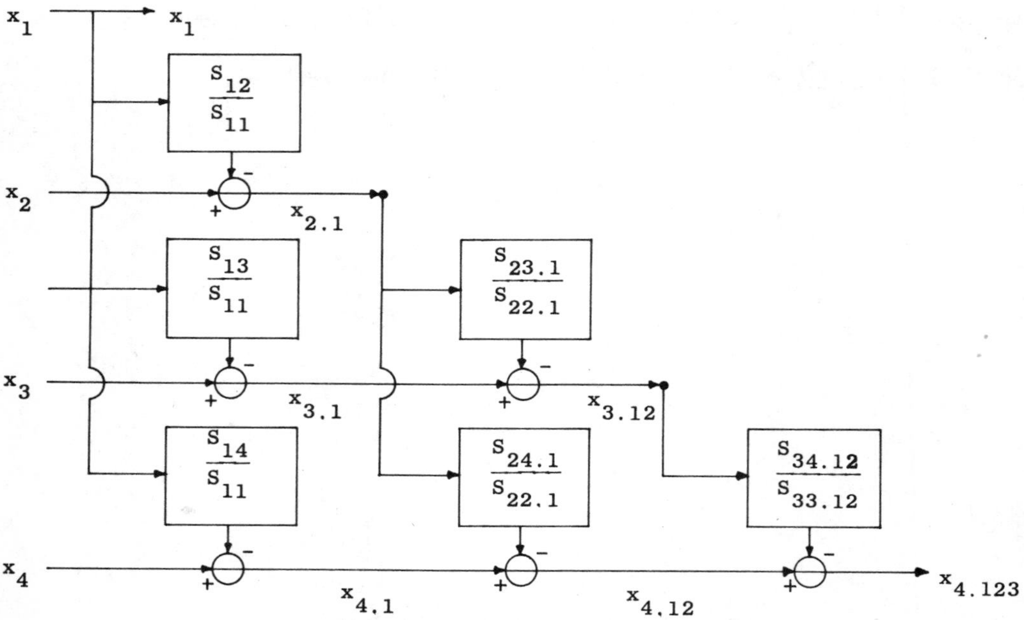

FIGURE 6. Ordered Conditioned Records from Arbitrary Records (These results extend to any number of records)

For a three input/output model, the record designated as $x_4(t)$ should be replaced by $y(t)$ and the conditioned record $x_{4.123}(t)$ should be replaced by n(t). The spectral quantity S_{14} becomes S_{1y}, $S_{24.1}$ becomes $S_{2y.1}$ and $S_{34.12}$ becomes $S_{3y.12}$ such that the bottom row in Figure 6 now represents the optimum linear systems L_1, L_2 and L_3 of Figure 5.

CONDITIONED SPECTRAL DENSITY FUNCTIONS

Conditioned spectral density functions are given by the iterative computational formulas shown in Figures 7 and 8. General results for any number of records were previously derived in References [1, 2].

$$S_{2y.1} = S_{2y} - \left(\frac{S_{1y}}{S_{11}} \right) S_{21}$$

$$S_{22.1} = S_{22} \quad (1 - \gamma^2_{12})$$

$$S_{3y.12} = S_{3y.1} - \left(\frac{S_{2y.1}}{S_{22.1}} \right) S_{32.1}$$

$$S_{33.12} = S_{33.1} \ (1 - \gamma^2_{23.1})$$

FIGURE 7

Formulas for Conditioned Spectral Density Functions

Observe that the formula for $S_{2y.1}$ includes $S_{22.1}$ and $S_{yy.1}$ as special cases. This formula also gives $S_{23.1}$, $S_{24.1}$ and $S_{y2.1} = S^*_{2y.1}$. Similarly, the formula for $S_{3y.12}$ includes $S_{34.12}$, $S_{33.12}$ and $S_{yy.12}$ as special cases. If there were four inputs to consider, then the formula for $S_{4y.123}$ (not shown but directly extended) would include $S_{yy.123}$ (which is shown) as a special case. When there are only three inputs, as assumed here, the term

$S_{4y.123}$ does not exist and the term $S_{yy.123} = S_{nn}$, the spectral density function of the output noise n(t).

$$S_{yy.1} = S_{yy} (1 - \gamma_{1y}^2)$$

$$S_{yy.12} = S_{yy} (1 - \gamma_{1y}^2)(1 - \gamma_{2y.1}^2)$$

$$S_{yy.123} = S_{yy} (1 - \gamma_{1y}^2)(1 - \gamma_{2y.1}^2)(1 - \gamma_{3y.12}^2)$$

FIGURE 8

Formulas for Conditioned Output Spectral Density Functions

PARTIAL AND MULTIPLE COHERENCE FUNCTIONS

Definitions for partial coherence functions are stated in Figure 9. Specific formulas follow by substituting the particular conditioned spectral density functions from Figures 7 and 8. Note that partial coherence functions are

$$\gamma_{1y}^2 = \frac{|S_{1y}|^2}{S_{11}\ S_{yy}} = \gamma_{y1}^2$$

$$\gamma_{2y.1}^2 = \frac{|S_{2y.1}|^2}{S_{22.1}\ S_{yy.1}} = \gamma_{y2.1}^2$$

$$\gamma_{3y.12}^2 = \frac{|S_{3y.12}|^2}{S_{33.12}\ S_{yy.12}} = \gamma_{y3.12}^2$$

FIGURE 9

Partial Coherence Functions

really ordinary coherence functions of conditioned variables and hence are bounded between zero and unity.

A general definition for the multiple coherence function is given at the top of Figure 10 which applies both to arbitrary inputs as in Figure 1 and to ordered conditioned inputs as in Figure 4. Formulas for cases of one, two or three inputs are listed in Figure 11. The general relation for any number of inputs was previously derived in Reference [1].

$$\gamma^2_{y.x} = \frac{S_{yy} - S_{nn}}{S_{yy}}$$

$$0 \leq \gamma^2_{y.x} \leq 1$$

For three inputs

$$\gamma^2_{y.x} = \frac{S_{yy} - S_{yy.123}}{S_{yy}}$$

FIGURE 10

Multiple Coherence Functions

Single Input $\gamma^2_{y.x} = 1 - (1 - \gamma^2_{1y}) = \gamma^2_{1y}$

Two Inputs $\gamma^2_{y.x} = 1 - \left[(1 - \gamma^2_{1y})(1 - \gamma^2_{2y.1})\right]$

Three Inputs $\gamma^2_{y.x} = 1 - \left[(1 - \gamma^2_{1y})(1 - \gamma^2_{2y.1})(1 - \gamma^2_{3y.12})\right]$

For 3 uncoherent inputs, $\gamma^2_{y.x} = \gamma^2_{1y} + \gamma^2_{2y} + \gamma^2_{3y}$

FIGURE 11

Formulas for Multiple Coherence Functions

DECOMPOSITION OF THREE INPUT/OUTPUT MODEL

The preceding analysis yields the frequency domain composition of the conditioned three input/output model as shown in Figure 12 where the $\{L_i\}$ are the optimum frequency response functions of Figure 5. It follows that the total output spectral density function S_{yy} is decomposed here into four distinct terms where:

$\gamma_{1y}^2 S_{yy}$ = spectral output at y(t) due to linear effects of $x_1(t)$ as $x_1(t)$ passes through L_1

$\gamma_{2y.1}^2 S_{yy.1}$ = spectral output at y(t) due to linear effects of $x_{2.1}(t)$ as $x_{2.1}(t)$ passes through L_2

$\gamma_{3y.12}^2 S_{yy.12}$ = spectral output at y(t) due to linear effects of $x_{3.12}(t)$ as $x_{3.12}(t)$ passes through L_3

$S_{yy.123} = S_{nn}$ = spectral output at y(t) due to unknown independent terms n(t).

This procedure can be extended to any number of inputs.

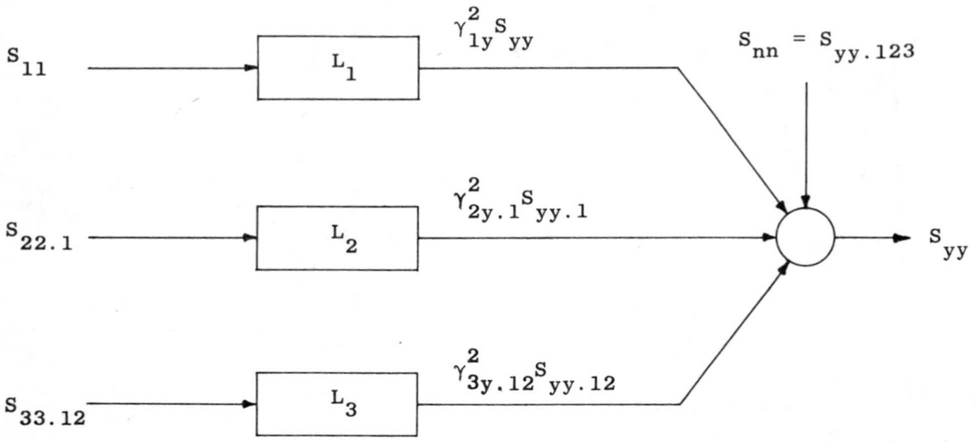

FIGURE 12. Decomposition of Three Input/Output Model
(Similar results follow for any number of inputs)

CONCLUSION

This paper has outlined some procedures to follow to analyze and predict useful frequency relationships in measured multiple input/output data. Many important engineering applications can be solved with the help of these techniques. To interpret results properly, since the data are statistical in nature and available records are limited in both duration and number, these applications also require appropriate statistical error analysis of any measured results. The author is fortunate to have had opportunities to extend and apply these ideas in recent consulting work and wishes to thank people who have supported these matters.

REFERENCES

1 J.S. Bendat, Solutions for the multiple input/output problem. J. of Sound and Vib. 44(3), (1976).

2 C.J. Dodds and J.D. Robson, Partial coherence in multivariate random processes. J. of Sound and Vib. 42(2), (1975).

3 J.S. Bendat and A.G. Piersol, Random Data: Analysis and Measurement Procedures, New York: John Wiley and Sons, Inc. (1971).

Dr. Julius S. Bendat, J.S. Bendat Company, 833 Moraga Drive, Los Angeles, California 90049, U.S.A.

DISCUSSION

In his introduction Dr. Bendat stated that these methods could be used for such application as sound source identification in electric motors. When this was questioned (Clarkson) because of the periodicity of the forces, Dr. Bendat agreed that the methods do not work for periodic forcing functions. In such cases it is necessary to remove the periodic inputs first before carrying out the analysis. The methods are for stationary random data, any periodicity will give a delta function in the spectra.

However, it is not necessary to know how many independent random inputs are present. The measurements of the coherency between the inputs will determine the degree of independence between the inputs. If it turns out that the coherency of one pair of inputs is unity for all frequency then one input is directly dependent on the other and should be eliminated from the analysis. It is giving redundant information. For example, if in the case of an automobile travelling along a road, the forces at the rear wheels are the same as those on the front wheels except for a time delay then they cannot be used as independent inputs. The coherence function between forces on each side will be close to one and therefore the problem must be treated as a two input problem rather than four.

Dr. Holmes quoted the recent work of Chung and Crocker (J. Acoustical Soc. Amer., 58, 1975) in which the not unsurprising result that the sources of sound in a diesel engine are the explosions within the cylinder is obtained. He said that in this case they had assumed that there may be some unmeasured uncorrelated inputs but that there were no unmeasured correlated inputs. He asked if the technique detects the presence of correlated unmeasured inputs. Dr. Bendat agreed that the work of Chung and Crocker was an oversimplification. The techniques described in their paper will allow determination only of effects of various other uncorrelated inputs. It gives a quantitative estimate as a function of frequency of possible uncorrelated inputs which were not considered in the model and also non-linear effects. It does not give an estimate of the number of unmeasured correlated inputs. In fact, all results would be erroneous because an incorrect model was used in this case.

Professor Willems commented that this method was in some senses equivalent to the Gauss elimination method. Dr. Bendat agreed that there may be some similarity but said that the emphasis here is in relating the method to physical problems. It attempts to distinguish effects of one source by itself when the effects of all other possible sources have been removed in a systematic fashion.

When asked (Ottens) which one of the inputs had to be removed first, Dr. Bendat replied that this was a matter of judgement. In practice it was usually best to use the largest or dominant input as x_1 and then remove its effect to find the effect of the other inputs. However it is important to be aware of the estimation problem because the method is only going to yield estimates of the quantities of interest. These estimates will be affected by the quality of the instrumentation and the skill of the user. Some results from recent error analysis for the functions H and L are now available (J. Sound and Vibration. 49, 1976).

Professor Akaike suggested that the time domain approach such as he had described (Chapter 18) might have some advantages particularly when considering such effects as nonstationarity. This method would allow a better mathematical interpretation of the results by such measures as maximum likelihood. When damping is present or when there is feedback the time domain method should give better results but the method is more difficult. It may be that the mathematical difficulties will become overwhelming when there are several inputs and outputs. In this case it may be better to go into the frequency domain. Thus the two methods can be seen as complementary with time domain being preferable for cases with a single or small number of inputs and the frequency domain preferable for a larger number of inputs.

Dr. Bendat agreed they were theoretically related but stated it was much more significant to obtain results as a function of frequency rather than over all frequencies. He pointed out that in many practical applications the time domain method was not so convenient for physical insight and the interpretation of results. Engineers preferred to think in terms of frequency response. Dr. Akaike replied that this may be so but the final results could always be converted back into the frequency domain for the engineering interpretation. Dr. Bendat stated that he does not recommend this indirect approach.

Professor Robson explained that in his work (paper number 17) the motivation to remove a dominant input had arisen from a practical problem where the results were coloured by one input. This pointed to the need for relationships between multiple inputs and outputs such as Dr. Bendat had presented. When asked (Robson) about the extension to multiple outputs, Dr. Bendat agreed that in considering only one output he had produced a building block for the more general problem of multiple inputs, multiple paths and multiple outputs but no feedback is included. There are some experimental results being obtained at the moment on such structures as North Sea Oil platforms and in other applications.

J D ROBSON and C J DODDS

Normal co-ordinates and residual spectra in the analysis of random vibration response

INTRODUCTION

The use of normal coordinates to uncouple the contributions of the several normal modes in the response of lightly damped systems has been a well established technique of vibration analysis since the time of Rayleigh. Such analysis reflects the separate contribution of the various normal modes, and has proved particularly well adapted to spectral response analysis of random vibration [1], [2].

The concepts of partial coherence and residual spectra, which depend on the conditioning of one variable to eliminate all coherence with another, have a less venerable history but again are well enough established. They are described, for example, in [3] and [4], and a recent paper by the present authors [5] has made available a rather simpler treatment, both of derivation and manipulation, for multivariate processes. (A still more recent paper by Bendat [6] has spelled out the implications for multi-excitation response).

The two techniques have similarities in that both depend on the separation of a signal into constituents, uncoupled from each other in the one case and coherent or uncoherent in the other. It seems not impossible that there exists some loose form of analogy, which might be uncovered with advantage. It seems even more likely that if the two techniques could be employed in combination, the combined advantages of both would be very powerful indeed.

It is the purpose of this paper therefore to examine the relationship between the two techniques, and the extent to which they can be used to advantage in combination.

The form of the paper is controlled by this purpose. There must be some account of normal coordinates and normal modes, the spectral response relations expressed in terms of them, and the coherence properties of the various variables involved. There must be some account of residual spectra and the technique of conditioning, at least to the extent that is required by the later arguments of the paper. This preparation will make possible a derivation of the formal relationships involving normal coordinates and residual spectra in combination. We shall then be able to examine the possibility of practical application, particularly in the processing of multivariate response signals.

NORMAL COORDINATES

General Relationships

Any m coordinates x_r of a lightly-damped n freedom system can be expressed conveniently in matrix form as

$$\underset{\sim}{x} = \begin{bmatrix} x_1 \\ x_2 \\ \vdots \\ x_m \end{bmatrix}$$

If the n normal coordinates ξ_j are also denoted by

$$\underset{\sim}{\xi} = \begin{bmatrix} \xi_1 \\ \xi_2 \\ \vdots \\ \xi_n \end{bmatrix}$$

we may write

$$\mathbf{x} = \mathbf{d}\,\boldsymbol{\xi}\ , \tag{1}$$

where $\mathbf{d}$ is of order mxn, having columns with the form of the respective normal modes of the system.

The ξ_j will be related to their corresponding generalised forces Ξ_j by the relationship

$$\boldsymbol{\xi} = \boldsymbol{\alpha}\,\boldsymbol{\Xi}\ , \tag{2}$$

where because of uncoupled nature of the normal coordinates the n x n matrix $\boldsymbol{\alpha}$ is diagonal.

Finally if there are actual forces X_r corresponding to each of the x_r we write

$$\boldsymbol{\Xi} = \mathbf{d}'\,\mathbf{X}\ , \tag{3}$$

where $\mathbf{d}'$ is the transpose of $\mathbf{d}$. There is no loss of generality by assuming that all coordinates x_r have corresponding forces X_r: if there are response points where no force acts we can take the corresponding elements of $\mathbf{X}$ to be zero, but we must include suitable response coordinates at all points at which forces are applied.

The matrices $\mathbf{x}$, $\boldsymbol{\xi}$, $\boldsymbol{\Xi}$, $\mathbf{X}$ introduced above can be taken to represent realisations of multivariate stationary random processes and for these the following spectral response relationships may be derived in the usual way (See [2]):

$$\mathbf{S}^{x} = \mathbf{d}\,\mathbf{S}^{\xi}\,\mathbf{d}' \tag{4}$$

$$\mathbf{S}^{\xi} = \boldsymbol{\alpha}^{*}\,\mathbf{S}^{\Xi}\,\boldsymbol{\alpha} \tag{5}$$

$$\mathbf{S}^{\Xi} = \mathbf{d}'\,\mathbf{S}^{X}\,\mathbf{d} \tag{6}$$

Here $\mathbf{S}^{x}$ and $\mathbf{S}^{X}$ are of order m x m: $\mathbf{S}^{\xi}$ and $\mathbf{S}^{\Xi}$ are of order n x n.

We shall also find it necessary to use a further set of response relationships - those that involve the cross spectral densities between the various pairs of variables x_r, ξ_j, etc. These may be derived simply enough using the technique of [7] for example, and are of the form

$$\boldsymbol{S}^{x\xi} = \boldsymbol{d}\ \boldsymbol{S}^{\xi} \tag{7}$$

$$\boldsymbol{S}^{\xi\zeta} = \boldsymbol{\alpha}^*\ \boldsymbol{S}^{\zeta} \tag{8}$$

$$\boldsymbol{S}^{\zeta X} = \boldsymbol{d}'\ \boldsymbol{S}^{X} \tag{9}$$

The left hand sides of these are respectively of order m x n, n x n, n x m.

It is sometimes desirable to consider relationships reciprocal to those stated above, and this can easily enough be done by the use of inverse matrices. For example if

$$\boldsymbol{k} = \boldsymbol{d}^{-1} \tag{10}$$

we may write

$$\boldsymbol{\xi} = \boldsymbol{k}\ \boldsymbol{x}$$

and so

$$\boldsymbol{S}^{\xi x} = \boldsymbol{k}\ \boldsymbol{S}^{x} \tag{11}$$

Such inversion presents no difficulty if $m = n$, or even if $m > n$: if $m < n$ it is of course not possible.

Special Cases

It is helpful to consider two special cases, each of some practical interest. These are first, the case where response arises from a single exciting force, and second, the case where all exciting forces are completely without coherence.

Single force Suppose that all elements of the $\boldsymbol{X}$ matrix, apart from X_m are zero: then in the $\boldsymbol{S}^X$ matrix, only the element S^X_{mm} will be other than zero. Expanding the $\boldsymbol{S}^{\zeta}$ matrix, using (6) it can be shown that all elements are proportional to S^X_{mm}, and that, in particular

$$S^{\Xi}_{rs} = d_{mr}\, d_{ms}\, S^X_{mm} \tag{12}$$

From (12) it follows that

$$\gamma^{\Xi}_{rs} = 1. \tag{13}$$

It will be shown later that in all cases

$$\underset{\sim}{\gamma}^{\xi} = \underset{\sim}{\gamma}^{\Xi} \tag{14}$$

so that in this case

$$\gamma^{\xi}_{rs} = 1\ , \quad \text{also.} \tag{15}$$

<u>Uncoherent excitations</u> Where the excitations of the $\underset{\sim}{X}$ matrix are uncoherent, the $\underset{\sim}{S}^X$ matrix becomes diagonal, and a considerable simplification in the response relationships might be expected. Unfortunately the $\underset{\sim}{S}^{\Xi}$ matrix is not simple in form, each element including contributions from all elements of $\underset{\sim}{X}$. We obtain, in fact,

$$S^{\Xi}_{rs} = d_{r1}d_{s1}S^X_{11} + d_{r2}d_{s2}S^X_{22} + \ldots + d_{rm}d_{sm}S^X_{mm}\ . \tag{16}$$

There is indeed no simplicity about the coherence of the elements of $\underset{\sim}{\Xi}$ - and so those of $\underset{\sim}{\xi}$. We then obtain

$$(\gamma^{\xi}_{rs})^2 = (\gamma^{\Xi}_{rs})^2 =$$

$$\frac{(d_{r1}d_{s1}S^X_{11}+d_{r2}d_{s2}S^X_{22}+\ldots+d_{rm}d_{sm}S^X_{mm})^2}{(d^2_{r1}S^X_{11}+d^2_{r2}S^X_{22}+\ldots+d^2_{rm}S^X_{mm})(d^2_{s1}S^X_{11}+d^2_{s2}S^X_{22}+\ldots+d^2_{sm}S^X_{mm})} \tag{17}$$

so that there is in general some coherence between all the elements of $\underset{\sim}{\xi}$.

There will be zero coherence between any ξ_r, ξ_s only if

$$d_{r_1}d_{s_1}S^X_{22} + d_{r_2}d_{s_2}S^X_{22} + \ldots + d_{r_m}d_{s_m}S^X_{mm} = 0 \tag{18}$$

and this would require the elements of $\underset{\sim}{X}$ to be applied in a very special way.

It may be noted that the adoption of normal coordinates, far from simplifying the problem, has here eliminated what simplicity there was. This stresses the already obvious fact that decoupling the decoherence are very different matters.

Coherences of normal coordinates

In considering the special cases above we have seen that complete coherence of responses is to be expected where there is a single exciting force, but that other simplicities in excitation are unlikely to be helpful. It is enlightening to consider coherence of the normal coordinates more generally.

We may first establish the equality of coherences between elements of the $\underset{\sim}{\xi}$ matrix and those of the $\underset{\sim}{\Xi}$ matrix. From (5) we have

$$S^{\xi}_{rr} = \alpha^*_r\alpha_r S^{\Xi}_{rr}; \quad S^{\xi}_{rs} = \alpha^*_r\alpha_s S^{\Xi}_{rs}; \quad S^{\xi}_{sr} = \alpha^*_s\alpha_r S^{\Xi}_{sr}; \quad S^{\xi}_{ss} = \alpha^*_s\alpha_s S^{\Xi}_{ss} \tag{19}$$

so that

$$(\gamma^{\xi}_{rs})^2 = \frac{\alpha^*_r\alpha_s S^{\Xi}_{rs} \quad \alpha^*_s\alpha_r S^{\Xi}_{sr}}{\alpha^*_r\alpha_r S^{\Xi}_{rr} \quad \alpha^*_s\alpha_s S^{\Xi}_{ss}} = (\gamma^{\Xi}_{rs})^2 \tag{20}$$

It will simplify the exposition if we confine attention to the case $n = 2$; let us treat the case where we apply only two forces X_3, X_4 corresponding to displacements x_3, x_4, and measure two responses x_1, x_2. Then we may write

$$\left.\begin{aligned} x_1 &= d_{11}\,\xi_1 + d_{12}\,\xi_2 \\ x_2 &= d_{21}\,\xi_1 + d_{22}\,\xi_2 \end{aligned}\right\} \tag{21}$$

and

$$\left.\begin{aligned} \Xi_1 &= d_{31}X_3 + d_{41}X_4 \\ \Xi_2 &= d_{32}X_3 + d_{42}X_4 \end{aligned}\right\} \tag{22}$$

The corresponding spectral relationship is

$$
\begin{aligned}
S^{\Xi}_{11} &= d^2_{31}S^X_{33} + d_{31}d_{41}S^X_{34} + d_{41}d_{31}S^X_{43} + d^2_{41}S^X_{44} \\
S^{\Xi}_{12} &= d_{31}d_{32}S^X_{33} + d_{31}d_{42}S^X_{34} + d_{41}d_{32}S^X_{43} + d_{41}d_{42}S^X_{44} \\
S^{\Xi}_{21} &= d_{32}d_{31}S^X_{33} + d_{32}d_{41}S^X_{34} + d_{42}d_{31}S^X_{43} + d_{42}d_{41}S^X_{44} \\
S^{\Xi}_{22} &= d^2_{32}S^X_{33} + d_{32}d_{42}S^X_{34} + d_{42}d_{32}S^X_{43} + d_{42}d_{42}S^X_{44}
\end{aligned}
\qquad (23)
$$

It must now be clear that it will be profitless to expand γ^{Ξ}_{12}, even in this simple case, unless we have some special knowledge about the magnitude of the elements of $\underset{\sim}{d}$.

If a single force acts of course, or if the two are arranged such that a single mode predominates, the problem then becomes quite trivial, with $\gamma^2 = 1$.

It may indeed often be the case that coherence between normal coordinates is negligible: where the natural frequencies of two modes are well separated the cross spectral densities will often be small. Where natural frequencies are well separated it will usually be permissible to assume that a simple mode predominates close to any resonance frequency: the question of coherence with other modes will not then arise. But although we must take advantage of these favourable circumstances when they arise, the occurrence of near-coincident natural frequencies is common enough to make it necessary to have available analysis which can cope with them.

RESIDUAL SPECTRA

A considerable insight into the relationship of the multivariate response to multivariate excitation may be obtained by examination of residual spectral densities - the term spectral density being used here to cover all elements of the spectral density matrix, including cross spectral densities. Often a few components of excitation will exert a predominant influence on the responses, so that the complete response relationships underlying actual response data only become apparent when these predominating constituents have been eliminated. Techniques exist for the conditioning of the components of multivariate random processes with respect to one or more particular components, and the residual spectra applicable to the process after this conditioning have particular application to such responses. They can reveal relationships between response and excitation which were hidden before the conditioning was applied to remove unwanted coherences.

Residual Spectral Relationships

The process of conditioning has the effect of removing from any component of a random process whatever part of it is completely coherent with one other chosen component or, if desired, with several; the resulting spectral densities - direct or cross - are residual spectral densities. The technique of conditioning and the determination of residual spectral densities is fully described in reference [5] and only the main results will be given here.

Consider the multivariate process $\underset{\sim}{x}$ whose m elements x_r represent a single realisation of the process. These variables may be responses, or excitations, or a combination of both. Denote the spectral density matrix of $\underset{\sim}{x}$ by $\underset{\sim}{S}^x$, which will in general be an m x n non-singular hermitian matrix. (Any singularity in the matrix should be eliminated before proceeding.)

It is helpful at this stage to recognise that the matrix $\underset{\sim}{S}^x$ may be expanded in the form

$$\underset{\sim}{S}^x = \underset{\sim}{A}^* \underset{\sim}{A}' \tag{24}$$

where $\underset{\sim}{A}$ is the lower triangular matrix given by

$$\underset{\sim}{A} = \begin{bmatrix} A_{11} & 0 & \cdots & 0 \\ A_{21} & A_{22} & & 0 \\ \vdots & & \ddots & \\ A_{m1} & A_{m2} & & A_{mm} \end{bmatrix} \tag{25}$$

The great advantage of the transformation (24) is that the various residual spectral densities can be expressed very simply in terms of the elements of $\underset{\sim}{A}$. Thus if we use the notation $S_{rs.pq}$ to denote the cross spectral density of x_r and x_s conditioned with respect to x_p and x_q - with the obvious extensions to higher or lower orders of conditioning - it can be shown that with $r \geq s$:

$$\left.\begin{aligned} S^x_{rs} &= \sum_{q=1}^{s} A^*_{rq} A_{sq} \qquad r,s > 1 \\ S^x_{rs.1} &= \sum_{q=2}^{s} A^*_{rq} A_{sq} \qquad r,s > 2 \\ S^x_{rs.12} &= \sum_{q=3}^{s} A^*_{rq} A_{sq} \qquad r,s > 3 \end{aligned}\right\} \tag{26}$$

etc.

Equations (26) not only enable the elements of the $\underset{\sim}{A}$ matrix to be defined - using the first of (26) - but then give immediately the various orders of residual spectral densities. It may be seen moreover that the series expression for S^x_{rs} contains those for $S^x_{rs.1}$, $S^x_{rs.12}$, etc., so each term of this series expression has a physical significance in terms of coherence: the removal of each term in turn eliminates all coherence with one more component.

We shall be concerned in this paper only with first-order conditioning - eliminating coherence only with one variable - and so shall be concerned only with the first two equations of (26). Using the first of these to establish

$\underset{\sim}{A}$ in terms of $\underset{\sim}{S}^x$, the second leads to the result

$$S^x_{rs.q} = S^x_{rs} - \frac{S^x_{rq} S^x_{qs}}{S^x_{qq}} \tag{27}$$

or in particular to the results

$$\left.\begin{aligned} S^x_{22.1} &= S^x_{22} - \frac{S^x_{21} S^x_{12}}{S^x_{11}} = S^x_{22} \left[1 - (\gamma^x_{12})^2\right] \\ S^x_{23.1} &= S^x_{23} - \frac{S^x_{21} S^x_{13}}{S^x_{11}} \end{aligned}\right\} \tag{28}$$

etc.

FORMAL RELATIONSHIPS

It must be obvious from the beginning that the splitting of response into uncoupled responses in the various modes, and into responses uncoherent with one or more other variables, are operations very different in kind. Yet there are loose similarities and we should first dispose of any suggestion that there is an analytical analogy. We must then establish the formal results in which residual spectral densities and normal coordinates appear in combination: these include particularly the expression of residual spectral densities of one or more response variables in terms of the spectral densities of their component normal coordinates, and expressions for response spectral densities conditioned with respect to normal coordinates. Most results are easily expressed in terms of normal coordinate spectral densities: it will be desirable to include some discussion of the problem of inversion - that is expression in terms of the basic response spectral densities.

Analysis will be kept general as far as possible: wherever it is desirable for reasons of clarity or simplicity to restrict attention to a two-response or two-modal system this will be done. The implications as to loss of generality will be examined in due course.

The Lack of Analogy

Normal coordinates $\boldsymbol{\xi}$ can be defined by

$$\mathbf{x} = \mathbf{d}\,\boldsymbol{\xi}\,, \tag{29}$$

where the columns of $\mathbf{d}$ are the normal modes of the system; the elements of $\boldsymbol{\xi}$ then appear uncoupled in the equations of motion of a lightly-damped system. It is perhaps natural to ask whether we can define uncoherent coordinates $\boldsymbol{\eta}$ such that if

$$\mathbf{x} = \mathbf{e}\,\boldsymbol{\eta} \tag{30}$$

then the elements of $\boldsymbol{\eta}$ are uncoherent.

It is more convenient to work in terms of $\mathbf{g} = \mathbf{e}^{-1}$; then

$$\boldsymbol{\eta} = \mathbf{g}\,\mathbf{x}\,, \tag{31}$$

and we would want the corresponding spectral density matrix, given by

$$\mathbf{S}^{\eta} = \mathbf{g}^{*}\,\mathbf{S}^{x}\,\mathbf{g}' \tag{32}$$

to be diagonal at all frequencies.

Expansion of (32) shows immediately that this cannot be done. Allocation of the elements of $\mathbf{g}$ so as to cause the off-diagonal elements of $\mathbf{S}^{\eta}$ to be zero automatically eliminates the diagonal elements also. (If the elements of $\mathbf{g}$ are made frequency-dependent it is certainly possible, but that is another matter).

Residual Spectra in Terms of Normal Coordinates

If a system has responses

$$\underset{\sim}{x} = \begin{bmatrix} x_1 \\ x_2 \\ \vdots \\ x_m \end{bmatrix}$$

and normal coordinates

$$\underset{\sim}{\xi} = \begin{bmatrix} \xi_1 \\ \xi_2 \\ \vdots \\ \xi_n \end{bmatrix} , \quad \text{then}$$

$$\underset{\sim}{x} = \underset{\sim}{d} \, \underset{\sim}{\xi} \tag{33}$$

and d is of order m x n. It follows therefore that

$$\underset{\sim}{S}^{x} = \underset{\sim}{d} \, \underset{\sim}{S}^{\xi} \, \underset{\sim}{d}' \,. \tag{34}$$

From (34) we may expand particular elements of $\underset{\sim}{S}^{x}$ in terms of the elements of $\underset{\sim}{S}^{\xi}$; it will be convenient here to restrict attention to the case n = 2, and we then obtain

$$\left.\begin{aligned} S^{x}_{11} &= d_{11}d_{11}S^{\xi}_{11} + d_{11}d_{12}S^{\xi}_{12} + d_{12}d_{11}S^{\xi}_{21} + d_{12}d_{12}S^{\xi}_{22} \\ S^{x}_{12} &= d_{11}d_{21}S^{\xi}_{11} + d_{11}d_{22}S^{\xi}_{12} + d_{12}d_{21}S^{\xi}_{21} + d_{12}d_{22}S^{\xi}_{22} \end{aligned}\right\} \tag{35}$$

etc.

Substitution from (35) in the known results (28) leads to

$$\left.\begin{aligned} S^{x}_{22.1} &= (d_{11}d_{22} - d_{12}d_{21})^2 \; \frac{S^{\xi}_{11}S^{\xi}_{22} - S^{\xi}_{12}S^{\xi}_{21}}{S^{x}_{11}} \\ S^{x}_{23.1} &= (d_{11}d_{22} - d_{12}d_{21})(d_{11}d_{32} - d_{12}d_{31}) \; \frac{S^{\xi}_{11}S^{\xi}_{22} - S^{\xi}_{12}S^{\xi}_{21}}{S^{x}_{11}} \end{aligned}\right\} \tag{36}$$

etc.

It will often happen that over a restricted range of frequency a particular modal contribution will predominate. If for example the contribution of ξ_1 predominates over a certain range the equations of (38) will be very much simplified over that range and become

$$\left.\begin{aligned} S^{x}_{22.1} &= \left(d_{22} - \frac{d_{12}d_{21}}{d_{11}}\right)^2 S^{\xi}_{22}\left[1 - (\gamma^{\xi}_{12})^2\right] \\ S^{x}_{23.1} &= \left(d_{22} - \frac{d_{12}d_{21}}{d_{11}}\right)\left(d_{32} - \frac{d_{12}d_{31}}{d_{11}}\right) S^{\xi}_{22}\left[1 - (\gamma^{\xi}_{12})^2\right] \end{aligned}\right\} \tag{37}$$

etc.

It will be noted that

$$S^{\xi}_{22}\left[1 - (\gamma^{\xi}_{12})^2\right] = S^{\xi}_{22.1} \tag{38}$$

It will be useful to record the general result corresponding to (36); this is

$$S^{x}_{rs.j} = S^{x}_{rs} - \frac{S^{x}_{rj}\,S^{x}_{js}}{S^{x}_{jj}}$$

$$= (d_{j2}d_{r1} - d_{j1}d_{r2})(d_{j2}d_{s1} - d_{j1}d_{s2})\frac{S^{\xi}_{11}S^{\xi}_{22} - S^{\xi}_{12}S^{\xi}_{21}}{S^{x}_{11}} . \tag{39}$$

Residual Spectral Densities of Normal Coordinates

Any set of normal coordinates $\underset{\sim}{\xi}$ forms a multivariate process, and is amenable to the usual rules for determining residual spectra. Thus we may write down for completeness:

$$S^{\xi}_{rs.q} = S^{\xi}_{rs} - \frac{S^{\xi}_{rq}\,S^{\xi}_{qs}}{S^{\xi}_{qq}} . \tag{40}$$

We may note that, as usual, in the special case where s = r we have

$$S^{\xi}_{rr.q} = S^{\xi}_{rr} - \frac{S^{\xi}_{rq} S^{\xi}_{qr}}{S^{\xi}_{qq}} = S^{\xi}_{rr}\left[1 - (\gamma^{\xi}_{rq})^2\right] \quad : \tag{41}$$

conversely,

$$S^{\xi}_{qq.r} = S^{\xi}_{qq} - \frac{S^{\xi}_{qr} S^{\xi}_{rq}}{S^{\xi}_{rr}} = S^{\xi}_{qq}\left[1 - (\gamma^{\xi}_{rq})^2\right] \quad . \tag{42}$$

Conditioning of Response with respect to a Normal Coordinate

Consider again the responses $\underset{\sim}{x}$ expressed in terms of the normal coordinates $\underset{\sim}{\xi}$, so that $\underset{\sim}{x} = \underset{\sim}{d}\,\underset{\sim}{\xi}$. It is of interest to consider the residual spectral densities of the elements of $\underset{\sim}{x}$, conditioned with respect to a particular normal coordinate. This will remove all coherence with the particular mode, though as all modes are in general to some extent coherent with all others the physical significance of this operation is not very clear: It should at least do something to reduce the content of the particular mode.

Conditioning with respect to the j th mode and adopting an obvious notation we have

$$S^{x}_{rs}\,.\,{}^{\xi}_{j} = S^{x}_{rs} - \frac{S^{x\xi}_{rj} S^{\xi x}_{js}}{S^{\xi}_{jj}} \quad ; \tag{43}$$

This may be expanded in terms of the elements of $\underset{\sim}{S}^{\xi}$ for any particular value of n, for we know that

$$\underset{\sim}{S}^{x\xi} = \underset{\sim}{d}\,\underset{\sim}{S}^{\xi} \quad . \tag{44}$$

It will again be more convenient here to restrict attention to the case of n = 2. We then have

$$\left.\begin{aligned} x_r &= d_{r1}\,\xi_1 + d_{r2}\,\xi_2 \\ x_s &= d_{s1}\,\xi_1 + d_{s2}\,\xi_2 \end{aligned}\right\} \tag{45}$$

and may construct the required spectral relations in the usual way to obtain, for example,

$$\left.\begin{aligned} S^{x}_{rs} &= d_{r1}d_{s1}S^{\xi}_{11} + d_{r1}d_{s2}S^{\xi}_{12} + d_{r2}d_{s1}S^{\xi}_{21} + d_{r2}d_{s2}S^{\xi}_{22} \\ S^{x\xi}_{r1} &= d_{r1}S^{\xi}_{11} + d_{r2}S^{\xi}_{21} \end{aligned}\right\} \tag{46}$$

etc.

Substituting from (46) in (43) we have relations of the following form:

$$\left.\begin{aligned} S^{xx.\xi}_{11\ 1} &= d^2_{12}\, S^{\xi}_{22}\left[1 - (\gamma^{\xi}_{12})^2\right] \\ S^{xx.\xi}_{12\ 1} &= d_{12}d_{22}S^{\xi}_{22}\left[1 - (\gamma^{\xi}_{12})^2\right] \\ S^{xx.\xi}_{22\ 1} &= d^2_{22}S^{\xi}_{22}\left[1 - (\gamma^{\xi}_{12})^2\right] \\ S^{xx.\xi}_{11\ 2} &= d^2_{11}S^{\xi}_{11}\left[1 - (\gamma^{\xi}_{12})^2\right] \\ S^{xx.\xi}_{12\ 2} &= d_{11}d_{21}S^{\xi}_{11}\left[1 - (\gamma^{\xi}_{12})^2\right] \\ S^{xx.\xi}_{22\ 2} &= d^2_{21}S^{\xi}_{11}\left[1 - (\gamma^{\xi}_{12})^2\right] \end{aligned}\right\} \tag{47}$$

etc.

From (47) we may note that

$$\left.\begin{aligned} \frac{S^{xx.\xi}_{11\ 1}}{d^2_{12}} &= \frac{S^{xx.\xi}_{12\ 1}}{d_{12}d_{22}} = \frac{S^{xx.\xi}_{21\ 1}}{d_{12}d_{22}} = \frac{S^{xx.\xi}_{21\ 1}}{d^2_{22}} \\ \frac{S^{xx.\xi}_{11\ 2}}{d^2_{11}} &= \frac{S^{xx.\xi}_{12\ 2}}{d_{11}d_{21}} = \frac{S^{xx.\xi}_{21\ 2}}{d_{11}d_{21}} = \frac{S^{xx.\xi}_{22\ 2}}{d^2_{21}} \end{aligned}\right\} \tag{48}$$

And from (47) and (40) we find that the terms of the two sets of equations of (48) are respectively $S^{\xi}_{22.1}$ and $S^{\xi}_{11.2}$.

Results in Terms of $\underset{\sim}{S}^x$

The results we have derived here are expressed in terms of the spectral densities of the normal coordinates. While such results may be enlightening, their application to practical problems would obviously be increased if they could be expressed instead in terms of the spectral densities of quantities which can actually be measured. We should therefore consider this possibility.

If we again have

$$\underset{\sim}{x} = \underset{\sim}{d}\, \underset{\sim}{\xi} \tag{49}$$

in the usual way, inversion can be achieved formally by premultiplying throughout by $\underset{\sim}{e} = \underset{\sim}{d}^{-1}$. This can be done even if $m \neq n$, provided of course that $m > n$. Then

$$\underset{\sim}{\xi} = \underset{\sim}{d}^{-1}\, \underset{\sim}{d}\, \underset{\sim}{\xi} = \underset{\sim}{e}\, \underset{\sim}{x} \tag{50}$$

and provided that the matrix $\underset{\sim}{d}$ is known, there is no difficulty here. The spectral densities of the normal coordinates are expressed in terms of the spectral densities of the actual responses by

$$\underset{\sim}{S}^{\xi} = \underset{\sim}{e}\, \underset{\sim}{S}^{x}\, \underset{\sim}{e}' \tag{51}$$

But if the $\underset{\sim}{d}$ matrix is not known, as will be the case in many practical problems, there is little virtue in the application of (51) to the results previously obtained. No such application will be attempted here, therefore; if our 'combined' relationships are to have value in the treatment of actual response records this must be discernable from the results as they stand.

POSSIBLE APPLICATIONS

It is the object of this paper to investigate the usefulness of the two concepts of normal coordinates and residual spectra applied in combination. Here we should certainly have in mind the insight which any analytical result brings, but we shall be particularly concerned with possible applications to the processing of response records, which is probably the principal reason for interest in the concept of residual spectra. Clearly the results obtained in the previous section must form the basis of any such application, and we must consider them first with a view to their direct application. After this we must consider other ways in which these results may prove useful.

Possibility of Direct Application

Our present situation differs from the more usual one where a practical need exists and it is required to set up analysis which will resolve it. Here the analysis is already developed, and we are investigating its applicability to practical problems within a certain field. This field is in fact quite narrowly defined, because the incorporation of residual spectra strongly suggests a concern with multivariate response records. The use of normal coordinates is only permissible in the field of lightly damped linear systems - effectively structures. So we may have in mind the processing of multivariate records of the response of practical structures under random excitation.

It is appropriate now to consider the results in the last section with this end in view, in particular equations (39) and (47): it may be helpful to reproduce here equations typical of each, thus

$$S^{x}_{22.1} = (d_{11}d_{22} - d_{12}d_{21})^2 \frac{S^{\xi}_{11}\, S^{\xi}_{22} - S^{\xi}_{12}S^{\xi}_{21}}{S^{x}_{11}} , \tag{52}$$

$$S^{xx.\xi}_{22.1} = d^2_{22}\, S^{\xi}_{22} \left[1 - (\gamma^{\xi}_{12})^2\right] , \tag{53}$$

Each of these is applicable as it stands only to a two-freedom system, but this will be sufficient for our present purpose. There will be no difficulty in extending the analysis to more general systems if this should prove to be necessary.

Equation (52) gives a typical residual spectral density in terms of the spectral densities of the normal coordinates. (The denominator on the right-hand side has been left unexpanded for brevity). Although in its complete form it is difficult to find any application for this it is a surprisingly simple result, considering the complexity of the analysis which leads to it, and it does indicate the dependence of the residual spectral densities on the coherence of the normal coordinates. Clearly with full coherence between the normal coordinates the residual spectral densities would be zero. We have seen too that where a single mode predominates - as for example at frequencies close to a system's resonance frequencies - the problem is much simplified. If the contribution of ξ_1 predominates, equation (52) becomes

$$S^{x}_{22.1} = (d_{22} - \frac{d_{12}\, d_{21}}{d_{11}})^2\, S^{\xi}_{22.1} \tag{54}$$

and this is closely related to $S^{xx.\xi}_{22.1}$ by (53).

Equation (53) gives a simple expression for a quantity which is surely of some practical interest - a response spectral density conditioned with respect to a single mode. But this can neither be measured directly, nor computed from measured spectra: the right-hand side of (53) could only be written in terms of the elements of $\underset{\sim}{S}^{x}$ if the normal modes were already known. However the close relationship in special circumstances between $S^{x}_{22.1}$ and $S^{xx.\xi}_{22.1}$ as revealed by (53) and (54) is of some interest.

It may well be that (52) and (53) have direct application to problems that cannot yet be envisaged. It seems more realistic however to accept that there is no obvious direct application, and to seek for an indirect approach to the problem which is of some current interest - the elimination of a single modal contribution from response records taken from a structure in service, where there is a random excitation which is of some complexity and not measurable, and where the mode shapes are not known.

Residual Spectra of Structural Response

It will be helpful at this stage to look at response spectra taken from an actual structure in service. The structure is of some size and approximately symmetrical about two mutually perpendicular vertical planes; it vibrates due to the excitation of its own environment. Little is known in detail about the spectral content of this excitation but it is believed to be approximately flat over the frequency range of interest.

Response spectra were computed from data recorded by two transducers mounted at right angles to each other in the horizontal plane, parallel to the two planes of symmetry. Figures 1 and 2 show the spectral densities S_{AA} and S_{BB} of the responses x_A, x_B in the two directions. In both spectra peaks occur at frequencies of 1.4 Hz and 2.0 Hz, but no identification of modes is possible from this information as it is presented.

If however we condition x_A with respect to x_B we obtain the residual spectral density $S_{AA.B}$ which is plotted in Figure 3. We find that the 2.0 Hz peak has been completely removed; what remains is evidently only the expected response curve for a single normal mode with natural frequency 1.4 Hz. In this we have perhaps been more successful than we could reasonably have hoped from such an operation, but for our present purpose this result is of importance as indicating how effective the conditioning process can be in eliminating unwanted constituents of a signal.

A similar result is obtained if we condition x_B with respect to x_A and obtain the residual spectral density $S_{BB.A}$: this is plotted in Figure 4.

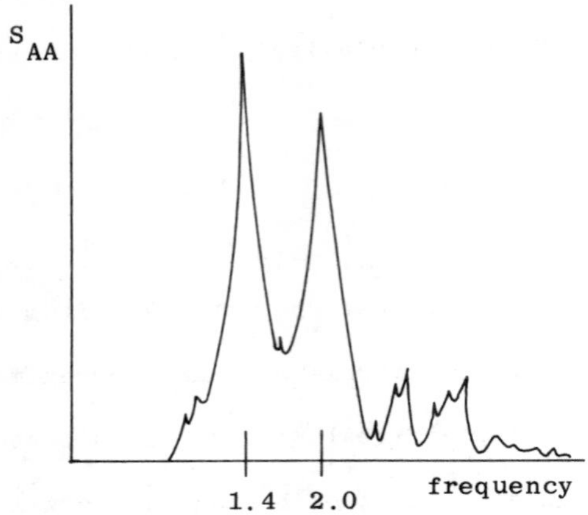

FIGURE 1

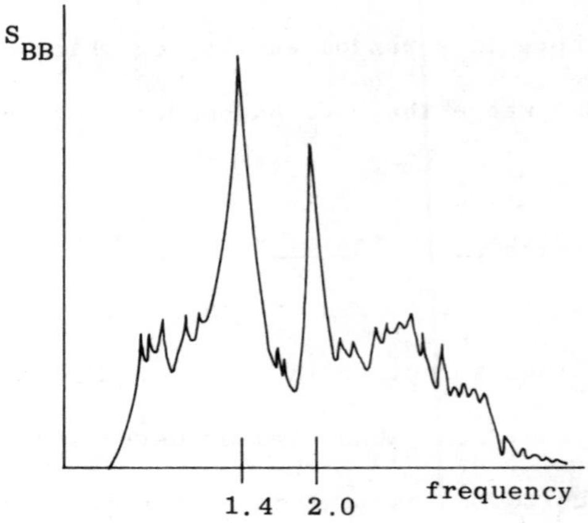

FIGURE 2

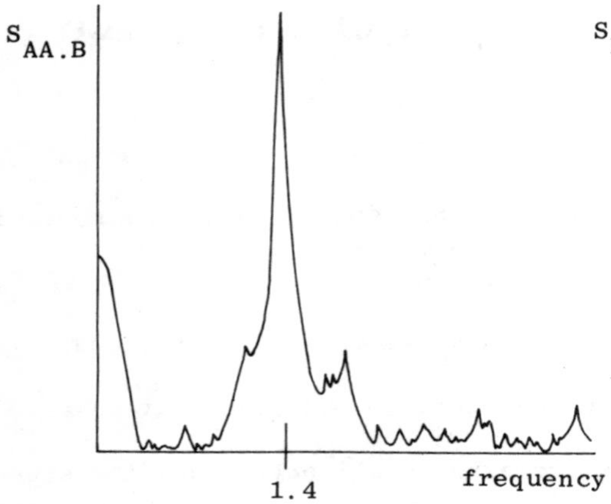

FIGURE 3

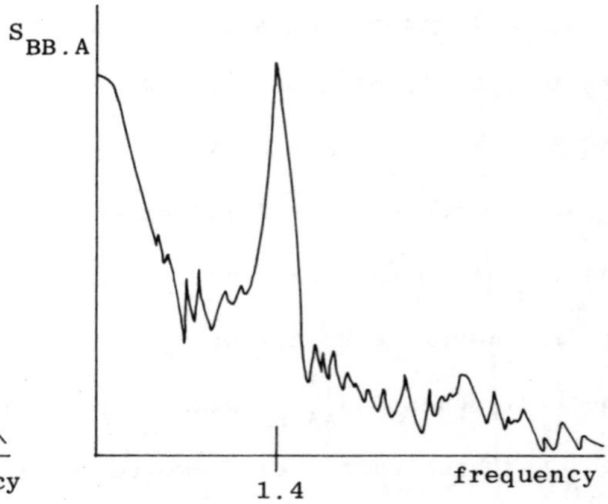

FIGURE 4

Structural Response Spectra and Residual Spectra

Again the 2.0 Hz mode has vanished: again the 1.4 Hz mode has survived.

The interpretation of these results is assisted by an investigation of the coherence of x_A and x_B which proves to be approximately zero at frequencies close to 1.4 Hz. It seems that there are similar (bending) modes of equal natural frequency in the two directions A and B and the responses in the two modes are uncoherent: evidently equation (18) is satisfied in this case. The mode whose natural frequency is 2.0 Hz is likely to be a torsional mode, and its contributions to the two responses therefore fully coherent. (The nature of these modes has recently been confirmed by beam-element simulation of the structure).

Proposed Iterative Technique

In the preceding section the operation of a straightforward conditioning technique isolated a single modal contribution. But this must be considered largely fortuitous, and we cannot rely on similar success in general. It would obviously be advantageous if we could remove a single modal contribution from our recorded spectra, yet the equations available to us offer no direct way of doing this.

A remedy may lie in the possibility of iterative application of the conditioning procedure, and in considering this we may be encouraged both by the effectiveness of the conditioning procedure, and - to some extent - the nature of the theoretical results.

Suppose that we have a number of simultaneously-taken response records $x_1, x_2, \ldots x_m$ from a given structure: suppose that we wish to eliminate from them the contribution of a single mode, the sth, say. We examine the spectra S^x_{rr} and identify the signal which is most coloured by the sth mode: suppose this to be x_1. We then condition all the remaining signals with respect to x_1, thus eliminating all constituents fully coherent with x_1, and so generate a new set of response signals $y_1, y_2, \ldots y_{m-1}$ which may be expected to be relatively deficient in their content of ξ_s. We then examine the spectra S^y_{rr} and pick out the signal most coloured by the sth mode: let this be y_1, and condition all the remaining y_r with respect to it. We then have m-2 conditioned signals of which the content of the sth mode should be considerably reduced.

The authors must admit that they have not yet found themselves in a position to try out this technique in practice. It seems promising however and worthy of a trial when a sufficient number of simultaneous recordings become available.

REFERENCES

1 A. Powell, On the fatigue of structures due to vibrations excited by random pressure fields. J. Acou.Soc. Am. 30, 1130. 1958.

2 J.D. Robson, The random vibration response of a system having many degrees of freedom. Aero Quart 17, 21. 1966.

3 J.M. Jenkins and D.G. Watts, Spectral Analysis, San Francisco: Holden Day. 1968.

4 J.S. Bendat and A.G. Piersol, Random Data Analysis and Measurement Procedures, New York: Wiley - Interscience, 1971.

5 C.J. Dodds and J.D. Robson, Partial coherence in multivariate random processes. Journal of Sound and Vibration 42, 243, 1975.

6 J.S. Bendat, Solutions for the multiple input/output problem. Journal of Sound and Vibration, 44, 311, 1976.

7 J.D. Robson, The derivation of multi variate spectral response relationships by means of the harmonic analogy. Journal of Sound and Vibration 40, 285, 1975.

Professor J.D. Robson, Rankine Professor of Mechanical Engineering,
Department of Mechanical Engineering, The University, Glasgow, G12 8QQ.
Dr. C. Dodds, Vibro Consultants Ltd., 1 Skye, East Kilbridge, Glasgow G74 2BX
Scotland.

DISCUSSION

Professor Clarkson pointed out that the result of the last process leading to Figures 3 and 4 appeared to introduce low frequency components into the spectra; he asked whether this was significant or a limitation of the procedure caused by poor signal-to-noise ratio in this frequency range? Professor Robson agreed that this was a worrying feature but said that this first program might need further refinement. He agreed that these first results were rather imperfect; they gave an impression rather than a clear indication.

The methods proposed by both Bendat and Robson showed great promise but it was considered (Clarkson) that the value would depend on the accuracy with which the data could be handled in the first place. Otherwise the various steps of addition, subtraction, multiplication and division would be manipulations of instrumentation noise. Dr. Bendat agreed that the value of the methods depended on the accuracy of the measurements; in particular the inputs must be defined accurately. He pointed out that they were evolving a technique which was conscious of instrumentation noise and actually taking it into account. Improvements could be obtained by using sufficient signal averaging. There were particular problems with highly correlated inputs to distributed systems. The measurement points must be chosen carefully in order to avoid some of the computations 'blowing up'. There are good applications providing it is possible to use good instrumentation, obtain enough data and use enough signal averaging.

In reply to a suggestion (Clarkson) that it might be useful for certain applications to express the relationships in the time domain Professor Robson agreed that this might be done.

The application of the results to transient data was questioned (Holmes). Professor Robson and Dr. Bendat agreed that the methods were not strictly applicable to non-stationary data. Depending on the nature of the transient it might be better to have the formulation in the time domain.

In answer to a further question Professor Robson pointed out that the generalised forces could be obtained easily; he thought that the fact that the coherency of the generalised coordinates was equal to the coherency of the generalised forces was a useful result. Asked (Lin) if the methods were more suited to cases where the modes were widely spaced in frequency Professor Robson replied that in such a case the methods were particularly simple to use. However, the advantage of this proposed procedure was the ability to eliminate one mode and hence the method was also suitable for the consideration of cases where resonances were close together.

Professor Crandall commented that this was the germ of an idea but he was groping for applications. Professor Robson replied that the methods should be general enough for different people to use them in their own way. He was concerned with problems where the inputs were unknown whereas Dr. Bendat was working on problems where the inputs were precisely defined. One purpose of this presentation was to expose the methods at a very early stage of development in order to attract the interest of different groups of workers.

H AKAIKE

Spectrum estimation through parametric model fitting

INTRODUCTION

The estimation of spectra forms a starting point of the analysis of a dynamic system operating within a stochastic environment. The conventional procedure of estimation of a spectrum is realized by applying some kind of smoothing operation to the periodogram defined as the Fourier transform of the sample autocovariance sequence. Although the procedure is simple and can be useful for many practical applications the decision on the choice of the smoothing operation poses a serious problem. When the smoothing is too extensive the details of the spectra will be wiped out and when the smoothing is not sufficient the reduction of statistical variability of the estimate will not be satisfactory.

The procedure of spectrum estimation to be discussed in this paper is based on the parametric model fitting procedure where a time domain model is used to express the observed time series as the output of a linear dynamic system driven by a white noise. Once the parameters of the system and the power of the white noise are specified the spectral characteristics of the output are completely determined. Thus the problem of estimation of a spectrum is essentially reduced to the problem of identification of a stochastic linear system.

TIME DOMAIN MODELS AND SPECTRA

The most general linear stochastic system defined with a finite number of parameters is the autoregressive moving average (AR-MA) model which is defined by

$$y(n)+B(1)y(n-1)+\ldots+B(M)y(n-M) = x(n)+A(1)x(n-1)+\ldots+A(L)x(n-L) \quad (1)$$

where x(n) is a white noise with zero-mean and x(n) is assumed to be independent of y(n-1), y(n-2), ... When y(n) is a stationary scalar process the power spectral density p(f) of y(n) is given by

$$p(f) = \frac{\left| 1 + \sum_{\ell=1}^{L} A(\ell)\exp(-i2\pi f\ell) \right|^2}{\left| 1 + \sum_{m=1}^{M} B(m)\exp(-i2\pi fm) \right|^2} \sigma^2 \quad (2)$$

where σ^2 denotes the variance of x(n). For a general p-dimensional vector process y(n), define A(f) and B(f) by

$$A(f) = I + \sum_{\ell=1}^{L} A(\ell)\exp(-i2\pi f\ell) \quad (3)$$

$$B(f) = I + \sum_{m=1}^{M} B(m)\exp(-i2\pi fm), \quad (4)$$

where I denotes a pxp identity matrix and $A(\ell)$ and B(m) are pxp matrices of the coefficients. The spectral density matrix P(f) of y(n) is defined by

$$P(f) = B(f)^{-1}A(f)CA(f)^*B(f)^{*-1}, \quad (5)$$

where C denotes the covariance matrix Ex(n)x(n)' and * denotes complex conjugate. The (ℓ,m)th element of P(f) defines the cross spectral density between the ℓth and mth components of y(n).

PARAMETER ESTIMATION

The procedure of determination of the parameters to be taken in this paper is to assume y(n) to be a Gaussian process and apply the method of maximum likelihood. For a series $\{y(n); n=1,2,\ldots,N\}$ of a finite number of observations the exact likelihood function takes a very complicated form but a simple and useful approximation is available when N is sufficiently large compared with the time constant of the model to be fitted.

Define the Fourier transform $Y(f)$ of $\{y(n), n=1,2,\ldots,N\}$ by

$$Y(f) = \frac{1}{\sqrt{N}} \sum_{n=1}^{N} \exp(-i2\pi fn)y(n) \tag{6}$$

Under the assumption of a Gaussian process the approximate log likelihood $L(A,B,C)$ of the AR-MA model (1) is given by

$$L(A,B,C) = -\frac{N}{2}\Big[p \log_e(2\pi) + \log_e(\det C) + \operatorname{tr}\{C^{-1} \int_{-\frac{1}{2}}^{\frac{1}{2}} X(f)X(f)^* \, df\}\Big] \tag{7}$$

where $X(f) = A(f)^{-1} B(f)Y(f)$ and $A(f)$ and $B(f)$ are defined by (3) and (4). The symbols det and tr denote determinant and trace, respectively. For a given set of coefficients $A(\ell)$ and $B(m)$ the C which maximizes $L(A,B,C)$ is given by

$$C(A,B) = \int_{-\frac{1}{2}}^{\frac{1}{2}} X(f)X(f)^* \, df. \tag{8}$$

The corresponding value of $L(A,B,C)$ is given by

$$L(A,B) = -\frac{N}{2} \log_e \{\det C(A,B)\} + K, \tag{9}$$

where K is a constant independent of $A(\ell)$ and $B(m)$. For the purpose of comparison of Gaussian models the value of K is immaterial and hereafter the maximum log likelihood will be defined as the value of $L(A,B)$, maximized with respect to $A(\ell)$ and $B(m)$, less K. The values of $A(\ell)$, $B(m)$ and C which maximize $L(A,B,C)$ will be denoted by $\hat{A}(\ell)$, $\hat{B}(m)$ and $\hat{C}$, respectively.

In Figure 1 are illustrated the estimates of the power spectrum of a simulated record of car vibration. The estimates were obtained by fitting AR models with successively increasing orders. By comparing the estimates with the theoretical spectrum it becomes obvious that the decision on the order is crucial in making a parametric model fitting procedure practically useful.

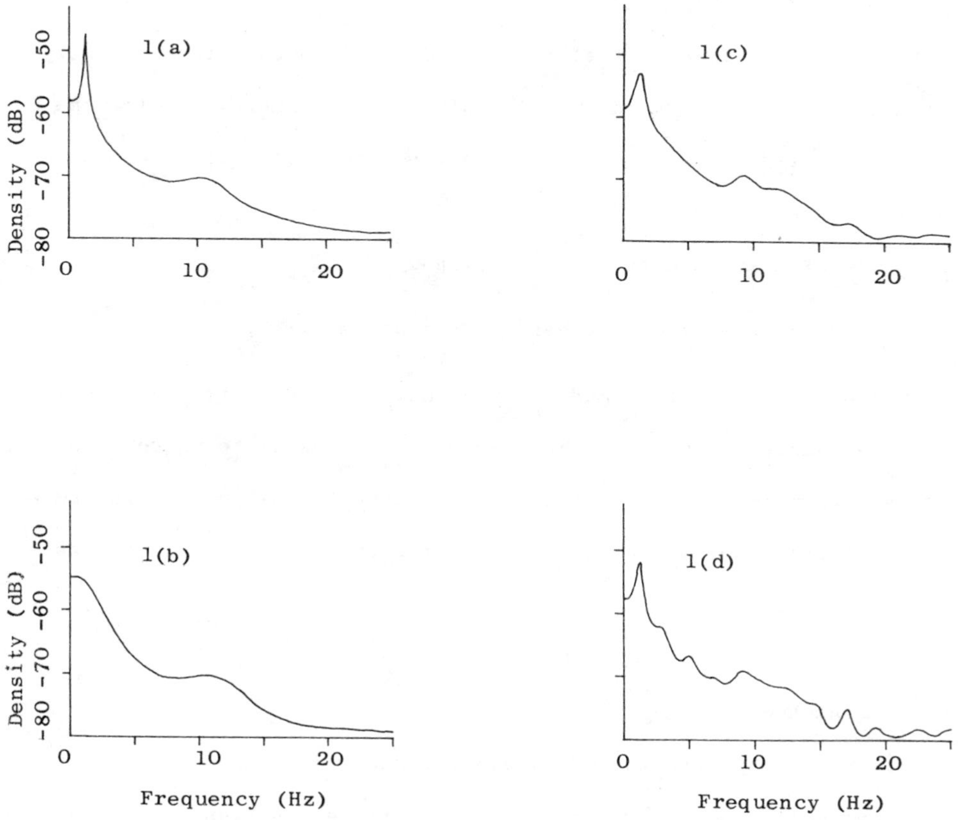

1(a) Theoretical Spectrum
AR 5 MA 4

1(b)-(d) Estimated Spectra
Data length N = 900
(b) AR 6 (c) AR 18
(d) AR 24

Figure 1 Effect of changing orders in AR model fitting

AIC CRITERION OF FIT

As a measure of badness of fit of a model obtained by the method of maximum likelihood an information criterion AIC is defined by

$$AIC = (-2) \log_e (\text{maximum likelihood}) + 2(\text{number of free parameters}), \tag{10}$$

where the number of free parameters denotes the number of independently adjusted parameters within the model. AIC is an estimate of twice the neg-entropy of the true structure with respect to the fitted model and a low value of AIC means a good fit.

For a Gaussian AR-MA model AIC is defined by

$$AIC = N \log_e(\det C) + 2(\text{number of free parameters}), \tag{11}$$

The values of AIC given in Figure 1 show that the model with the minimum AIC gives a reasonable estimate of the spectrum. The estimate which gives the minimum AIC within a limited number of estimates is called the minimum AIC estimate (MAICE). For the discussions of the use of AIC in statistical model identification, see [1,2].

NUMERICAL EXAMPLES

An example of the application of parametric model fitting by AIC for the spectrum estimation of car vibration is illustrated in Figure 2. The AR-MA models are the MAICE's within the AR-MA and AR models fitted to the data. On the right column are illustrated the estimates obtained by smoothing the periodogram by the Hanning window described by Blackman and Tukey [3]. The inherent difficulty in choosing the maximum lag, or the number of sample autocovariances used for the spectrum estimation, is quite apparent.

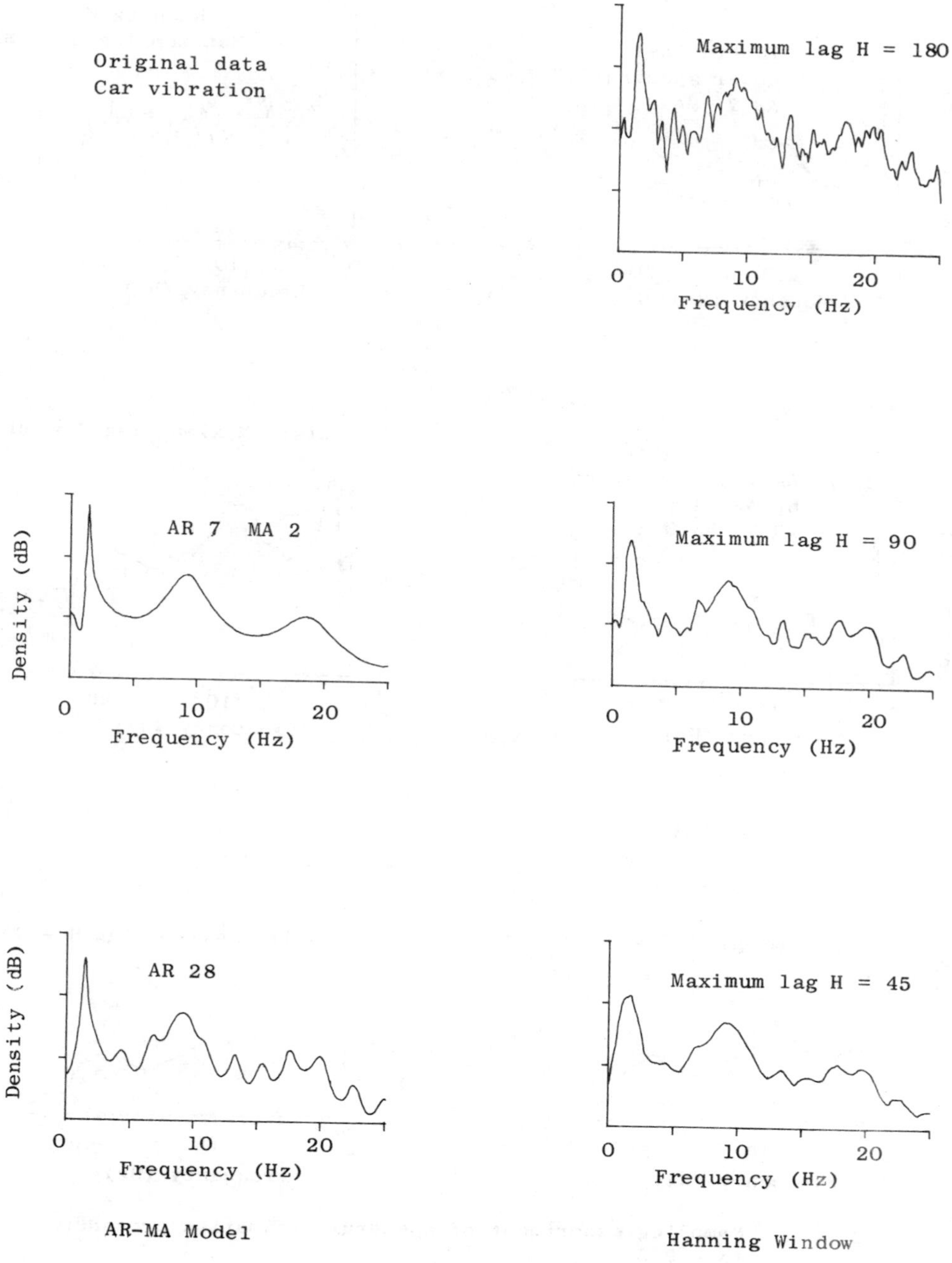

Figure 2 Spectrum estimation of car vibration (N = 900)

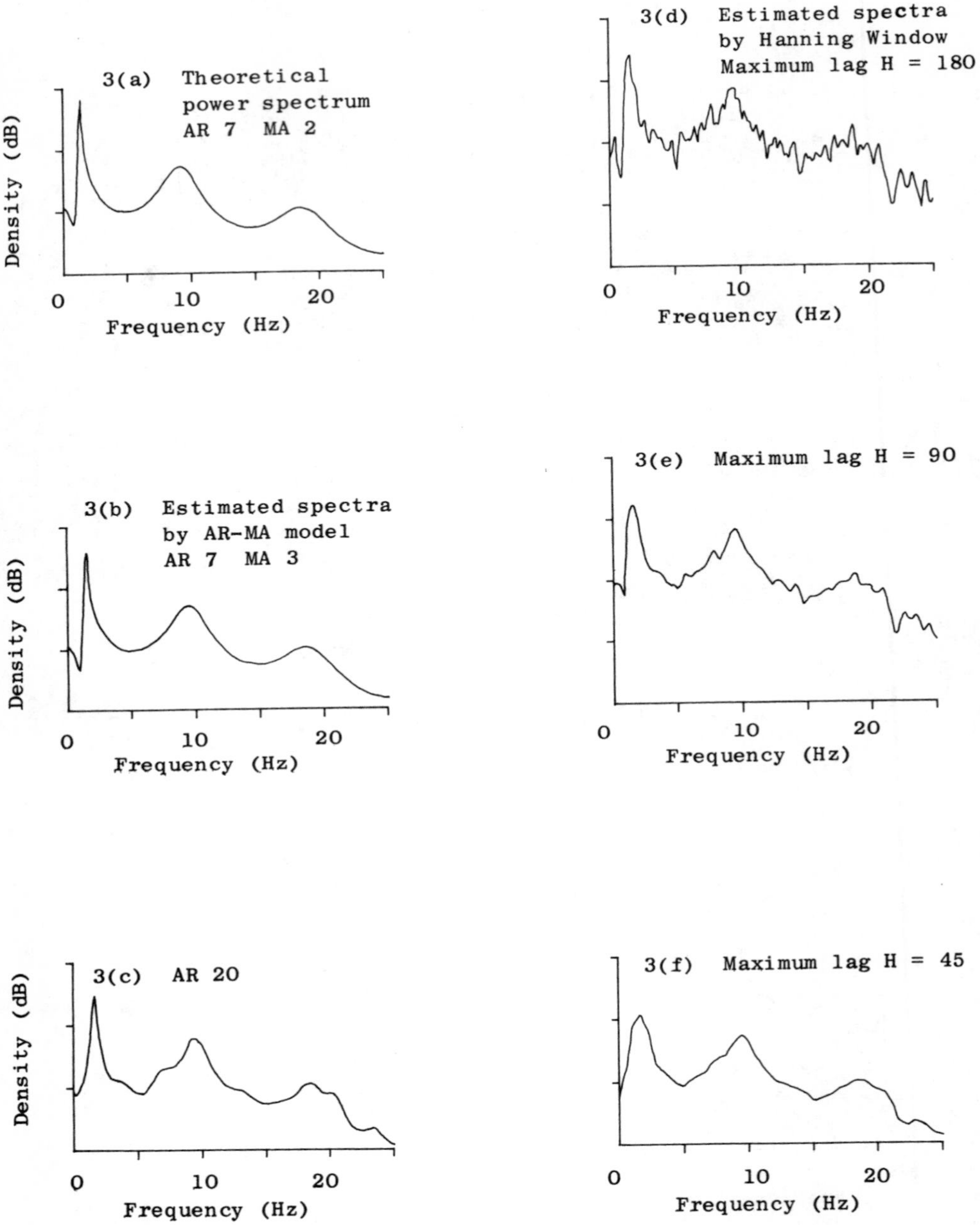

Figure 3 Sampling experiment of spectrum estimation procedures (N = 900)

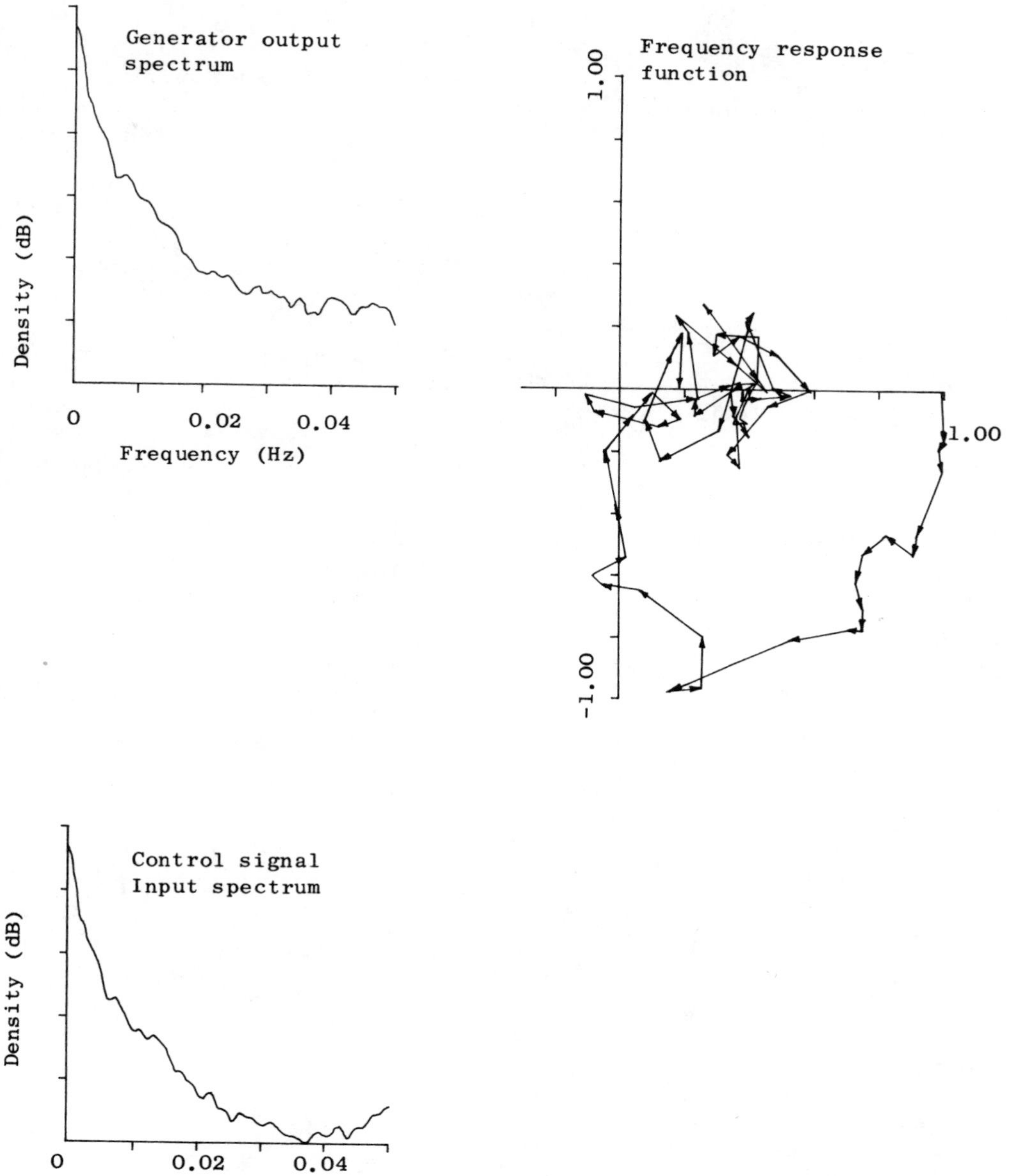

Figure 4 Frequency response analysis by spectrum windowing (N = 1600)

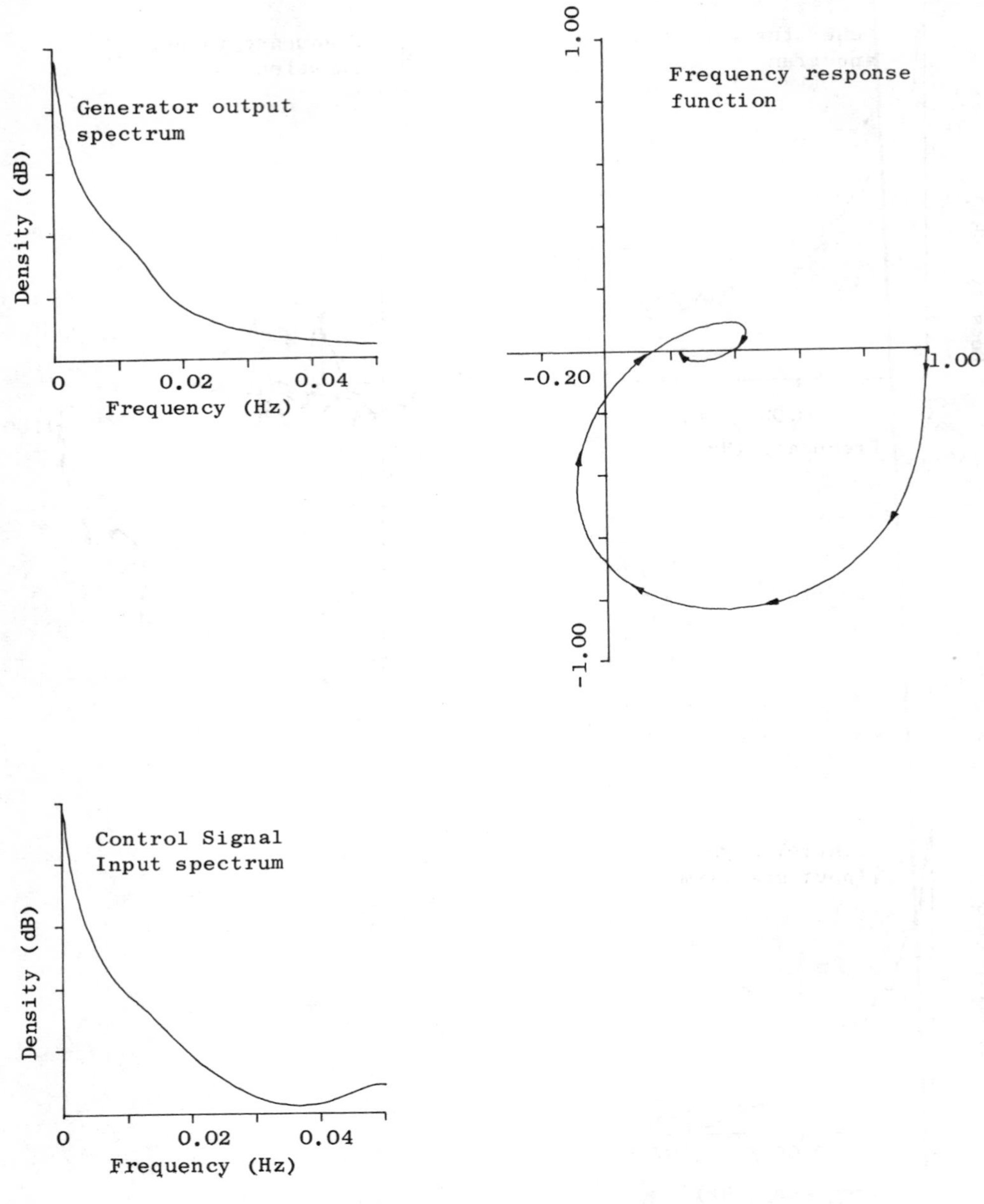

Figure 5 Frequency response analysis by AR model fitting (N = 1600)

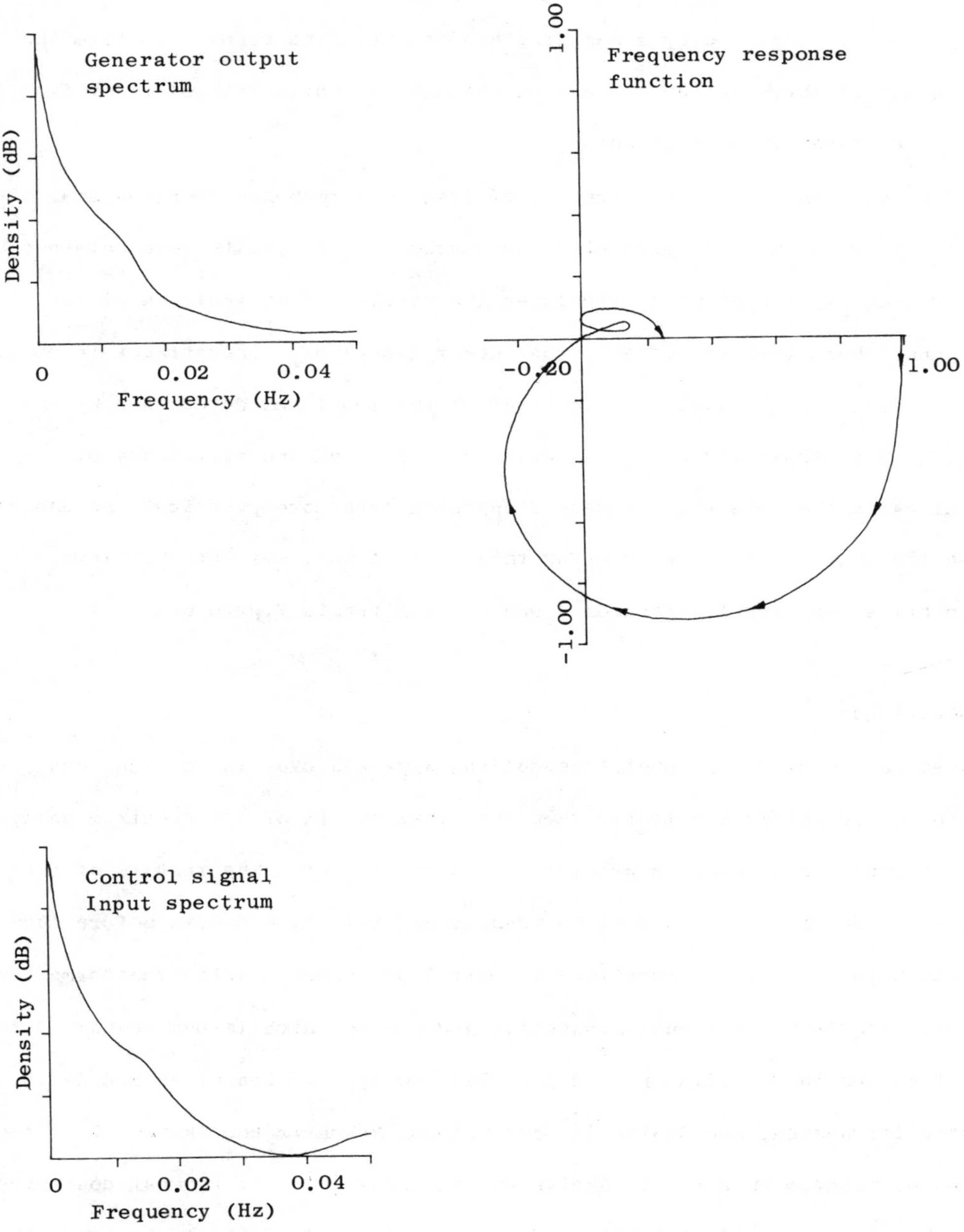

Figure 6 Theoretical spectra corresponding to Figures 4 and 5

In Figure 3 are illustrated the results corresponding to those of Figure 2, which were obtained by a sampling experiment. The results confirm the advantage of the spectrum estimation through the parametric model fitting over the conventional procedure.

Figures 4 and 5 show the results of frequency response function analysis by the conventional and parametric approaches. The results were obtained by a sampling experiment which simulates the results of an analysis of the response characteristic of an actual power generator. The parametric model used here is an AR model and the order of the model was determined by the MAICE. The simplicity of the required decision and the efficiency of the final estimates make the parametric approach much more practical and useful than the conventional approach for this type of multi-variate problems. Theoretical spectra for Figures 4 and 5 are given in Figure 6.

DISCUSSIONS

The advantage of the parametric modelling approach over the conventional approach becomes more definite when the final object of the spectrum analysis is the implementation of a predictor or a controller. The results of the frequency domain analysis must be transformed into time domain before they are used to develop a prediction or control procedure. This transformation usually requires additional subjective judgements which is unnecessary for the time domain modelling procedure. For the application of AR models to controller design, see Akaike [4] and Ootomo, Nakagawa and Akaike [5]. A computer program is given in Akaike and Nakagawa [6]. It is most convenient for the assessment of statistical characteristics of the estimates that the time domain model provides a convenient framework for the necessary sampling experiment.

The most significant disadvantage of the conventional procedure is the complete lack of the objective method of determination of the maximum lag or the window width. In spite of this disadvantage the conventional procedure is rather free from the problem of bias usually inherent in the model fitting procedure and can be useful as a supplementary procedure to the time domain modelling. The AR model has an efficient procedure of fitting and at present will be the most practically useful in many applications. The AR-MA model obviously produces better results. The only problem with this model is the difficulty of the maximum likelihood computation. Since the basic difficulty of identifiability of a multivariate AR-MA model is given a complete solution [7] and a practical procedure of the maximum likelihood computation is already developed [8,9] it is hoped that the model will eventually become the standard for most of the practical applications.

CONSISTENCY OF ORDER DETERMINATION

One of the recurring questions on the performance of MAICE in fitting AR and AR-MA models is the consistency of the order determination of the models. Since the definition of AIC is based on the asymptotic distribution of the sample covariances it is obvious that the behaviour of AIC will exhibit a stationary distribution as the data length N is increased indefinitely, when real AR or AR-MA process data are treated and models with orders equal to or higher than the true order are fitted. This observation shows clearly that the order of MAICE cannot be a consistent estimate of the true order even when the original process has a clearly defined finite order.

Although the order is not consistently estimated the model chosen by MAICE provides a consistent estimate of the structure of the true process. This is obvious from the consistency of the sample autocovariances and in fact a precise mathematical proof of this fact has been given by Shibata [10] for the case of AR model fitting.

It is rather easy to modify AIC so that MAICE would give a consistent estimate of the order when applied to a set of data from a real finite order process. For scalar processes, a reasonable modification of AIC with this property is given by

$$\text{BIC} = (N-k)\log_e\{S(k)\cdot \frac{N}{N-k}\} + k\log_e\{(S(0)-S(k))\frac{N}{k}\}, \tag{12}$$

where N is the number of observations, S(k) denotes the maximum likelihood estimate of the variance σ^2 of the white noise x(n) and k = L+M, the number of the coefficients in (1). The motivation for the modified information criterion BIC will be discussed elsewhere [11].

In practical applications of the model fitting procedure it is quite probable that the actual process demands an infinite order model. This is the case when an AR model is fitted to an AR-MA process with non-trivial MA part. In such situations, by asking the consistency of the order determination the procedure tends to pick a model of excessively low order. Experimentally it has been observed that the difference of the predictive ability of the models obtained by the original and modified information criteria is very small, yet it seems that the original AIC would be more useful for ordinary exploratory analyses and BIC will be more suitable for routine applications where the consistency of results in repeated applications is of more importance than the details of the spectra. Comparison of the spectra obtained by using AIC and BIC would certainly be quite useful even in the case of an exploratory analysis of a single set of data. Much further experience should be accumulated to make the most of these information criteria in the fitting of parametric time domain models.

CONCLUSION

The feasibility of spectrum estimation through parametric model fitting in the time domain is discussed and demonstrated by numerical examples. The crucial problem of order determination finds a solution by using the concept of AIC and MAICE. The advantage of the parametric modelling over the conventional periodogram smoothing is clearly demonstrated by the numerical examples. With the improvement of the efficiency of the maximum likelihood computation the parametric modelling approach will eventually replace the conventional periodogram smoothing approach. Although the time domain modelling procedure by MAICE is already sufficiently useful to warrant its general use, further accumulation of experience of the comparative use of the information criteria AIC and BIC will certainly lead to a better understanding of the statistical characteristics of estimates of the time domain models.

ACKNOWLEDGEMENT

The author is grateful to Mr H. Nakamura of Kyushu Electric Company for allowing the use of the original record of the power generator.

REFERENCES

1 H. Akaike — A new look at the statistical model identification. IEEE Trans.Automat. Contr. AC-19 716-723, (1974).

2 H. Akaike — Information theory and an extension of the maximum likelihood principle. 2nd International Symposium on Information Theory, Akademiai Kiado, Budapest, 267-281, (1973).

3 R.B. Blackman and J.W. Tukey — The measurement of power spectra. Dover, New York (1958).

4 H. Akaike Autoregressive model fitting for control. Ann. Inst. Statist. Math. 23 163-180, (1971).

5 T. Ootomo, T. Nakagawa and H. Akaike Statistical approach to computer control of cement rotary kilns. Automatica 8, 35-48, (1972).

6 H. Akaike and T. Nakagawa Statistical Analysis and Control of Dynamic Systems (In Japanese, with a list of computer program packages for AR model fitting) Saiensu-sha Tokyo (1972).

7 H. Akaike Markovian representation of stochastic processes and its application to the analysis of autoregressive moving-average processes. Ann. Inst. Statist. Math. 26. 363-387 (1974).

8 H. Akaike, E. Arahata and T. Ozaki A time series analysis and control program package (1). Computer Science Monographs No.5, Inst. Statist. Math., TIMSAC-74 (1975).

9 H. Akaike Canonical correlation analysis of time series and the use of an information criterion. System Identification: Advances and Case Studies. Academic Press.

10 R. Shibata Selection of the order of an autoregressive model by Akaike's information criterion. Biometrika 63, 117-126, (1976).

11 H. Akaike On entropy maximization principle. Proc. Symposium on Application of Statistics, Dayton, Ohio, June, (1976).

Professor H. Akaike, The Institute of Statistical Mathematics, 4-6-7 Minami-Azabu Minato-Ku, Tokyo, Japan.

Professor Kohei Otsu, The Tokyo University of Mercantile Marine, 2-1 Echujima, Fukagawa, Koto-ku, Tokyo 135.

Mr. G. Kitagawa, The Institute of Statistical Mathematics, 4-6-7 Minami-Azabu Minato-ku, Tokyo 106.

DISCUSSION

In the discussion it was emphasized that there were no 'a priori' assumptions made concerning the form of the spectrum built into the estimation procedure. In developing the AIC criterion some assumptions are made, however, in relation to the asymptotic behaviour of the maximum likelihood estimate, which are commonly accepted. It was noted though, (Mark, Holmes) that one should use any specific advance knowledge (in the choice of AR-MA model) for example by emphasising the importance of parameters like damping ratios and natural frequencies characterising spectra arising from vibrating systems.

It was agreed (Clarkson) that data from a phenomenon having a multimode character (e.g. car vibration) might pose problems for the time domain procedure presented. In fact the results produced would probably be a smoothed version of the true spectrum which might tend to explain the results in terms of an (over) simple model. The author indicated that the procedure on such data had been to filter the data and apply the method to band-limited realizations. In relation to the frequency response estimation of Figures 5 and 6 where in Figure 5 the estimated frequency response function had lost (Holmes) the 'loop' of Figure 6, it was pointed out that this was accounted for by the input power being very low at these frequencies.

With reference to the confidence intervals that may be computed when using the Blackman-Tukey approach (Mark) the author said that confidence intervals concerning the estimate could easily be obtained for this procedure if one assumed that the AR-MA model was correct since then the asymptotic behaviour of the coefficients is known. The problem was in the determination of the correct order of the model.

Following comments (Bendat) that no mention had been made of the widely used frequency domain methods developed in the last decade, the author felt that there had been no significant improvement in statistical estimation theory in the frequency domain during this period even though there had been considerable improvements in computational procedures. Also the time domain approach ensured the availability of all the measures usually required, e.g. mean square response, cross spectra, coherency. However it was agreed that an advantage of the frequency domain approach is that it is model free and that a parametric model might ignore some portion of the true spectrum, so that in some circumstances one might use both procedures to help obtain the best results.

S H CRANDALL

Structured response patterns due to wide-band random excitation

INTRODUCTION

When lightly damped uniform continuous structures are excited by wide-band stationary random forces applied at isolated points, the spatial distributions of the mean square velocity responses exhibit surprisingly regular patterns if the number of modes responding is sufficiently large. This was demonstrated for taut strings and rectangular plates in [1] and for circular plates in [2]. Similar results have also been obtained for rectangular membranes [3] and for square plates, equilateral triangular plates and certain trapezoidal shaped plates [4]. In all these cases, if the excitation includes enough modes (enough is of the order of a dozen or so for a one-dimensional structure, or a hundred or more for a two-dimensional structure), the response mean square velocity is substantially uniform over most of the structure with certain spots (on one-dimensional structures) or lanes (on two-dimensional structures) in which there is enhanced response. Depending on the nature of the boundary conditions, lanes along the boundaries may have either enhanced or diminished response. In the case of plates and membranes the lanes of enhanced response can be observed by sprinkling salt grains over the surface [1]. When the broad band excitation level is gradually increased the salt grains in the lanes of enhanced response vibrate most actively and migrate away toward areas of lesser response leaving cleared strips behind them. In Figure 1 the location of the lanes of enhanced response are sketched for plates of various shapes. In each case the location of the driving point is indicated by a small circle. The relative enhancement of

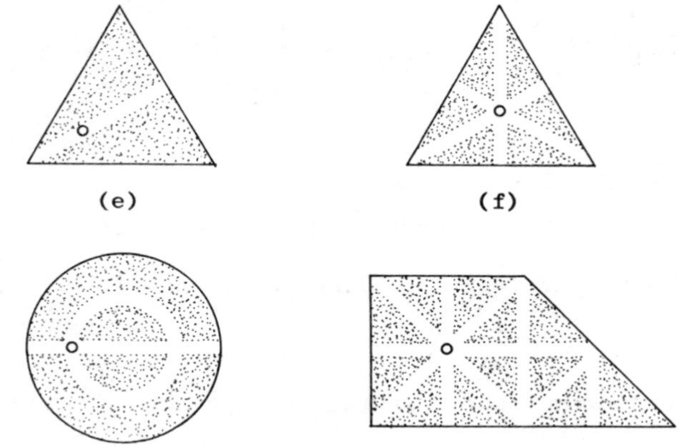

(a) (b) (c) (d) (e) (f) (g) (h)

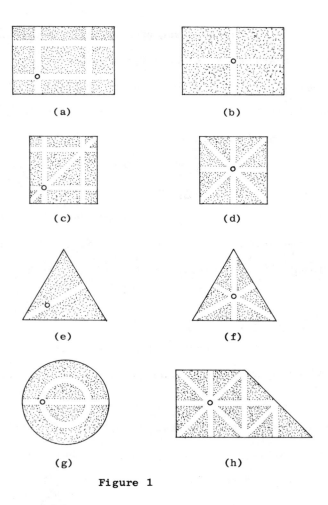

Figure 1

the mean square velocity response in the lanes as compared with the nearly uniform background response has been predicted analytically for all of the shapes in Figure 1 except (h). Direct measurement has confirmed the predictions in cases (a), (b) and (g). In [2] and [3] mean square displacement and mean square acceleration distributions were obtained in addition to mean square velocity distributions.

In the present paper attention will be limited to the mean square velocity response. The main purpose is to document the image-sum method for calculating the response distribution pattern. This method may be used to estimate the absolute magnitude of the mean square response at all points of systems for which a complete set of image sources can be established. For this purpose the image-sum method is an interesting alternative to the modal-sum method described earlier [1]. A major advantage of the image-sum procedure is the ease with which it can be used to predict the location and relative enhancement of the lanes of enhanced response. The image-sum method is not however as general as the modal-sum procedure because not all shapes (e.g., circular shapes) allow an image-sum expansion. General results for strings, beams, membranes and plates will be stated, although for simplicity of exposition the detailed development is limited to the case of the taut string.

CONDUCTANCE PARAMETERS

The velocity admittance Y of a specified point on a structure due to a harmonic force applied at another specified point generally is complex; i.e.,

$$Y = G + iB \tag{1}$$

where G is the conductance and B is the susceptance. It will be convenient to employ two conductance parameters G_0 and G_∞ to characterize the finite uniform structures under discussion. The parameter G_0 is the driving point conductance that would be obtained if the structure were perfectly rigid and free of any supports, and it was undergoing single degree of freedom transverse oscillations under the influence of inertia, viscous damping and a transverse exciting force. In the case of a string or beam of length L with a viscous damping coefficient c_1, force per unit length/unit velocity, the rigid body conductance is

$$G_0 = \frac{1}{c_1 L} \tag{2}$$

while for a membrane or plate of area A and viscous damping coefficient c_2, force per unit area/unit velocity, the rigid body conductance is

$$G_0 = \frac{1}{c_2 A} \tag{3}$$

The parameter G_∞ is the driving point conductance that would be obtained if the structure were totally undamped and infinite in extent; i.e., G_∞ is the radiation conductance for point excitation of an infinite undamped structure having the same local inertia and stiffness properties as the actual structure. For a string with uniform tension T_0 and mass per unit length μ_1 the radiation conductance is

$$G_\infty = \frac{1}{2(T_0\mu_1)^{\frac{1}{2}}} \tag{4}$$

For a Bernoulli-Euler beam with uniform flexural modulus EI and mass per unit length μ_1 the radiation conductance is

$$G_\infty = \frac{1}{4(EI\mu_1^3\,\omega^2)^{\frac{1}{4}}} \tag{5}$$

where ω is the circular frequency. For a membrane with uniform surface tension T_1 and mass per unit area μ_2 the radiation conductance is

$$G_\infty = \frac{\omega}{4T_1} \tag{6}$$

For a thin plate with uniform flexural modulus D and mass per unit area μ_2 the radiation conductance is

$$G_\infty = \frac{1}{8(D\mu_2)^{\frac{1}{2}}} \tag{7}$$

Note that for the string and for the plate the radiation conductance is independent of frequency. When we deal with a band-limited white noise excitation with uniform spectral density extending from $\omega = 0$ out to a cut-off frequency ω_c it will be convenient to introduce a frequency-averaged conductance

$$\bar{G}_\infty = \frac{1}{\omega_c}\int_0^{\omega_c} G_\infty \, d\omega \tag{8}$$

For the string and the plate $\bar{G}_\infty = G_\infty$ while for the beam

$$\bar{G}_\infty = \frac{1}{2(EI\mu_1^3\,\omega_c^2)^{\frac{1}{4}}} \tag{9}$$

and for the membrane

$$\bar{G}_\infty = \frac{\omega_c}{8T_1} \tag{10}$$

Note that both G_0 and G_∞ have been defined so as to be independent of the location of the driving point.

It turns out that under wide-band excitation the effective rms velocity admittance for lightly damped structures, except in restricted lanes of enhanced or diminished response, is simply the geometric mean of G_0 and $\bar{G}_\infty$. For this result to hold it is necessary for $\bar{G}_\infty$ to be small in comparison to G_0. The requirement that $\bar{G}_\infty/G_0 << 1$ is essentially a quantification of the term 'light damping'. Further insight into this requirement is obtained by translating it into a requirement on the vibration modes of the structure.

MODAL DENSITY AND NOISE BANDWIDTH

For a uniform string of length L with fixed ends the number of modes with natural frequencies smaller than ω is the largest integer smaller than

$$N(\omega) = \frac{L\omega}{\pi}\left(\frac{\mu_1}{T_0}\right)^{\frac{1}{2}} \tag{11}$$

The modal density if

$$n = \frac{dN}{d\omega} = \frac{L}{\pi}\left(\frac{\mu_1}{T_0}\right)^{\frac{1}{2}} \tag{12}$$

and the modal separation or modal spacing is

$$\Delta\omega = \frac{1}{n} = \frac{\pi}{L}\left(\frac{T_0}{\mu_1}\right)^{\frac{1}{2}} \tag{13}$$

Since the modes of a uniform string with fixed ends are equally spaced on the frequency axis the modal density and modal spacing are independent of frequency. For other boundary conditions this may no longer be precisely so but the results (12) and (13) become asymptotically correct for large ω. For a beam of length L with simply supported ends the number of modes with natural frequencies smaller than ω is the largest integer smaller than

$$N(\omega) = \frac{L}{\pi}\left(\frac{\mu_1\omega^2}{EI}\right)^{\frac{1}{4}} \tag{14}$$

The modal density

$$n = \frac{dN}{d\omega} = \frac{L}{2\pi}\left(\frac{\mu_1}{EI\omega^2}\right)^{\frac{1}{4}} \tag{15}$$

is a function of frequency. Its average value over the frequency range from $\omega = 0$ to $\omega = \omega_c$ is

$$\bar{n} = \frac{1}{\omega_c}\int_0^{\omega} n d\omega = \frac{L}{\pi}\left(\frac{\mu_1}{EI\omega_c^2}\right)^{\frac{1}{4}} \tag{16}$$

and an average or effective modal spacing can be defined as

$$\overline{\Delta\omega} = \frac{1}{\bar{n}} = \frac{\pi}{L}\left(\frac{EI\omega_c^2}{\mu_1}\right)^{\frac{1}{4}} \tag{17}$$

For other boundary conditions these results are not strictly correct but become asymptotically correct for large ω and large ω_c. For both the string and the beam, if the uniform viscous damping coefficient is c_1, the noise bandwidth [5] of all modes is

$$b = \frac{\pi}{2}\frac{c_1}{\mu_1} \tag{18}$$

For a membrane of area A the precise number of modes with natural frequencies smaller than ω is an irregular staircase function depending on the precise shape of the membrane and on the precise nature of the boundary conditions. Asymptotically for large ω

$$N(\omega) = \frac{A}{4\pi} \frac{\mu_2\omega^2}{T_1} \tag{19}$$

independently of these details [6]. The asymptotic modal density

$$n = \frac{dN}{d\omega} = \frac{A}{2\pi} \frac{\mu_2\omega}{T_1} \tag{20}$$

has the average value

$$\bar{n} = \frac{A}{4\pi} \frac{\mu_2\omega_c}{T_1} \tag{21}$$

over the frequency range from $\omega = 0$ to $\omega = \omega_c$. The effective average modal spacing is

$$\overline{\Delta\omega} = \frac{1}{\bar{n}} \frac{4\pi}{A} \frac{T_1}{\mu_2\omega_c} \tag{22}$$

Similarly, for a plate of area A the asymptotic number of modes with natural frequencies less than ω is [6]

$$N(\omega) = \frac{A}{4\pi} \left(\frac{\mu_2\omega^2}{D}\right)^{\frac{1}{2}} \tag{23}$$

The asymptotic modal density

$$n = \frac{dN}{d\omega} = \frac{A}{4\pi} \left(\frac{\mu_2}{D}\right)^{\frac{1}{2}} \tag{24}$$

and effective average modal spacing

$$\Delta\omega = \frac{1}{n} = \frac{4\pi}{A} \left(\frac{D}{\mu_2}\right)^{\frac{1}{2}} \tag{25}$$

are independent of frequency. For both the membrane and the plate, if the uniform viscous damping coefficient is c_2, the noise bandwidth of all modes is

$$b = \frac{\pi}{2} \frac{c_2}{\mu_2} \tag{26}$$

For all four types of structures it is possible to relate the conductance ratio $\bar{G}_\infty/G_0$ to the modal overlap ratio $b/\overline{\Delta\omega}$. In every case

$$\frac{\bar{G}_\infty}{G_0} = \frac{b}{\overline{\Delta\omega}} \tag{27}$$

For the string (27) follows from equations 2, 4, 13 and 18. For the beam (27) follows from equations 2, 9, 17 and 18. For the membrane (27) follows from equations 3, 10, 22 and 26 and for the plate (27) follows from equations 3, 7, 25 and 26. The light damping requirement $\bar{G}_\infty/G_0 << 1$ thus translates into the requirement that there be little modal overlap. On the average the modal separation should be much larger than the noise bandwidth of a single mode.

The underlying identity which is responsible for (27) is the proportionality of asymptotic modal density to radiation conductance. If m is the total mass of the structure ($m = \mu_1 L$ or $m = \mu_2 A$) then in all four cases

$$n(\omega) = \frac{dN}{d\omega} = \frac{2}{\pi} \; m \; G_\infty(\omega) \tag{28}$$

This follows from (4) and (12) for the string, from (5) and (15) for the beam, from (6) and (20) for the membrane, and from (7) and (24) for the plate. The relation (28) has been noted previously in connection with statistical energy analysis [7, 8].

MODAL SUM PROCEDURE

If a uniform string of length L with fixed ends is excited by a stationary wide-band random force applied at x = a the stationary mean square velocity response at a general location x is given [1] by the double sum

$$E\left[v^2\right] = \sum_{j=1}^{\infty} \sum_{k=1}^{\infty} \psi_j(x) \; \psi_k(x) \; \psi_j(a) \; \psi_k(a) \; I_{jk} \tag{29}$$

where the natural modes are

$$\psi_j(x) = \sqrt{2} \sin j\pi x/L \tag{30}$$

and the symbol I_{jk} stands for the modal interaction integral

$$I_{jk} = \int_{-\infty}^{\infty} H_j(-\omega)\ S_f(\omega)\ H_k(\omega)\ d\omega \tag{31}$$

where $H_j(\omega)$ is the frequency response of the j-mode velocity and $S_f(\omega)$ is the (two-sided, circular frequency) mean square spectral density of the exciting force. When $G_\infty/G_0 << 1$ there is negligible modal overlap and $I_{jk} \simeq 0$ for $j \neq k$. If the excitation is band-limited white noise with $S_f(\omega) = S_0$ for $|\omega| < \omega_c$ and $S_f(\omega) = 0$ for $\omega_c < |\omega|$, with $N(\omega_c) = N_c$, the infinite double sum (29) can be approximated by the finite single sum [1]

$$E\left[v^2\right] \simeq \sum_{j=1}^{N_c} \psi_j^2(x)\ \psi_j^2(a)\ I_{jj}$$

$$= \frac{\pi S_0}{\mu_1 c_1 L^2} \sum_{j=1}^{N_c} \psi_j^2(x)\ \psi_j^2(a) \tag{32}$$

$$= E\left[f^2\right]\ G_0 G_\infty \frac{1}{N_c} \sum_{j=1}^{N_c} \psi_j^2(x)\ \psi_j^2(a)$$

where $E\left[f^2\right] = 2\ S_0\omega_c$ is the mean square driving force and equations 2, 4 and 11 have been used to obtain the final form. The function

$$F_N(x;a) = \frac{1}{N} \sum_{j=1}^{N} \psi_j^2(x)\ \psi_j^2(a) \tag{33}$$

is displayed in Figure 2 for the case N = 20, a = L/3. The function $F_N(x;a)$ is symmetric about the string centre, x = L/2. For large N, the magnitude of $F_N(x;a)$ fluctuates in the neighbourhood of unity for all x except for values of x in the vicinity of the end points (x = 0 and x = L), the drive point x = a, and the image point x = L-a. At the latter two points the asymptotic

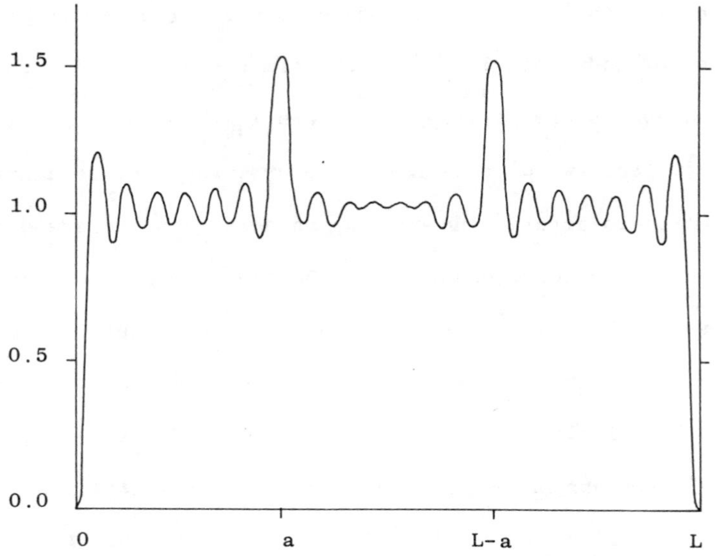

Figure 2

limit of the locally enhanced value of F_N is 1.5. The width of the zones of enhanced response are of order L/N. Except in these exceptional zones we then have the asymptotic result

$$\frac{E\left[v^2\right]}{E\left[f^2\right]} \simeq G_0\, G_\infty \tag{34}$$

or that the effective rms admittance of the string is the geometric mean between the rigid body conductance G_0 and the radiation conductance G_∞. A similar result also applies to uniform beams, membranes [3] and plates [1]. See equation (57). For two-dimensional structures the number of modes required for a given degree of approach to the asymptotic limit is of the order of the square of the number of modes required for a one-dimensional structure; e.g., the response pattern developed on a plate with 400 modes would have roughly the same degree of definition as the response pattern developed on a string with 20 modes.

The justification for passing from the double sum (29) to the single sum (32) was absence of modal overlap. For uniform one-dimensional structures the natural frequencies are well-separated and $I_{jk} \simeq 0$ for $j \neq k$ is assured by light damping. For two-dimensional structures the precise distribution of natural frequencies is erratic and depends in detail on the shape of the structure and on the boundary conditions. On the average the asymptotic modal density is maintained in the high-frequency range but instances of repeated frequencies or near equality of frequencies can occur. If these occurrences are relatively rare in the totality of modes entering into a sum of the form (29), the approximation provided by (32) is still useful. If, however, due to some special symmetry of the structure there is a systematic occurrence of repeated roots it will be necessary to add to (32) the contributions of those terms in (29) for which $\omega_j = \omega_k$. A good example of this is provided by the square plate. Here modes with p nodal lines parallel to x and q nodal lines parallel to y have the same frequencies as modes with q nodal lines parallel to x and p nodal lines parallel to y. If only the terms corresponding to (32) are summed the pattern predicted for a square plate is the same tic-tac-toe pattern as that for an arbitrary rectangular plate (see Figure 1a). This result is essentially correct for most points of driving-force application. The additional terms for $\omega_j = \omega_k$ contribute little. When however, the driving-force is applied along a diagonal of the square, there appears a diagonal lane of enhanced response (see Figure 1c) which is not predicted by (32) but is represented by the additional terms from (29) for which $\omega_j = \omega_k$. When the driving point is moved to the centre of the square (see Figure 1d) both diagonals of the union-jack pattern are represented by these additional terms.

IMAGE SUM PROCEDURE

The image sum procedure is described here for the case of a point-excited taut string. The summation can be performed in the time domain or in the frequency domain. The time domain is considered first. The basic idea here is that the stationary random force process applied at $x = a$ can be modelled by a sequence of pulses of fixed shape, but random amplitudes, occurring at random times. The response velocity at a general location x can also be modelled by a similar random sequence. Furthermore the statistical properties of the response depend in a straight forward way on the solution of the deterministic problem of finding the response time history for a single application of the excitation pulse. The method of images is employed to solve this deterministic problem.

Let p(t) be the fixed excitation pulse shape. The stationary random force process is then represented [9] by the sum

$$f(t) = \sum_i a_i \, p(t-t_i) \tag{35}$$

where the a_i are independent random variables with mean zero and variance σ_a^2 and the t_i are Poisson distributed random times with expected arrival rate ν. The auto correlation function of f(t) is [9]

$$R_f(\tau) = E\left[f(t)\ f(t+\tau)\right] = \nu\sigma_a^2 \int_{-\infty}^{\infty} p(t)\ p(t+\tau)dt \tag{36}$$

and the mean square spectral density of f(t) is

$$S_f(\omega) = \frac{1}{2\pi} \int_{-\infty}^{\infty} R_f(\tau)\ e^{-i\omega\tau}\ d\tau = \frac{\nu\sigma_a^2}{2\pi}\ |P(\omega)|^2 \tag{37}$$

where $P(\omega)$ is the Fourier transform of p(t).

$$P(\omega) = \int_{\infty}^{\infty} p(t)\ e^{-i\omega t}dt \tag{38}$$

For example, if

$$p(t) = \left[\frac{2S_0}{\pi\nu\sigma_a^2}\right]^{\frac{1}{2}} \frac{\sin \omega_c t}{t} \tag{39}$$

then, according to equation 36,

$$R_f(\tau) = 2S_0 \frac{\sin \omega_c \tau}{\tau} \tag{40}$$

and according to equation 37,

$$S_f(\omega) = \begin{cases} S_0, & |\omega| < \omega_c \\ 0, & \omega_c < |\omega| \end{cases} \tag{41}$$

Note that the pulse (39) has a single high peak at $t = 0$ with duration of order $2\pi/\omega_c$ surrounded by symmetric decaying oscillations.

If the velocity response at a location x due to a single excitation pulse p(t) applied at x = a is q(t) then the stationary random velocity response at x due to the random force process (35) applied at x = a is

$$v(t) = \sum_i q_i\ a(t-t_i) \tag{42}$$

The auto correlation function and spectral density of v(t) may be obtained from formulas corresponding to equations 36 and 37. In particular, the mean square velocity is

$$E\left[v^2\right] = \nu\sigma_a^2 \int_{-\infty}^{\infty} q^2(t)\ dt \tag{43}$$

The method of images is used to determine the response q(t) to the pulse p(t). Instead of a finite string of length L and a single excitation point we consider an infinite string with an infinite set of image excitation points as indicated in Figure 3. To model fixed boundary points at x = 0 and x = L the senses of the image sources must alternate as shown. At each

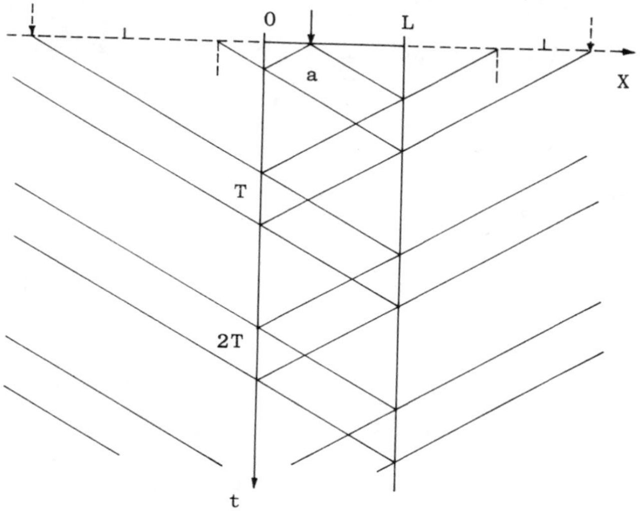

Figure 3

excitation point $x_p = 2nL \pm a$ the force pulse $\pm$ p(t) would generate diverging velocity waves $\pm G_\infty p(t-|x-x_p|/c)$ where $c = (T_0/\mu_1)^{\frac{1}{2}}$ is the string propogation velocity, if there were no damping. The trajectories of the pulse peaks are indicated by the sloping lines in Figure 3. At most locations x with $0 < x < L$ there are four peaks (two positive and two negative) during each fundamental period $T = 2L(\mu_1/T_0)^{\frac{1}{2}}$.

The effect of light damping can be incorporated approximately by assuming that during the first period T there is no attenuation at all of the four pulses arriving at a location x within the finite string, and that during the second period all four new pulses have an attenuation corresponding to a travel path of 2L, and that during the third period all four new pulses have an attenuation corresponding to a travel path 4L, etc. During the first period T the contribution to the integral in equation 43 from four pulses will be very nearly

$$4 \int_{-\infty}^{\infty} \left[G_\infty \, p(t)\right]^2 dt \tag{44}$$

if the decaying tails of the pulse p(t) die out rapidly in a time that is short compared to T. During the second period the contribution to the integral in equation 43 will be very nearly equal to

$$4 \left[e^{-2\alpha L} \right]^2 \int_{\infty}^{\infty} \left[G_\infty \, p(t) \right]^2 dt \tag{45}$$

where α is the attenuation constant

$$\alpha = \frac{c_1}{2\mu_1} \left[\frac{\mu_1}{T_0} \right]^{\frac{1}{2}} \tag{46}$$

Each succeeding period's contribution will be multiplied by an additional factor of exp $\{-\ 4\alpha L\}$ so that the total contribution to the integral in (43) is obtained by multiplying the first period contribution (44) by the sum of the series

$$\sum_{n=0}^{\infty} \exp\{-4n\alpha L\} = \frac{1}{1 - \exp\{-4\alpha L\}}$$

$$\simeq \frac{1}{4\alpha L} = \frac{2\left(T_0\mu_1\right)^{\frac{1}{2}}}{4c_1 L} = \frac{G_0}{4G_\infty} \tag{47}$$

where the light damping approximation has again been made and the conductances (2) and (4) have been introduced. The mean square velocity response (43), at a location where the four pulses arrive separately during a period, is thus given by

$$E|v^2| = G_0 G_\infty \nu\sigma_a^2 \int_{\infty}^{\infty} p^2(t)\, dt = G_0 G_\infty\, E\left[f^2\right] \tag{48}$$

which is equivalent to the modal sum result (33). The result (48) is independent of the precise shape of the excitation spectral density function. The underlying pulse shape p(t) must however be sufficiently narrow that it is meaningful to consider the response to consist of four nonoverlapping pulses during each fundamental period. For example, in the case of the pulse (39) corresponding to band-limited white noise with cut-off frequency ω_c the pulse duration $2\pi/\omega_c$ must be short in comparison with the fundamental period $t = 2\pi/\Delta\omega$ where $\Delta\omega$ is the modal spacing (13). This is equivalent to the requirement that the number of modes excited $N_c = \omega_c/\Delta\omega$ should be large.

The image sum procedure provides a clear insight into the location and relative response enhancement of the zones of enhanced response. In Figure 3 it is seen that during a single period T, four separate pulses arrive at most locations x, however, at the driving point x = a and at its image point x = L-a, only three pulses arrive in the same period. Two of the pulses, one positive and one negative, have the same magnitude as the pulses at other locations, but the third pulse which results from a superposition of two identical pulses has double magnitude. Instead of the contribution (44) from the first period T to the integral in (43), at these exceptional points the corresponding contribution would be

$$2\int_{-\infty}^{\infty} \left[G_\infty\ p(t)\right]^2 dt + \int_{\infty}^{\infty} \left[2G_\infty\ p(t)\right]^2 dt \qquad (49)$$

which is 1.5 times as large as (44). The same ratio applies to every period and hence to the total mean square velocity response (43). Enhancement of response thus occurs at those locations where pulses from more than one image source arrive simultaneously with the same sense. The relative enhancement can be quickly estimated by assigning unit magnitude to the contribution to the mean square of each single arrival, 4 units to each simultaneous arrival from two sources, 9 units to each simultaneous arrival from 3 sources, etc., and comparing the sums of these units over a fundamental period for points having enhanced response with the corresponding sums for ordinary points.

For example, the relative enhancement, at the drive point of the string above, would be estimated as follows

$$\frac{1 + 1 + 4}{1 + 1 + 1 + 1} = \frac{3}{2} \tag{50}$$

IMAGE SUM IN THE FREQUENCY DOMAIN

The image sum procedure in the time domain is relatively straightforward for non-dispersive systems such as the string. There does not however appear to be a simple way to extend the time domain calculation to dispersive systems. A frequency domain procedure which can be readily applied to dispersive systems is described below. For simplicity of exposition the same example of a uniform taut string of length L excited by a wide-band stationary random force process f(t) applied at x = a is considered. The mean square velocity at a general location x is obtained by using the method of images. The finite string with a single source is replaced by an infinite string acted on by the set of image sources indicated in Figure 3. Except for alternating signs all sources exert the same wide-band random force with mean square spectral density $S_f(\omega)$. Each source generates a pair of diverging velocity waves with mean square spectral density $G_\infty^2 S_f$ right at the source. After travelling a distance r the velocity spectral density will be attenuated by the factor

$$\left[\exp\{-\alpha r\}\right]^2 = \exp\left\{-\frac{c_1 L}{2(\mu_1 T_0)^{\frac{1}{2}}}\frac{2r}{L}\right\}$$

$$= \exp\left\{-\frac{G_\infty}{G_0}\frac{2r}{L}\right\} \tag{51}$$

The mean square velocity at x due to a single source at distance r is

$$\int_0^\infty 2\, S_f(\omega)\, G_\infty^2 \exp\left\{-\frac{G_\infty}{G_0}\frac{2r}{L}\right\} d\omega \tag{52}$$

This velocity process will be a wide-band process; i.e., a process with very short correlation time. The velocity waves arriving from other sources will be of the same type. The resultant velocity at a point is obtained by summing the velocities of the entire set of waves arriving from all the image sources. At most locations x, the source distances are all different from one another and each arriving wave is essentially uncorrelated with every other wave. At the special points $x = a$ and $x = L - a$ there are sets of equidistant sources which give rise to pairs of arriving waves having perfect correlation.

In the uncorrelated case the resultant mean square velocity from all sources is obtained by simply adding the mean square contributions of each single source. Once it has been agreed that all correlation will be neglected, it no longer is necessary to account for the precise distance to each source in estimating the total sum. What is needed is a systematic summing procedure which accounts for all sources and which assigns nominally correct attenuation factors to each source. For example, the sources can be arranged in groups of four with the same attenuation distance assigned to each source in a group, as was done in the preceding section. Alternatively the discrete sum can be approximated by a continuous integral. This latter approach is particularly effective for two-dimensional structures. Applying this approach to the string, we have the contribution (52) from a single source at distance r from the receiver point. The number of sources on both sides of the receiver is 2/L per unit length on one side. The resultant mean square velocity from all sources is then given by

$$E\left[v^2\right] = \int_0^\infty \frac{2dr}{L} \int_0^\infty 2S_f(\omega)\ G_\infty^2 \exp\left\{-\frac{G_\infty}{G_0}\frac{2r}{L}\right\} d\omega \tag{53}$$

at any point where there is no appreciable correlation between waves arriving from different image sources. The integration in (53) is performed by inverting the order of integration. Integration over r yields

$$E\left[v^2\right] = \int_0^\infty 2\ S_f(\omega)\ G_0\ G_\infty\ d\omega \tag{54}$$

In the case of the string where the conductances are independent of frequency, (54) implies

$$E\left[v^2\right] = G_0\ G_\infty\ E\left[f^2\right] \tag{55}$$

independently of the precise shape of $S_f(\omega)$, as was noted in connection with (48). The result (55) gives the substantially uniform mean square response which forms the background from which the zones of enhanced response stand out.

The zones of enhanced response are centred at the special points x=a and x=L-a where families of pairs of perfectly correlated waves exist. For example at x = L-a the distances to the negative sources at x = -2(n-1)L-a are exactly equal to the distances to the negative sources at x = 2nL-a for n = 1,2,... Thus only half of the waves arriving at x = L-a are single strength uncorrelated waves. The other half combine in pairs to form a family of double strength local responses which are uncorrelated with each other or with the single strength waves. The relative enhancement of the mean square response, in comparison with the background, can be estimated by assigning unit amplitude to the sum (53) when all waves are single strength and uncorrelated and by taking account of what fractions of the original number of contributions have single strength, double strength, etc. In the present case, half of the original number have single strength while the other half reduces to one quarter of the original number of uncorrelated contributions but each contribution now has double strength. The relative enhancement may then be estimated as

$$\frac{1}{2}\,(1^2) + \frac{1}{4}\,(2^2) = \frac{3}{2} \tag{56}$$

Image-summation in the frequency domain can thus be used to locate zones or lanes of enhanced response and to estimate the relative enhancement of response.

The result (54) derived here for the string may be extended to the uniform beam, membrane and plate. It is necessary to neglect the near field wave response, and (in the two-dimensional cases) to use asymptotic approximations for the far-field radiation. In the cases of the beam and the membrane, where G_∞ is frequency dependent, it is not possible to pass directly to (55) for arbitrary excitation spectral density. If, however, $S_f(\omega)$ is taken to be the band-limited white spectrum (41) the integration of (54) leads to

$$E\left[v^2\right] = G_0\;\bar{G}_\infty\;E\left[f^2\right] \tag{57}$$

where $\bar{G}_\infty$ is the frequency averaged conductance (8). Thus, for all four structures under band-limited white noise excitation at a point, the effective rms velocity admittance at points outside the exceptional zones or lanes of enhanced or diminished response is essentially uniform and equal to the geometric mean between the rigid-body conductance G_0 and the frequency averaged radiation conductance $\bar{G}_\infty$. The location and relative enhancement of the zones or lanes of enhanced response can often be determined from simple geometric analyses based on image summation procedures.

CONNECTION WITH STATISTICAL ENERGY ANALYSIS

The final result (57) can be interpreted as an approximation to one of the basic principles of statistical energy analysis [10]. When (57) is rearranged in the form

$$\bar{G}_\infty \, E\left[f^2\right] = E\left[v^2\right]/G_0 \tag{58}$$

the left-hand side can be interpreted as the expected power flow into an infinite structure from the external excitation, while the right-hand side represents the expected rate of energy dissipation by the finite structure due to the viscous damping mechanism under the condition that all points on the structure have the same mean square velocity $E\left[v^2\right]$. Under wide-band excitation, the power flow into an infinite structure is nearly the same as into a finite structure, and, as we have seen, the mean square velocity is nearly equal to $E\left[v^2\right]$ over most of the structure (with the exception of the zones or lanes of enhanced or diminished response). Thus (58) is very nearly a simple power balance statement applied to the finite structure.

ACKNOWLEDGEMENT

This research was supported by a Grant from the National Science Foundation.

REFERENCES

1 S.H. Crandall and L.E. Wittig, Dynamic response of structures Proc. Symp. Pergamon Press, 55 Chladni's pattern for random vibrations of a plate, (1972).

2 S.H. Crandall, Proc. 7th U.S. Natl. Congr. Appl. Mech. ASME, 131 Wide-band random vibration of structures (1974).

3 L.E. Young, On the dynamic response of a damped membrane to a point stochastic load. MIT Acoustics and Vibration Lab. Rep. 81678-2 (1975).

4 S.S. Lee, Lanes of intensified response in structures excited by wide-band random excitation. Thesis proposal Department of Mechanical Engineering MIT (1975).

5 S.H. Crandall, Random Vibration Vol. 2, MIT Press, 48 (1963).

6 R. Courant and D. Hilbert, Methods of Mathematical Physics Vol. 1, Interscience Publishers, Chap. 6 (1953).

7 R.H. Lyon, Statistical analysis of power-injection and response in structures and rooms. Journ. Acoust. Soc. Amer. 45, 545 (1969).

8 L. Cremer, M. Heckl and E.E. Ungar, Structure-Borne Sound, Springer Verlag, 295 (1973).

9 S.O. Rice, Mathematical analysis of random noise. Bell System Tech. Journ. 23, 282; 24, 46 (1944-45).

10 R.H. Lyon, Statistical energy analysis of dynamical systems. MIT Press (1975).

Professor S.H. Crandall, Ford Professor of Engineering, Massachusetts Institute of Technology, Department of Mechanical Engineering, Cambridge, Massachusetts 02139.

DISCUSSION

Professor Frolov commented on the development of the special random theory from one-dimension to two-dimensions and asked if it is possible to extend these ideas to three-dimensions for application to shell structures. Professor Crandall replied that there are great experimental difficulties with shells (the salt will not stay on!) and therefore it is necessary to use transducers. This is a very laborious process but the limited work to date indicates that it may be possible to find order in rectangular shallow shells. In principle the analytical methods work well. Bolotin and Elishakoff have done work on shells but they were not looking for order in the patterns of surface motion. In reply to a second question (Frolov) about multipoint excitation Professor Crandall reported that he had just presented a paper* in which it is shown that if there are two uncorrelated point sources on a rectangular plate the resultant pattern is the superposition of the two individual patterns. There are some special cases where correlated forces produce very marked patterns but the general superposition requires uncorrelated forces.

Replying to a question on the effect of arbitrary shape (Kozin) Professor Crandall said that the pattern disappeared when the shape deviated from rectangular. The specific example given was when one side of the plate was cut off at an angle of 10°. The effect of other boundary conditions (Elishakoff) such as degree of fixing, damping etc. on the pattern for a

---

* S.H. Crandall and A.P. Kulvets, Source Correlation Effects on Structural Response, to appear, Proceedings of the Symposium on Applications of Statistics, June 14 - 18, 1976, Wright State University, Dayton, Ohio.

rectangular plate was minimal. There was essentially very little change of the interior pattern when one of the four edges was unsupported or when rubber damping was added along one side. Fully clamped boundaries have not been tried but this change is not expected to make any difference. The effect of damping (Shiehlen) was not significant for damping less than 5% of critical. For a heavily damped structure such as a membrane the pattern can be identified close to the driver but disappears further away. Professor Crandall repeated that the most significant effect was the shape of the boundary.

I ELISHAKOFF

Flutter and random vibrations in plates

INTRODUCTION

This study considers the interaction of flutter and random vibrations of plates, excited by pressure fluctuations in a turbulent boundary layer. The theory of stochastic stability of elastic bodies in supersonic flow, developed in reference [1] , is used. The external load is composed of pressure fluctuations in the turbulent boundary layer, and pressure perturbations associated with the plate deformations according to the 'piston theory'. In describing the probabilistic behaviour of the external pressure, the cross-spectral density of the latter is assumed as an exponentially damped cosine wave.

A bibliography of probabilistic interaction of structures and supersonic flow is given in reference [1]. A comprehensive review of deterministic flutter problems may be found in Dowell's survey [3].

THEORETICAL ANALYSIS

Let the vibrations of a plate be governed by the following differential equation (1)

$$D \frac{\partial^4 w}{\partial x^4} + 2\varepsilon_o m \frac{\partial w}{\partial t} + gmU \frac{\partial w}{\partial x} + m \frac{\partial^2 w}{\partial t^2} = q(x,t) \tag{1}$$

where

$$m = \rho h, \quad \varepsilon_o = \varepsilon + 0.5g, \quad g = \frac{\kappa p_\infty}{mc_\infty}, \quad D = \frac{E}{12(1-\nu^2)} \tag{2}$$

w(x,t) is the plate displacement, D - the stiffness modulus, h - thickness,

ν - Poisson's ratio, E - Young's modulus, p_∞ c_∞ - the unperturbed pressure and unperturbed velocity, respectively, κ - the polytropic exponent and U the flow velocity. We confine ourselves to clamped plates for which the boundary conditions are:

$$w = \frac{\partial w}{\partial x} = 0, \quad \text{for } x = 0,\ x = a \tag{3}$$

The mode shapes are given by the functions

$$\begin{aligned} \psi_\beta(x) &= \sin\lambda_\beta x - \sinh\lambda_\beta x + A_\beta(\cos\lambda_\beta x - \cosh\lambda_\beta x) \\ A_\beta &= (\cos\lambda_\beta a - \cosh\lambda_\beta a)(\sin\lambda_\beta a + \sinh\lambda_\beta a)^{-1} \end{aligned} \tag{4}$$

The eigenvalues are the roots of

$$1 - (\cos \lambda_p a)(\cosh \lambda_p a) = 0 \tag{5}$$

and have the approximate values

$$\begin{aligned} &\lambda_1 a = 4.7300,\ \lambda_2 a = 7.8532,\ \lambda_3 a = 10.9956,\ \lambda_4 a = 14.1371, \\ &\lambda_5 a = 17.2787,\ \lambda_6 a = 20.4203,\ \lambda_\beta a \cong (\beta + 0.5)\pi, \text{ for } \beta \geq 7 \end{aligned} \tag{6}$$

The natural frequencies are

$$\omega_\beta^2 = \lambda_\beta^2 D/m \tag{7}$$

The fluctuating components of the external force q(x,t) and of the plate displacement w (x,t), are expressed as

$$q(x,t) = \sum_{\alpha=1}^{N} \int_{-\infty}^{\infty} Q_\alpha(\omega)\psi_\alpha(x)\exp(i\omega t)d\omega \tag{8}$$

$$w(x,t) = \sum_{\alpha=1}^{N} \int_{-\infty}^{\infty} W\ (\omega)\psi_\alpha(x)\exp(i\omega t)d\omega \tag{9}$$

where $Q_\alpha(\omega)$ and $W_\alpha(\omega)$ are random complex spectra and N is the number of modes taken into account.

Following reference [1], we use Galerkin's procedure to arrive at a set of linear equations

$$\begin{bmatrix} a_{11} & a_{12} & a_{13} & \cdots & a_{1N} \\ a_{21} & a_{22} & a_{23} & \cdots & a_{2N} \\ a_{31} & a_{32} & a_{33} & \cdots & a_{3N} \\ \cdots & & & & \\ a_{N1} & a_{N2} & a_{N3} & \cdots & a_{NN} \end{bmatrix} \begin{bmatrix} W_1 \\ W_2 \\ W_3 \\ \cdot\cdot \\ W_N \end{bmatrix} = \frac{1}{m} \begin{bmatrix} Q_1 \\ Q_2 \\ Q_3 \\ \cdot\cdot \\ Q_N \end{bmatrix} \tag{10}$$

where

$$a_{jk} = (\omega_j^2 - \omega^2 + 2\varepsilon_0 i\omega)\ \delta_{jk} + gUb_{jk} \tag{11}$$

$$b_{jk} = \left(\int_0^a \frac{d\psi_k}{dx}\ \psi_j dx\right)\left(\int_0^a \psi_j^2\ dx\right)^{-1}\ ,j,k = 1,2,3,\ldots,N. \tag{12}$$

For the system parameters, where the matrix $A \equiv [a_{jk}]$ is nonsingular, the solution of set (10) is

$$W_\beta = \sum_{=1}^{N} (A^{-1})_{\beta\delta}\ Q_\delta(\omega)/m \tag{13}$$

$(A^{-1})_{\beta\delta}$ being elements of the inverse matrix A^{-1}. The cross-spectral densities of the plate displacements are then given by

$$S_w(x,x',\omega) = \sum_\alpha \sum_\beta \sum_\gamma \sum_\delta (A^{-1})^*_{\alpha\gamma}\ (A^{-1})_{\beta\delta} S_{\gamma\delta} \psi_\alpha(x)\psi_\beta(x')/m^2 \tag{14}$$

The cross-spectral density of the displacement at point x is given by

$$S_w(x,x,\omega) = S_{w,I} + S_{w,II} \tag{15}$$

where

$$S_{w,I} = \sum_\gamma S_{\gamma\gamma} \left| \sum_\alpha (A^{-1})_{\alpha\gamma}\psi_\alpha \right|^2/m^2 \tag{16}$$

$$S_{w,II} = 2\mathrm{Re} \sum_{\substack{\gamma\ \delta \\ \gamma<\delta}} \sum S_{\gamma\delta} \sum_\alpha (A^{-1})^*_{\alpha\gamma}\psi_\alpha\ \sum_\beta (A^{-1})_{\beta\delta}\psi_\beta/m^2 \tag{17}$$

In equations (16) - (17) $S_{\gamma\delta}$ denote the cross-spectral densities of the generalized forces, defined as

$$S_{\gamma\delta} = \left(\int_0^a \psi_\delta^2 dx \int_0^a \psi_\delta^2 dx\right)^{-1} \int_0^a\int_0^a S_q(x,x',\omega)\psi_\gamma(x)\psi_\delta(x')dxdx' \qquad (18)$$

where $S_q(x,x',\omega)$ is the cross-spectral density of the external forces

$$S_q(x,x',\omega) = \frac{1}{2\pi}\int_{-\infty}^{\infty} \langle q^*(x,t)q(x',t+\tau)\rangle \exp(-i\omega\tau)d\tau \qquad (19)$$

The mean-square stability condition (Equation 46 of Reference [1]) takes the form

$$\sum_\alpha \sum_\beta \sum_\gamma \sum_\delta \psi_\alpha(x)\psi_\beta(x) \int_{-\infty}^{\infty} (A^{-1})^*_{\alpha\gamma}(A^{-1})_{\beta\delta}S_{\gamma\delta}d\omega < k^2 \qquad (20)$$

When the flow velocity is such that the matrix A is singular and the critical frequency is non-zero, we have flutter-type instability of the plate. Consequently, equation (14) is valid only in the range $U < U_{flutter}$.

If the deformational pressure perturbations are not taken into account, the third term in equation (1) vanishes and ε_0 is replaced by ε. Then

$$(A^{-1})_{\alpha\gamma} = \delta_{\alpha\gamma}/a_{\alpha\alpha}, \qquad a_{\alpha\alpha} = \omega_\alpha^2 - \omega^2 + 2\varepsilon i\omega \qquad (21)$$

$\delta_{\alpha\gamma}$ being Kronecker's delta.

The mean-square stability condition in the sense S_∞ (for $k \to \infty$) is then satisfied for all flow velocities U, and the approximate solution obtained by reduction coincides with the exact solution. This case is widely discussed in literature (see, for example, Reference [4]).

$$S_w(x,x',\omega) = \sum_{\alpha=1}^{\infty}\sum_{\beta=1}^{\infty} S_{\alpha\beta}\psi_\alpha(x)\psi_\beta(x')/(a^*_{\alpha\alpha}\, a_{\beta\beta}) \qquad (22)$$

We consider now how the number of modes taken into account influences probabilistic response characteristics.

One-Mode Approximation

For this case, the determinant of coefficients of the spectral equation is

$$\det A(\omega) = \omega_1^2 - \omega^2 + 2\varepsilon_o i\omega + gUb_{11} \tag{23}$$

The critical velocity is given by

$$\mathrm{Re}\left[\det A\ (\omega)\right] \equiv \omega_1^2 - \omega^2 + gUb_{11} = 0,\ \mathrm{Im}\left[\det A(\omega)\right] \equiv 2\varepsilon_o i\omega = 0 \tag{24}$$

The total damping coefficient ε_o differs from zero, and the latter equation holds only for $\omega = 0$, i.e. flutter-type instability is apparently impossible. At the same time $\mathrm{Re}\left[\det A(\omega)\right] = 0$, because of $b_{11} = 0$, is also impossible. It is thus seen that the one-mode approximation, which is often used in random vibration analysis, does not cover complete physical picture and actually leads to the incorrect conclusion of non-existence of flutter-type instability when the component of pressure perturbation associated with the plate deformations is taken into account.

Two-Mode Approximation

The spectral equations (1) take the form

$$\begin{aligned} a_{11}W_1 + a_{12}W_2 &= \frac{1}{m} Q_1 \\ a_{21}W_1 + a_{22}W_2 &= \frac{1}{m} Q_2 \end{aligned} \tag{25}$$

The determinant of the system becomes:

$$\det A(\omega) = (\omega_1^2 - \omega^2 + gUb_{11} + 2\varepsilon_o i\omega)(\omega_2^2 - \omega^2 + gUb_{22} + 2\varepsilon_o i\omega) - g^2U^2b_{12}b_{21} \tag{26}$$

The real and imaginary parts of this determinant are

$$\mathrm{Re}\left[\det A(\omega)\right] = c_o\omega^4 - c_2\omega^2 + c_4,\quad \mathrm{Im}\left[\det A(\omega)\right] = -c_1\omega^3 + c_3\omega \tag{27}$$

$$c_0 = 1,\ c_1 = 4\varepsilon_0,\ c_2 = \omega_1^2 + \omega_2^2 + gU(b_{11} + b_{22}) + 4\varepsilon_0^2$$

$$c_3 = 2\varepsilon_0(c_2 - 4\varepsilon_0^2),\ c_4 = (\omega_1^2 + gUb_{11})(\omega_2^2 + gUb_{22}) - g^2U^2b_{12}b_{21} \tag{28}$$

To clarify the incompatibility condition for equations (24), we seek the common root of $\mathrm{Re}[\det A(\omega)] = 0$, $\mathrm{Im}\ [\det A(\omega)] = 0$. From the latter we have

$$\omega^2_{1,\text{critical}} = 0,\ \omega^2_{2,\text{critical}} = \frac{c_3}{c_1} = \frac{1}{2}[\omega_1^2 + \omega_2^2 + gU(b_{11} + b_{22})] \tag{29}$$

for which

$$\mathrm{Re}[\det A(\omega_{1,\text{critical}})] = \omega_1^2\omega_2^2 - g^2U^2b_{12}b_{21} \tag{30}$$

$$\mathrm{Re}[\det A(\omega_{2,\text{critical}})] = [-c_3(c_1c_2 - c_0c_3) + c_4c_1^2]c_1^{-2} \tag{31}$$

The critical flow velocities $U_{\text{divergence}}$ and U_{flutter} are given by the following conditions, respectively

$$\omega_1^2\omega_2^2 - g^2U^2b_{12}b_{21} = 0 \tag{32}$$

$$c_3(c_1c_2 - c_0c_3) + c_4c_1^2 = 0 \tag{33}$$

It may be shown that $b_{12}b_{21} < 0$. Consequently, equation (32) has no real roots and divergence-type (static) instability is ruled out in this approximation also.

The critical flutter velocity is obtained from (33)

$$U_{\text{flutter}} = \{-[(\omega_2^2 - \omega_1^2)^2 + 8\varepsilon_0^2(\omega_1^2 + \omega_2^2)]/(b_{12}b_{21})\}^{1/2}/2g \tag{34}$$

For $U < U_{\text{flutter}}$ the matrix A is non-singular and the spectral equations have the following solution

$$W_1(\omega) = \frac{a_{22}}{\Delta}\ Q_1 - \frac{a_{12}}{\Delta}\ Q_2\ , \tag{35}$$

$$W_2(\omega) = \frac{a_{11}}{\Delta}\ Q_2 - \frac{a_{21}}{\Delta}\ Q_1 \tag{36}$$

the relationship (11) determining a_{jk}. Utilizing the Hermetian property of the cross-spectral densities of the generalized forces, the root mean-square value may be formulated as

$$m^2\langle w^2 (x,t)\rangle = \int_{-\infty}^{\infty} \left[S_{11}\psi_{11} + S_{22}\psi_{22} - 2\mathrm{Re}(S_{12}\psi_{12})\right]\Delta^{-2}d\omega \tag{37}$$

where

$$\begin{aligned}
\psi_{11} &= |a_{22}|^2\psi_1^2 + |a_{21}|^2\psi_2^2 + 2\mathrm{Re}(a_{21}^*a_{22})\psi_1\psi_2 \\
\psi_{22} &= |a_{12}|^2\psi_1^2 + |a_{11}|^2\psi_2^2 + 2\mathrm{Re}(a_{11}^*a_{12})\psi_1\psi_2 \\
\psi_{12} &= a_{22}^*a_{12}\psi_1^2 + a_{21}^*a_{11}\psi_2^2 + (a_{21}^*a_{12} + a_{22}^*a_{11})\psi_1\psi_2
\end{aligned} \tag{38}$$

Three-Mode Approximation

The elements of the 3 x 3 matrix A are obtained from (11) with b_{jk} given by (12). The real and imaginary parts of the determinant are, in turn, given by

$$\mathrm{Re}\left[\det A(\omega)\right] = c_6 - c_4\omega^2 + c_2\omega^2 - c_0\omega^6 \tag{39}$$

$$\mathrm{Im}\left[\det A(\omega)\right] = (c_5 - c_3\omega^2 + c_1\omega^4)\omega \tag{40}$$

where

$$\begin{aligned}
&c_0 = 1,\ c_1 = 6\varepsilon_0,\ c_2 = \Omega_1^2 + \Omega_2^2 + \Omega_3^2 + 12\varepsilon_0^2,\ c_3 = 4\varepsilon_0(\Omega_1^2 + \Omega_2^2 + \Omega_3^2) \\
&+ 8\varepsilon_0^2,\ c_4 = \Omega_1^2\Omega_2^2 + \Omega_1^2\Omega_3^2 + \Omega_2^2\Omega_3^2 + 4\varepsilon_0^2(\Omega_1^2 + \Omega_2^2 + \Omega_3^2) - g^2U^2(b_{12}b_{21} \\
&+b_{13}b_{31}+b_{23}b_{32}),\ c_5 = 2\varepsilon_0(\Omega_1^2\Omega_2^2 + \Omega_1^2\Omega_3^2 + \Omega_2^2\Omega_3^2) - 2\varepsilon_0 g^2U^2 \\
&(b_{12}b_{21}+b_{13}b_{31}+b_{23}b_{32}),\ c_6 = \Omega_1^2\Omega_2^2\Omega_3^2 - g^2U^2(\Omega_3^2b_{12}b_{21} + \Omega_2^2b_{13}b_{31} \\
&+ \Omega_1^2b_{23}b_{32}) + g^3U^3(b_{12}b_{23}b_{31} + b_{13}b_{21}b_{23}), \\
&\Omega_j^2 = \omega_j^2 + gUb_{jj},\quad j = 1,2,3,
\end{aligned} \tag{41}$$

For determining the critical frequency and velocity, we again seek the common root of Re[det A(ω)] = 0 and Im[det A(ω)] = 0. From the latter equation, we have

$$\omega_{1,\text{critical}} = 0, \quad \omega^2_{(2,3),\text{critical}} = \frac{1}{2c_1}\left[c_3 \pm (c_3^2 - 4c_1c_5)^{1/2}\right] \tag{42}$$

The requirement that ω_{critical} be a root of Re[det A(ω)] = 0, leads to the following equation for critical velocity

$$c_6 = 0 \tag{43}$$

$$\left[c_5(c_4c_3 - c_2c_5) + c_6(2c_1c_5 - c_3^2)\right](c_1c_2 - c_0c_3)$$

$$+ (c_1c_4 - c_0c_5)\left[c_1c_3c_6 - c_5(c_1c_4 - c_0c_5)\right] - c_1^3c_6^2 = 0 \tag{44}$$

It is readily shown that the latter condition derives from the requirement that the principal Hurwitz determinant for the polynomial

$$\Delta(s) = c_0s^6 + c_1s^5 + c_2s^4 + c_3s^3 + c_4s^2 + c_5s + s_6 \tag{45}$$

be zero.

Equation (43) is equivalent to

$$\omega_1^2\omega_2^2\omega_3^2 - g^2U^2(b_{12}b_{21}\omega_3^2 + b_{13}b_{31}\omega_2^2 + b_{23}b_{32}\omega_1^2)$$

$$+ g^3U^3(b_{12}b_{23}b_{31} + b_{13}b_{32}b_{21}) = 0 \tag{46}$$

It is readily seen that

$$b_{12}b_{23}b_{31} + b_{13}b_{32}b_{21} \equiv 0,$$

$$b_{12}b_{21}\omega_3^2 + b_{13}b_{31}\omega_2^2 + b_{23}b_{32}\omega_1^2 \leq 0 \tag{47}$$

and divergence-type instability is again ruled out.

Equation (44) yields the critical flutter velocity. The random vibration in the pre-flutter range $U < U_{\text{flutter}}$ is analysed in complete analogy to the proceeding section.

Four-Mode Approximation

The polynomial (38) of Reference [1] reads, for N = 4

$$\Delta(s) = c_0 s^8 + c_1 s^7 + c_2 s^6 + c_3 s^5 + c_4 s^4 + c_5 s^3 + c_6 s^2 + c_7 s + c_8 \tag{48}$$

its coefficients being

$$c_0 = 1,\ c_1 = 8\varepsilon_0,\ c_2 = 24\varepsilon_0^2 M_1,\ c_3 = 32\varepsilon_0^3 + 6M_1\varepsilon_0,\ c_4 = 16\varepsilon_0^4 + 12\varepsilon_0^2 M_1 + M_2,\ c_5 = 8\varepsilon_0^3 M_1 + 4\varepsilon_0 M_2,\ c_6 = 4\varepsilon_0^2 M_2 + M_3,\ c_7 = 2\varepsilon_0 M_3,\ c_8 = M_4 \tag{49}$$

where

$$M_1 = \omega_1^2 + \omega_2^2 + \omega_3^3 + \omega_4^2,\ M_2 = \omega_1^2\omega_2^2 + \omega_2^2\omega_3^2 + \omega_3^2\omega_4^2 + \omega_1^2\omega_4^2 - g^2U^2(b_{12}b_{21} + b_{23}b_{32} + b_{34}b_{43} + b_{14}b_{41} + b_{24}b_{42} + b_{13}b_{31})$$

$$M_3 = \begin{bmatrix} \omega_1^2 & gUb_{12} & 0 \\ gUb_{21} & \omega_2^2 & gUb_{23} \\ 0 & gUb_{32} & \omega_3^2 \end{bmatrix} + \begin{bmatrix} \omega_1^2 & 0 & gUb_{14} \\ 0 & \omega_3^2 & gUb_{34} \\ gUb_{41} & gUb_{41} & \omega_4^2 \end{bmatrix} + \begin{bmatrix} \omega_1^2 & gUb_{12} & gUb_{14} \\ gU_{12} & \omega_2^2 & 0 \\ gUb_{41} & 0 & \omega_4^2 \end{bmatrix} \begin{bmatrix} \omega_2^2 & gUb_{23} & 0 \\ gUb_{32} & \omega_3^2 & gUb_{34} \\ 0 & gUb_{43} & \omega_4^2 \end{bmatrix} \tag{50}$$

$$M_4 = \begin{bmatrix} \omega_1^2 & gUb_{12} & 0 & gUb_{14} \\ gUb_{21} & \omega_2^2 & gUb_{23} & 0 \\ 0 & gUb_{32} & \omega_2^2 & gUb_{34} \\ gUb_{41} & 0 & gUb_{43} & \omega_3^2 \end{bmatrix}$$

The critical flutter velocity is obtained from the requirement that the principal Hurwitz determinant be zero:

$$\begin{bmatrix} c_1 & c_o & 0 & 0 & 0 & 0 & 0 \\ c_3 & c_2 & c_1 & c_o & 0 & 0 & 0 \\ c_5 & c_4 & c_3 & c_2 & c_1 & c_0 & 0 \\ c_7 & c_6 & c_5 & c_4 & c_3 & c_2 & c_1 \\ 0 & c_8 & c_7 & c_6 & c_5 & c_4 & c_3 \\ 0 & 0 & 0 & c_8 & c_7 & c_6 & c_5 \\ 0 & 0 & 0 & 0 & 0 & c_8 & c_7 \end{bmatrix} = 0 \qquad (51)$$

Descartes' rule of signs, applied to the expression for M_4

$$M_4 = g^4U^4(b_{12}b_{34} - b_{14}b_{23})^2$$

$$g^2U^2(\omega_1^2\omega_2^2b_{34}^2 + \omega_1^2\omega_4^2b_{23}^2 + \omega_3^2\omega_4^2b_{12}^2 + \omega_2^2\omega_3^2b_{14}^2) + \omega_1^2\omega_2^2\omega_3^2\omega_4^2 \qquad (52)$$

indicates that $M_4 = 0$ has no real roots. Divergence-type instability is thus again ruled out, and only flutter type instability is relevant. The subsequent procedure is as in the two-mode approximation.

Higher approximations are realized in a similar manner.

EXAMPLE AND DISCUSSION

The numberical results were obtained on an IBM 360/165 computer at the Technion. Given plate data and flow parameters were: $a = 0.55m$, $h = 1.5 \times 10^{-3} m$, $\rho = 2{,}700\ kg/m^3$, $E = 7.05 \times 10^{10}\ kg/m.sec^2$, $\nu = 0.3$, $\varepsilon = 0.1$, $c_\infty = 303.3\ m/sec$ (995 ft/sec), $\rho_\infty = 0.45915\ kg/m^3$ (0.00089 slug/ ft^3), altitude = 30,000 ft. The cross-spectral density of the external forces was given as

$$S_q = \sigma_q^2 \psi_q(\omega) \exp\left[\left(-\frac{0{-}1\omega a}{U_c} + 0.265\frac{a}{\delta_o}\kappa\right)|\xi' - \xi| - \frac{i\omega a}{U_c}(\xi' - \xi)\right] \quad (53)$$

where σ_q^2 is the mean-square fluctuation in the turbulent boundary layer, $\psi_q(\omega)$ - the frequency spectral density, ω - frequency, U_c - convection velocity, $\xi' = x'/a$ and $\xi = x/a$ - the nondimensional plate coordinates, δ_o - thickness of the boundary layer. The nondimensional thickness of the boundary layer was taken as $\delta_o/a = 0.32$. For $\kappa = 0$ and $\kappa = 1$, Equation (53) reduces, respectively, to Wilby's and Crocker's approximations References [5,6]. For $\kappa = 1$, it also reduces to

$$S_q = \sigma_q^2 \psi_q(\omega) \exp\left[-\left(\frac{0.1\omega a}{U_c} + \frac{0.037}{\Delta_1^*}\right)|\xi' - \xi| - \frac{i\omega a}{U_c}(\xi' - \xi)\right] \quad (54)$$

as given in References [7] and [8], where $\Delta_1^* = 0.04468$. Note that the non-dimensional cross-spectral density $S_q \sigma_q^{-2} \psi_q^{-1}$ which may be obtained from formula (53) represents probably good empirical expression for the normalized cross-spectral density at supersonic as well as subsonic speeds [6,9]. The root mean-square pressure of a boundary-layer pressure field is about 0.003q for the Mach numbers $2 \leq M \leq 5$, where q is the dynamic pressure of the free flow [10]. The frequency spectral density $\psi_q(\omega)$ may be taken, for example, from Reference [11]. The convective velocity was taken constant as $U_{conv.} = 0.95U$ being the free stream velocity [12]. In Reference [2], inter alia, the cross-spectral densities of the generalized forces $S_{\gamma\delta}$ are calculated - on the basis of exact [4] and approximate mode shapes, according to Bolotin's dynamic edge effect method Reference [13] - as a function of the non-dimensional frequency $\bar{\bar{\omega}} = \omega a/U_c$. The two-mode approximation yields the flutter Mach number $M^{(2)}_{flutter} = 1.545$, the three-mode approximation - $M^{(3)}_{flutter} = 1.976$ and the four-mode approximation (which may be regarded in practice as an exact solution (see Reference [14]) - $M^{(4)}_{flutter} = 2.08$.

The nondimensional cross spectral density of the maximal normal stresses

$$\bar{S} = \frac{m^2 h^4}{36D^2} \frac{S_\sigma(\omega)}{\sigma_q^2 \psi_q(\omega)} \tag{55}$$

was calculated as function of the nondimensional frequency

$$\bar{\omega} = \omega a^2 (m/D)^{\frac{1}{2}} \tag{56}$$

the relationship between $\bar{\omega}$ and $\bar{\bar{\omega}}$ being given by

$$\bar{\omega} = z\bar{\bar{\omega}}, \quad z = \omega_c \; a/U_c \tag{57}$$

where

$$\omega_c = U_c^2 \; (m/D)^{1/2} \tag{58}$$

is the coincidence frequency. The following nondimensional quantities were also used

$$\bar{\omega}_\alpha = \omega_\alpha f, \; \bar{\varepsilon} = \varepsilon f, \; \bar{\varepsilon}_o = \varepsilon_o f, \; \bar{g} = gf, \; \bar{b}_{jk} = ab_{jk},$$

$$\bar{U} = Uf/a, \; \bar{\omega}_\alpha = (\lambda_\alpha a)^2, \; \bar{x} = x/a \tag{59}$$

with

$$f = a^2 (m/D)^{1/2} \tag{60}$$

and $\lambda_\alpha a$ as per equation (6).

In Figures 1 and 2, $\bar{S}_\sigma$ is plotted against $\bar{\omega}$ at the edge $\bar{x} = 0$ in the two-mode approximation. At M = 1.05 and M = 1.3, the plots are seen to contain

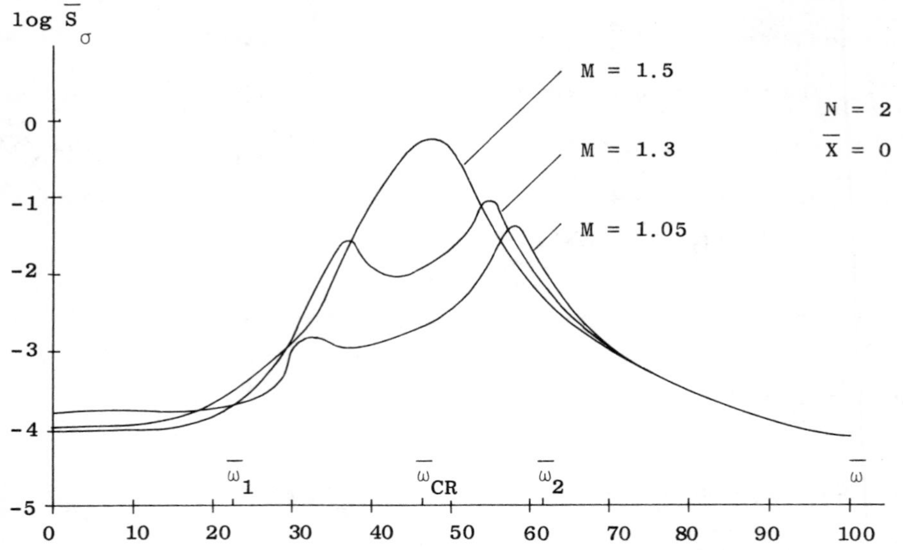

FIGURE 1

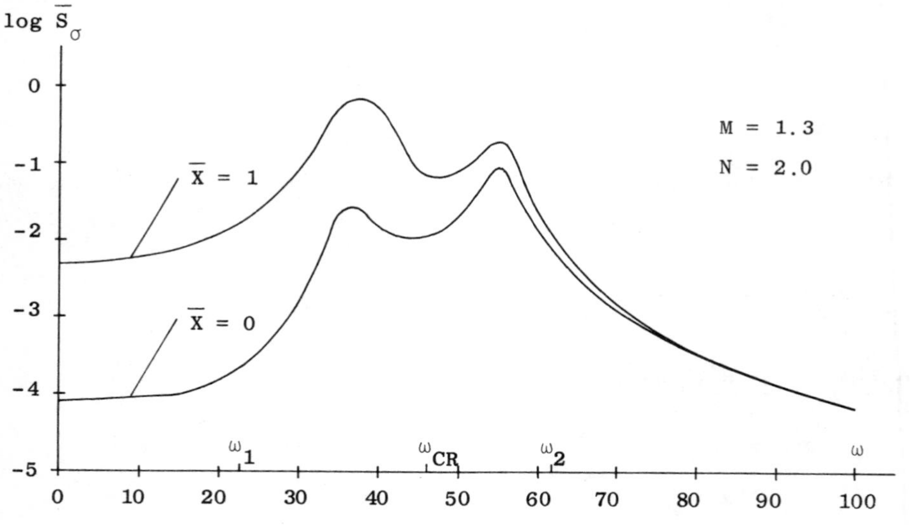

FIGURE 2

two distinct maxima; for M = 1.05 the first maximum appears at $\bar{\omega} = 32$, the first nondimensional eigenfrequency being $\bar{\omega}_1 = 22.37$; the second maximum - at $\bar{\omega} = 56$, the second nondimensional eigenfrequency being $\bar{\omega}_2 = 61.67$. As the Mach number increases and approaches the flutter level $M_{flutter}$, they degenerate to a single maximum close to the critical flutter frequency $\bar{\omega}_{CR} = 46.4$. At $M = M_{flutter}$, the cross-spectral density of the maximal normal stresses (obtainable formally from:

$$S_\sigma(x,x,\omega) = \frac{36D^2}{h^4 m^2} \sum_{\alpha,\beta,\gamma,\delta} (A^{-1})^*_{\alpha\gamma} (A^{-1})_{\beta\delta} S_{\gamma\delta} \psi''_\alpha(x) \psi''_\beta(x) \tag{61}$$

and valid only in the preflutter range) has a pole at $\bar{\omega}_{CR}$. Note that for the Mach numbers used in Figure 1 and 2 the 'piston theory' is inapplicable |14-15|. The figures are plotted only for the demonstration of the above mentioned degenerative tendency. An additional reason why these figures are only qualitative is that two mode approximation is not enough to describe accurately the random vibrations of plates in supersonic flow.

Figures 3 and 4 present the nondimensional cross-spectral densities of the maximal normal stresses in the four-mode approximation for M = 1.4, 1.6, 1.8. In the two-mode approximation, the two last Mach number fall within the post-flutter range, illustrating how approximations below the minimum safe number of modes N^* (in the present case, four) are of purely qualitative significance. Figure 4 shows four maxima, the first of them occurring considerably above the first eigenfrequency and the remaining three close to the corresponding eigenfrequencies; as the Mach number increases the first two maxima draw together and eventually merge, leaving a total of three maxima (as though the system had only three degrees of freedom!). This degenerative tendency may perhaps serve as an advance indication of possible flutter-type instability.

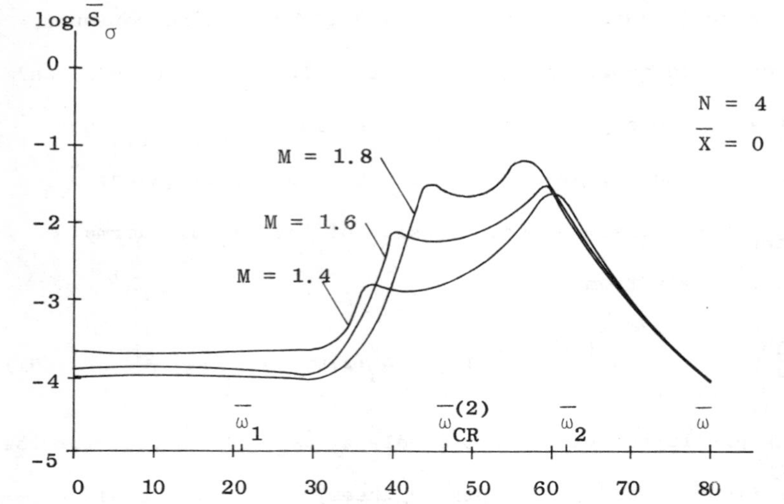

FIGURE 3

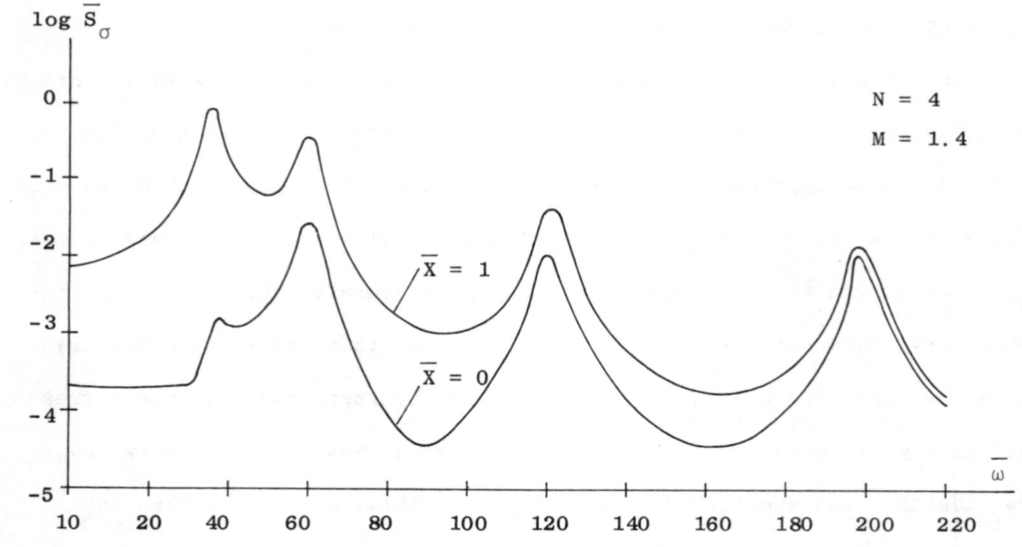

FIGURE 4

Figures 2 and 4 show that for the given free stream velocity, the RMS of the maxima at x = a:

$$D_\sigma(a) = \int_{-\infty}^{\infty} S_\sigma(a,a,\omega)\, d\omega \tag{62}$$

exceeds its counterpart $D_\sigma(0)$ at $x = 0$, provided the function $\psi_q(\omega)$ is sufficiently smooth. Figures 1 and 3 show that, moreover, in the four-mode approximation the cross-spectral density is insensitive to change of the free stream velocity in the interval outside the first two eigenfrequencies.

On comparing the true maximal stress-spectral density (TMSSD) with its partial counterpart (PMSSD), based on terms corresponding to like modes of vibration alone (joint acceptances) - it turned out that recourse to the PMSSD results in 40-45% underestimation at some frequencies. Consequently, when taking into account the deformational pressure perturbations, the terms containing cross-acceptances may not be neglected in calculating the probabilistic response.

The convergence analysis, undertaken in order to determine the minimum number of modes to be taken into account, showed that $N = N^*$ is still too small: at least twelve modes are necessary for reliable estimation of log $\bar{S}_\sigma$ in the frequency interval $\bar{\omega} \leq 500$. The logarithm is plotted in Figures 5 and 6 against $\bar{\omega}$ at twelve mode approximation. For $M = 1.2$ the first maximum appears at $\bar{\omega} = 32$ and the second at $\bar{\omega} = 60$, while for $M = 1.8$ the first maximum shifts to $\bar{\omega} = 44$, and the second to $\bar{\omega} = 59$. The RMS of the maxima at $x = a$ again exceeds that at $x = 0$ (Figure 6). Log $\bar{S}_\sigma$ is plotted against $\bar{\omega}$ in Figure 7 with the last term in Equation (11) formally taken as zero, i.e. with only part of the deformational pressure perturbations taken into account (according to formula (22)). As it may be seen from Figure 7, such an approximation leads to an underestimate of the TMSSD, especially at the low frequencies.

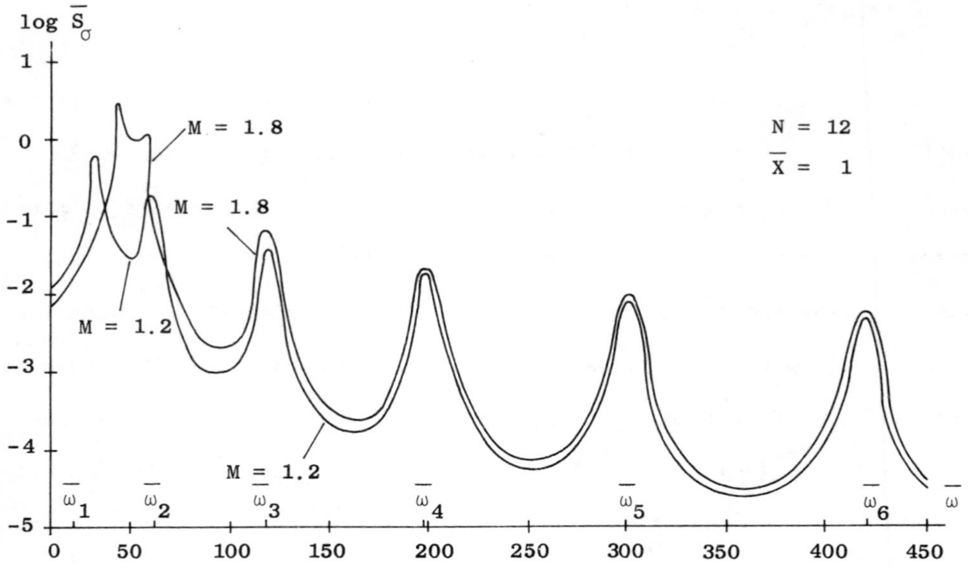

FIGURE 5

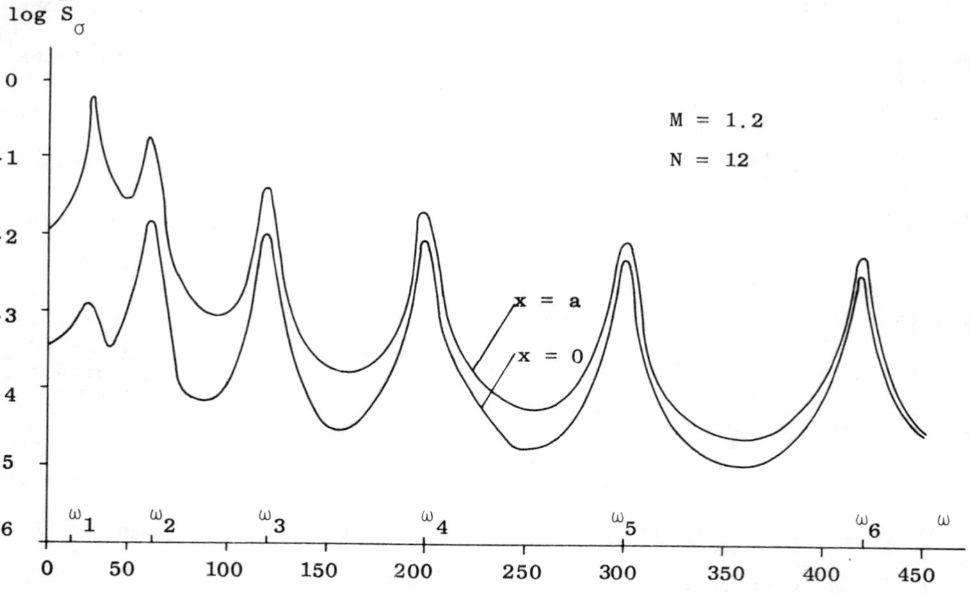

FIGURE 6

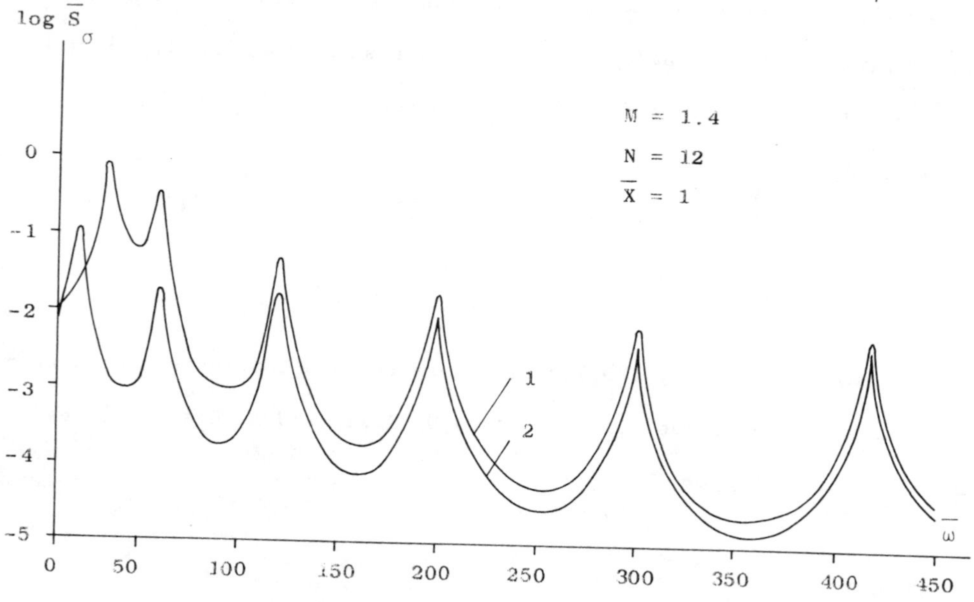

FIGURE 7

CONCLUSION

New analytical and numerical results on the interaction of flutter and random vibrations of plates are presented. It is shown that the proposed theory enables us to take flutter-type instability into account. When deformational pressure perturbations (according to the 'piston theory') are neglected, the one-mode approximation often used in the random vibration problems is inconsistent, and a certain minimum number of modes N^* (in this case, four) be included for safe determination of the critical flutter velocity and a satisfactory qualitative picture of probabilistic behaviour. Moreover, numerical analysis showed that at least twelve modes are needed for an accurate determination of the cross-spectral density of the maximal stresses. When the free-stream velocity is well removed from the critical flutter level, the cross-spectral density has the same number of maxima as the modes used; as the velocity increases, the first two maxima tend to draw together and eventually degenerate to a single maximum. It was shown that the RMS of the maximal stresses in a clamped plate is much smaller at $x = 0$ than at $x = a$, indicating that it is the latter section that should be considered first in the acoustic fatigue analysis.

ACKNOWLEDGEMENTS

The study was supported by the Technion Research and Development Foundation. The numerical analysis was carried out by H. Arkin and A. Ovadia at the Technion Computer Center.

REFERENCES

1 I. Elishakoff, Mean-square stability of elastic bodies in supersonic flow, Journal of Sound and Vibration 33 (1), 67-78 (1974).

2 I. Elishakoff, Statistical methods in vibrations. Part 1: Interaction of flutter and random vibration in plates. Technion-Israel Institute of Technology, Department of Mechanics, Report TDM 75-07, (March 1975).

3 E.M. Dowell, Panel Flutter: a review of aeroelastic stability of plates and shells, American Institute of Aeronautics and Astronautics Journal, 8, 385-399 (1970).

4 A. Powell, In Random Vibration (S.H. Crandall, ed.). On the response of structures to random pressures and to jet noise in particular, Wiley, New York, (1958). Chapter 8, pp. 187-229.

5 J.F. Wilby, In Jahbuch 1964 der Wissenschftlichen Gesselschaft für Luft-und Raumfahrt. Turbulent boundary layer pressure fluctuations and their effect on adjacent structures (1964).

6 M.J. Crocker, The response of a supersonic transport fuselage to boundary layer and to reverberant noise. Journal of Sound and Vibration 9, 6-20, (1969).

7 I. Elishakoff and B. Efimtsoff, Vibrations of unbounded plates in a random force field, Soviet Applied Mechanics, 8 (8), 103-107 (1972). (Consultants Bureau, New York).

8 W.J. Chyu and M.K. Au-Yang, Random response of rectangular panels to the pressure field beneath a turbulent boundary layer in subsonic flows, NASA TND-6970, (October 1972).

9 M.J. Crocker and R.W. White, Response of Lockheed L-2000 Supersonic Transport Fuselage to turbulent and to reverberant noise, Wyle Laboratories - Research Staff Consulting Report No. WCR 66-11 (1966).

10 Y.K. Lin, Probabilistic Theory of Structural Dynamics, McGraw-Hill Book Company, New York, p. 219 (1967).

11 P.M. Belcher, Prediction of boundary layer turbulence spectra and correlations for supersonic flight. Proceedings 5th International Congress on Acoustics, Liege (1965) Paper L54.

12 W.V. Speaker and C.M. Ailman, Spectra and space-time correlations of the fluctuating pressure at a wall beneath a supersonic turbulent boundary layer perturbed by steps and shock waves. NASA CR-486, (1966).

13 V.V. Bolotin and U.N. Moskalenko, Vibration of Plates, in Strength, Stability, Vibration (I. Birger and Ya. Panovko, Eds.) "Mashinostroenie" Publishing House, Moscow, (1968), pp. 370-416 (in Russian).

14 V.V. Bolotin, Yu. N. Novichkoff and Yu. Yu. Sheveiko, Theory of Aerohydroelasticity in Strength, Stability, Vibrations (I. Birger and Ya. Panovko, Eds.), "Mashinostroenie" Publishing House, Moscow, (1968), pp. 487-488 (in Russian).

15 H. Ashley and G. Zartarian, Piston theory - a new aerodynamic tool for the aeroelastician, Journal of the Aeronautical Sciences, 23 (12), 1109-1118, (1956).

Dr. I. Elishakoff, Department of Aeronautical Engineering, Technion - Israel Institute of Technology, Haifa, Israel.

DISCUSSION

Questioned about the form of the function S_q in equation (53) (Kozin) Dr. Elishakoff explained that this function is the coefficient of the cross-spectral density. The cross-spectral density is represented as a product of two functions, one of them being the coefficient of the spectral density in the longitudinal direction and the second the lateral direction.

Professor Clarkson asked if the empirical relationships used for the cross correlation and cross-spectral functions were based on subsonic results such as those of Wilby Reference [5]. Dr. Elishakoff replied that because the scale of the correlation function goes to infinity as ω goes to zero he took the correlation functions for supersonic results and used corrected coefficients in the expressions for cross-spectral density References [1, 9, 11, 12]. If noncorrected coefficients are used for computation of internal noise levels in stiffened cylindrical shell enclosing the acoustic medium, the results are about 10 dB higher than for corrected coefficients in the expressions for cross-spectral density (see: I. Elishakoff, 1971, Ph.D. Thesis, Department of Dynamics and Strength of Machines, Moscow Energy Institute: Vibration fields in cylindrical shells, excited by random forces. Chapter 5, Influence of supersonic flow on random vibration and noise in shells, pp. 202-237).

Questioned on the effect of nonlinearity (Clarkson) Dr. Elishakoff said that the nonlinear analysis had been started. The present work was trying to identify the number of terms which must be included in the linear analysis in preparation for extension to the nonlinear case.

K SOBCZYK and D B MACVEAN

Non-stationary random vibrations of systems travelling with variable velocity

INTRODUCTION

There are engineering problems in which the excitation of a system is caused by its traversal of a fixed profile. For example, a road vehicle receives a time-varying displacement excitation from the undulations of the road on which it runs. Often, the roughness of a profile is stochastically irregular and hence both the excitation and response of a travelling system are random.

Various simplified models of travelling vibratory systems with profile-imposed excitation can be considered. Many response problems associated with such models have been studied in the literature and significant practical information has been obtained. The following two main assumptions are usually made:

(a) the profile is a statistically homogeneous (stationary) random function;

(b) the traversal velocity of the system is constant.

Under the above assumptions the basic characteristics of the system response can be determined by use of standard methods of stochastic dynamics [1], [2], [3]. Usually the spectral method is used for calculation of stationary response.

There is no evidence to doubt stationarity of most real profiles (the existing measurements of real road surfaces - e.g. [4] - seem to justify assumption (a)). However the assumption concerning the uniformity of traversal velocity is evidently not strictly true and it is usually introduced for simplification of the analysis. Often, the actual velocity of travelling systems is variable. Moreover, when the variations are due to imperfections in speed control arrangements and some external factors, the traversal velocity can depart randomly from a constant nominal value. It is therefore of great practical significance to obtain information about the response of a system travelling with variable velocity. It is also of interest to clarify what kinds of probabilistic and statistical problems are produced by variable velocity. However, such a more sophisticated formulation of the problem leads to essential difficulties in the analysis. First of all, variable velocity generally introduces non-stationarity in the response.

Various problems of non-stationary response of physical systems have been considered in the literature and interesting results for some special problems have been obtained (e.g. [2], [5], [6], [7]). Non-stationarity generated by variable traversal velocity of a system has, however, its specific features and, as first indicated in paper [7], introduces specific difficulties.

In the present paper an effort has been made to provide a systematic approach to the problem of the response of a travelling system with variable traversal velocity. Attention is focussed on the non-stationary response generated by deterministically variable velocity, but also the formulation and some results concerned with randomly-varying traversal velocity are reported. A complete analysis of travelling systems encountered in practice (e.g. road vehicles) generally requires somewhat complicated models. However, to display the main features of the response associated with variable traversal velocity and to obtain quantitative results we employ here a tractable, simple model shown at Figure 1. The non-stationarity generated by variable velocity is clearly not of the separable type. The concept of evolutionary spectra also does not seem easily applicable. So, a direct

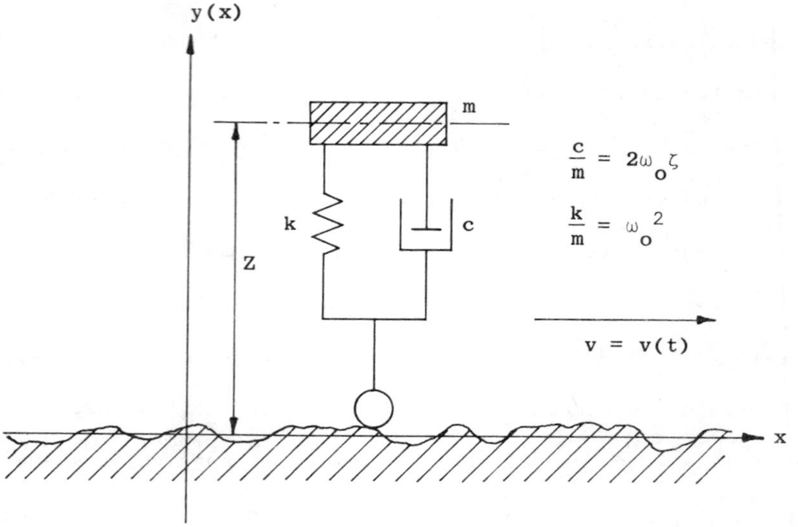

Figure 1 Vehicle model

time-domain approach is adopted in the analysis of the response with deterministically variable velocity. The problem with randomly-varying velocity can be analysed by use of different methods depending on the profile description

DETERMINISTICALLY VARYING VELOCITY

Formulation. General Analysis

Let a profile be described by the random function $y(x)$ of spatial coordinate x (horizontal distance). It will be assumed that $y(x)$ is statistically homogeneous (stationary) and Gaussian with $\langle y(x) \rangle = 0$ and correlation function $R_y(x_1, x_2) = R_y(x_2 - x_1) = R_y(\tilde{x})$.

If a system is traversing the profile in the x-direction the input to the system due to the profile irregularities can also be considered as a function $y(t)$ of time t. In general, $x = x(t)$ and the functions $y(x)$ and $y(t)$

are related by

$$y(x(t)) = y(t) \tag{1}$$

In the case of constant traversal velocity u, $x = x(t) = ut$ and simply $y(t) = y(ut)$. In this case

$$R_y(t_1, t_2) = \langle y(ut_1)y(ut_2)\rangle = R_y(u(t_2 - t_1)) \tag{2}$$

and process y(t) is also stationary; $R_y(t_1, t_2) = R_y(\tau)$, $\tau = t_2 - t_1$.

If the function $x = x(t)$ is an arbitrary (strictly increasing) deterministic function of t then

$$R_y(t_1, t_2) = \langle y(x(t_1))y(x(t_2))\rangle = R_y(x(t_2) - x(t_1)). \tag{3}$$

It is seen that in general the excitation process y(t) is non-stationary and Gaussian.

Let $t = 0$ be the initial instant of time at which the system begins to travel with variable velocity. Two different cases can be distinguished.

1. The system starts its traversal motion at $t = 0$ with velocity $v(0) = 0$ and with acceleration a(t); in this case function x(t) is defined as follows

$$\ddot{x}(t) = a(t), \qquad x(0) = \dot{x}(0) = 0 ,$$

that is

$$x(t) = \begin{cases} 0 , & t \leq 0 \\ x_1(t), & t > 0 \end{cases} \qquad v(t) = \begin{cases} 0 & t \leq 0 \\ \dot{x}_1(t), & t > 0 \end{cases} \tag{4}$$

where

$$x_1(t) = \int_0^t (t - \tau)\, a(\tau)\, d\tau , \qquad t > 0 . \tag{5}$$

2. The system travels with constant velocity u from the 'infinite' past and then at time $t = 0$ it begins to accelerate with acceleration a(t) i.e.

$$x(t) = \begin{cases} ut & t < 0 \\ ut + x_1(t), & t > 0 \end{cases} \qquad v(t) = \begin{cases} u & t < 0 \\ u + \dot{x}_1(t), & t > 0 \end{cases} \tag{6}$$

The function a(t) is usually such that $v(t) > 0$. Case 1 can be understood as a particular case of Case 2.

By virtue of (1) and (3) we have, respectively:

In Case 1:

$$y(t) = y(x_1(t)) , \qquad t > 0 \tag{7}$$

$$R_y(t_1, t_2) = R_y(x_1(t_2) - x_1(t_1)) , \qquad t_1, t_2 > 0;$$

In Case 2:

$$y(t) = \begin{cases} y(ut) , & t < 0 \\ y(ut + x_1(t)) , & t > 0; \end{cases} \tag{8}$$

$$R_y(t_1, t_2) = \begin{cases} R_y(u(t_2 - t_1) + x_1(t_2) - x_1(t_1)), & t_1, t_2 > 0 \\ R_y(u(t_2 - t_1) - x_1(t_1)), & t_1 > 0, \quad t_2 < 0 \\ R_y(u(t_2 - t_1) + x_1(t_2)) & t_1 < 0, \quad t_2 > 0 \\ R_y(u(t_2 - t_1)) & t_1, t_2 < 0 \end{cases} \tag{9}$$

When the acceleration a(t) tends to zero, formula (2) is obtained for all values of t_1, t_2. General formula (3) and relations (7), (9) show explicitly that the non-stationarity of the excitation y(t) is of a formal origin; the 'physical' randomness occurring in the problem (roughness of a profile) is stationary.

Let z(t) denote the vertical displacement of the system traversing the profile y(t). We wish to determine the statistical properties of the response process z(t) when the excitation has the correlation function given by (7) or (9). In what follows the process will be characterised by its mean square $\langle z^2(t) \rangle$ and the system will be idealised by a linear and single-degree of freedom model (Figure 1).

The governing equation for the vertical displacement z(t) of the mass of the given system is

$$\ddot{z}(t) + 2h\,\dot{z}(t) + \omega_o^2\, z(t) = 2h\,\dot{y}(t) + \omega_o^2\, y \tag{10}$$

where $h = \omega_o\zeta$; ω_o and ζ denote the natural frequency and damping ratio of the system, respectively. The initial conditions required in Case 1, are the following

$$z(0) = y(0), \qquad \dot{z}(0) = 0 \tag{11}$$

The response of the system and its mean square are respectively:

<u>In Case 1</u>:

$$z(t) = z_1(t) + y(0)\,\psi(t) \tag{12}$$

$$z_1(t) = \int_0^t p(\tau)\, y(t - \tau)\, d\tau \tag{13}$$

$$\psi(t) = e^{-ht}(\cos\lambda_o t - \frac{h}{\lambda_o}\sin\lambda_o t), \qquad \lambda_o = (\omega_o^2 - h^2)^{\frac{1}{2}} \tag{14}$$

$$p(\tau) = \frac{1}{\lambda_o}\, e^{-ht}\left[2h\lambda_o \cos\lambda_o\tau - (2h^2 - \omega_o^2)\sin\lambda_o\tau\right] \tag{15}$$

$$\langle z^2(t)\rangle = \langle z_1^2(t)\rangle + 2\psi(t)\,\langle y(0)\, z_1(t)\rangle + \psi^2(t)\,\langle y^2(0)\rangle \tag{16}$$

$$\langle z_1^2(t)\rangle = \int_0^t\!\!\int_0^t p(\tau_1)\, p(\tau_2)\, R_y(x_1(t-\tau_2) - x_1(t-\tau_1)\, d\tau_1\, d\tau_2 \tag{17}$$

$$\langle y(0)\, z_1(t)\rangle = \int_0^t p(\tau)\, R_y(x_1(t - \tau))\, d\tau \tag{18}$$

<u>In Case 2</u>:

$$z(t) = \int_0^\infty p(\tau)\, y(t - \tau)\, d\tau \tag{19}$$

$$\langle z^2(t)\rangle = \int_0^\infty\!\!\int_0^\infty p(\tau_1)\, p(\tau_2)\, R_y(t - \tau_1, t - \tau_2)\, d\tau_1\, d\tau_2 \tag{20}$$

where R_y occurring in (20) is given by formula (9). In the special case when acceleration tends to zero, formula (20) gives the classical expression for mean square of stationary response

$$\langle z^2 \rangle = \int_0^\infty \int_0^\infty p(\tau_1)\, p(\tau_2)\, R_y(u(\tau_2 - \tau_1))\, d\tau_1\, d\tau_2 = \text{const.} \tag{21}$$

The comparison of non-stationary response (20) with stationary (zero-acceleration) response (21) at a given velocity v_1 is obtained by comparing (20) with (21) making the velocity u in (21) equal to $v_1 = v(t_1) = u + \dot{x}(t_1)$. For determination of component stresses the relative displacement $w(t) = z(t) - y(t)$ should be considered. The mean square of this relative motion is therefore given by (in both cases, respectively)

$$\langle w^2(t) \rangle = \langle z^2(t) \rangle - 2\langle y(t)\; z(t) \rangle + \sigma_y^2 \tag{22}$$

where σ_y^2 is the variance (here also mean square) of the profile and $\langle y(t)\; z(t) \rangle$ is in both cases easily expressible by a single integral.

Remark: In some applications, for instance in road vehicle analysis, such quantities as the vertical velocity $\dot{z}(t)$ and the acceleration of vertical displacement $\ddot{z}(t)$ are also of practical interest and even more the joint characteristics of vector random process $\{z(t), \dot{z}(t), \ddot{z}(t)\}$ are required if some reliability problems are to be solved.

The intention of this study is to indicate the analytical technique for treatment of variable traversal velocity rather than to calculate all characteristics of practical interest. It is, however, worth noting that such quantities as $\langle \dot{z}^2(t) \rangle$ and $\langle \ddot{z}^2(t) \rangle$ can be calculated in a similar way using the following relations

$$\dot{z}(t) = \dot{z}_1(t) + y(0)\ \psi(t)$$

$$= \int_0^t p_1(\tau)\ y(t - \tau)\ d\tau + p(0)\ y(t) + y(0)\ \dot{\psi}(t)$$

$$p_1(\tau) = \frac{1}{\lambda_o} e^{-h\tau} \left[(2h^3 - \omega_o^2 h + 2h\lambda_o^2) \sin\lambda_o\tau - (4h^2 + \omega_o^2)\lambda_o \cos\lambda_o\tau\right]$$

$$\ddot{z}(t) = 2h \left[y(t) - z(t)\right] + \omega_o^2 \left[y(t) - z(t)\right]$$

and expressions (8), (9). Also, the problem of calculation of characteristics of the process $\{z(t), \dot{z}(t), \ddot{z}(t)\}$ is tractable due to Gaussianity of the response $z(t)$.

Response of a System with Uniform Traversal Acceleration

In order to obtain more concrete results a special form of the function $x = x(t)$ has to be selected. We shall consider in detail the case when the velocity of the system varies with constant acceleration a. In this case formula (5) gives

$$x_1(t) = \frac{1}{2} a\ t^2, \qquad t > 0 \tag{23}$$

and we have, respectively

In Case 1:

$$<z_1^2(t)> = \int_0^t\int_0^t p(\tau_1)\ p(\tau_2)\ R_y(at(\tau_1 - \tau_2) - \frac{1}{2} a(\tau_1^2 - \tau_2^2))\ d\tau_1\ d\tau_2 \tag{24}$$

$$<y(0)\ z_1(t)> = \int_0^t p(\tau)\ R_y(\frac{1}{2} a(t - \tau)^2)\ d\tau \tag{25}$$

In Case 2:

$$<z^2(t)> = I_1(t) + I_2(t) + I_3(t) + I_4(t) \tag{26}$$

$$I_1(t) = \int_0^t\int_0^t p(\tau_1)\, p(\tau_2)\, R_y((u + at)(\tau_1 + \tau_2) - \frac{1}{2} a(\tau_1^2 - \tau_2^2)\, d\tau_1\, d\tau_2 \tag{27}$$

$$I_2(t) = \int_t^\infty\int_0^t p(\tau_1)\, p(\tau_2)\, R_y(u(\tau_1 - \tau_2) - \frac{1}{2} a(t - \tau_1)^2)\, d\tau_1\, d\tau_2 \tag{28}$$

$$I_3(t) = \int_0^t\int_t^\infty p(\tau_1)\, p(\tau_2)\, R_y(u(\tau_1 - \tau_2) + \frac{1}{2} a(t - \tau_2)^2)\, d\tau_1\, d\tau_2 \tag{29}$$

$$I_4(t) = \int_t^\infty\int_t^\infty p(\tau_1)\, p(\tau_2)\, R_y(u(\tau_1 - \tau_2))\, d\tau_1\, d\tau_2 \tag{30}$$

Formula (26) and integrals (27) - (30) give the required mean square of the response in the general Case 2. For a given correlation function of the profile, integrals (27) - (30) can be evaluated numerically. Analytical calculations are rather involved mainly due to integrals $I_2(t)$ and $I_3(t)$. In what follows we will focus our attention on Case 1 and on formula (24).

To simplify integral (24) the following new variables r and s are introduced

$$r = \tau_1 - \tau_2, \qquad \tau_1 = \frac{1}{2}(r + s)$$
$$s = \tau_1 + \tau_2, \qquad \tau_2 = \frac{1}{2}(s - r \tag{31}$$

In terms of r and s the product of the unit response function (15) is

$$p(\tau_1)\, p(\tau_2) = p(\frac{r+s}{2})\, p(\frac{s-r}{2}) = \frac{1}{2\lambda_o^2}\, e^{-hs} \left[(c_1^2 + c_2^2) \cos\lambda_o r + (c_1^2 - c_2^2) \cos\lambda_o s - 2c_1 c_2 \sin\lambda_o s\right] \tag{32}$$

where

$$c_1 = 2h\lambda_o, \quad c_2 = 2h^2 - \omega_o^2$$

Formula (24) takes now the following form

$$<z_1^2(t)> = \frac{1}{4\lambda_o^2}\left[(c_1^2 + c_2^2)J_1 + (c_1^2 - c_2^2)J_2 - 2c_1c_2J_3\right] \tag{33}$$

where

$$J_1 = \iint_D e^{-hs} \cos\lambda_o r \, R_y(atr - \frac{a}{2} rs) \, dr \, ds \tag{34}$$

$$J_2 = \iint_D e^{-hs} \cos\lambda_o s \, R_y(atr - \frac{a}{2} rs) \, dr \, ds \tag{35}$$

$$J_3 = \iint_D e^{-hs} \sin\lambda_o s \, R_y(atr - \frac{a}{2} rs) \, dr \, ds \tag{36}$$

and D is the domain in the rs - plane corresponding to the domain $\{0 < \tau_1 < t,\ 0 < \tau_2 < t\}$ in the $\tau_1\tau_2$ - plane. Often, the correlation function used in estimation of experimental measurements contain a modulus sign, that is they are functions of $|\tilde{x}| = |x_2 - x_1|$. The domain D splits according to the solution of the inequalities

$$\begin{aligned} D^+ &: atr - \frac{a}{2} rs \geq 0 , \\ D^- &: atr - \frac{a}{2} rs < 0 , \end{aligned} \qquad D = D^+ \cup D^- \tag{37}$$

that is

$$D^+ : \begin{cases} 0 < r < t \\ r < s < -r + 2t \end{cases} \qquad D^- : \begin{cases} -t < r < 0 \\ -r < s < r + 2t \end{cases} \tag{38}$$

Particular Case

The correlation functions of real profiles can be approximated by different analytical expressions depending on the type of roughness. Here, we shall take the following expression

$$R_y(\tilde{x}) = \sigma_y^2 \, e^{-\alpha|\tilde{x}|} , \qquad \tilde{x} = x_2 - x_1 , \qquad \alpha > 0 \tag{39}$$

which, as was shown in [8], constitutes an adequate estimation of the correlation function of a wide class of real road profiles. Numerical values of σ_y and α characterising the vertical variation of road roughness and its radius of correlation are different for specific roads.

In this case the double integrals (34) - (36) can be integrated once analytically. The choice of s as first variable of integration in J_1 but r in J_2 and J_3 leads to the results of the form

$$\frac{1}{\sigma_y^2} J_1 = \int_0^t f_1(r) \cos\lambda_0 r \, dr$$

$$\frac{1}{\sigma_y^2} J_2 = \int_0^t \{[f_2(s) + k_1 f_3(s)] \cos\lambda_0 s + k_2 f_3(s) \sin\lambda_0 s\} \, ds \qquad (40)$$

$$\frac{1}{\sigma_y^2} J_3 = \int_0^t \{[f_2(s) - k_1 f_3(s)] \sin\lambda_0 s + k_2 f_3(s) \cos\lambda_0 s\} \, ds$$

where

$$f_1(r) = \frac{2}{br - h} \{\exp[-h(2t - r) - br^2] - \exp[-hr - br(2t - r)]\}$$

$$f_2(s) = \frac{2}{b(2t-s)} \exp(-hs) \{1 - \exp[-bs(2t - s)]\} \qquad (41)$$

$$f_3(s) = \frac{2}{bs} \exp[-h(2t - s)] [1 - \exp(-bs^2)] ,$$

$$k_1 = \cos 2\lambda_0 t, \quad k_2 = \sin 2\lambda_0 t, \quad b = \frac{\alpha a}{2} .$$

Formula (25) can be easily reduced to

$$\langle y(0) \, z_1(t)\rangle = \frac{\sigma^2}{\lambda_0} e^{-bt^2} \{ \int_0^t f_4(\tau) [c_1 \cos\lambda_0 \tau - c_2 \sin\lambda_0 \tau] \, d\tau\}$$

$$f_4(\tau) = \exp[-b\tau^2 + (2bt - h)\tau]. \qquad (42)$$

Integrals (40) and (42) were evaluated numerically for selected values of α and a.

For large values of t calculations were made for $\lambda_o t$ a multiple of π in order to enable use of Filon's integration technique. For comparison with the case of zero-acceleration the mean square of stationary response (21) was evaluated analytically (on the basis of equivalent spectral representation) with the resulting formula

$$\langle z^2(t)\rangle = \frac{\sigma_y^2}{2h} \frac{(4h^2 + \omega_o^2)\,\alpha u + 2h\omega_o^2}{\alpha^2 u^2 + 2h\alpha u + \omega_o^2} \tag{43}$$

The results for non-stationary, and stationary response are presented graphically in Figure 2. For small values of t a method based on Chebyshev expansions proposed in [11] was used in numerical integration because of the reliability of its error estimate. Also, by use of the same integration technique the mean $\langle y(t)\, z(t)\rangle$ occurring in (22) was calculated. The mean square of the relative displacement is shown in Figure 3.

RANDOMLY-VARYING VELOCITY

Sinusoidal Profile

In vehicle engineering useful design information has been obtained in the past using the assumption that the road profile is sinusoidal and that the traversal velocity is constant - c.f. [9]. It is interesting to study how these results are affected by random fluctuations of the velocity.

Let the traversal velocity of the system v(t) be a stationary and Gaussian random process with mean v(t) = u and autocorrelation function $\langle v(t_1)\, v(t_2)\rangle = R_v(\tau)$, $\tau = t_2 - t_1$. If v(t) is given, the space-time relation x = x(t) is defined by the stochastic integral

$$x = x(t) = \int_o^t v(t')\, dt' \tag{44}$$

Figure 2 Comparison of stationary and non-stationary mean square response

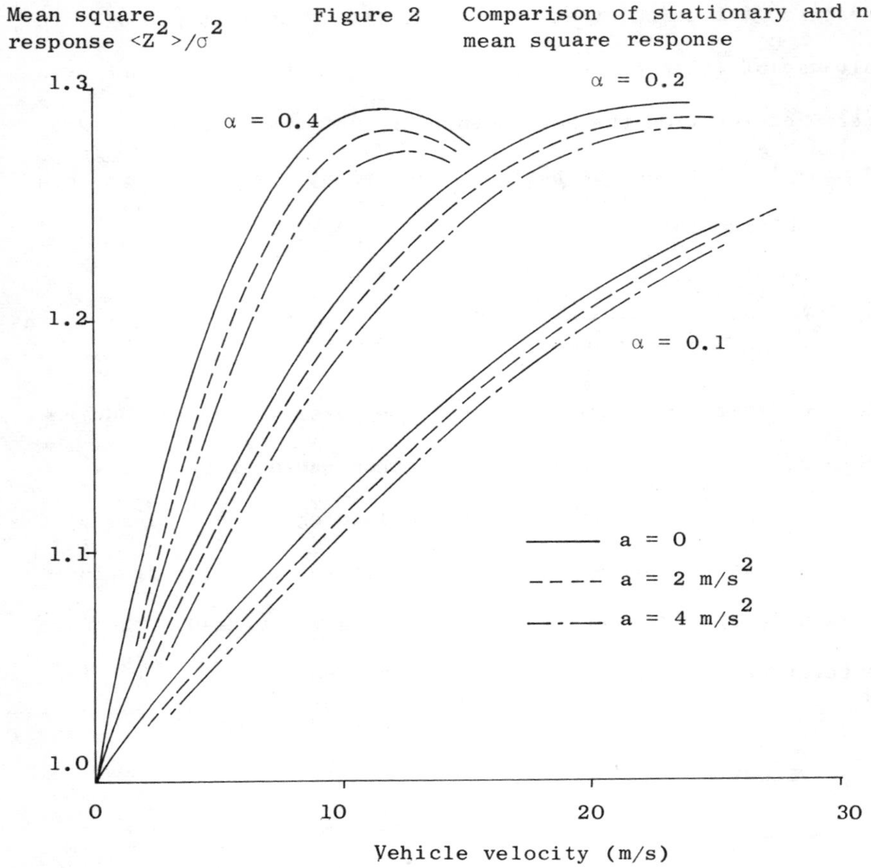

α: correlation parameter
a: constant value of vehicle acceleration

$$<x(t)> = ut, \quad R_x(t_1, t_2) = \int_0^{t_1}\int_0^{t_2} R_v(t'' - t')\, dt'\, dt'' \tag{45}$$

If the profile is described by a sinusoidal function $y(x) = A \sin q x$ of frequency q then, according to (1) the time-varying excitation of the system is

$$y(t) = A \sin q\, x(t) \tag{46}$$

where process x(t) and its characteristics are defined by (44) and (46).

It should be noted that, because of integration (44) and non-linear transformation (46), the excitation process y(t) is now both non-stationary and non-Gaussian.

The characteristics of the excitation $y(t)$ can be expressed in terms of characteristic functions of the Gaussian process $x(t)$, namely

$$\langle y(t)\rangle = A\langle \sin q\, x(t)\rangle = \text{Im}\, \phi_t(q) = Ae^{\frac{-q^2\sigma_x^2(t)}{2}} \sin q_1 t, \quad q_1 = uq \qquad (47)$$

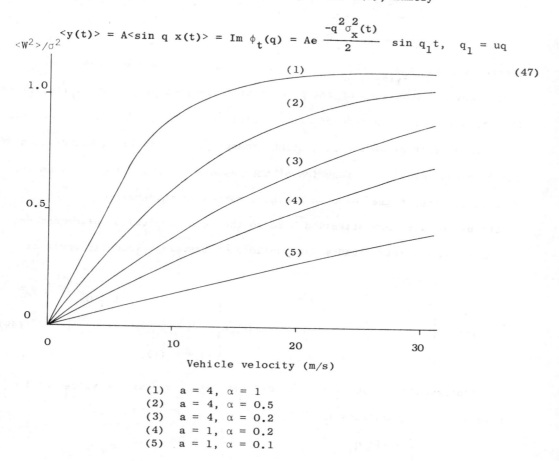

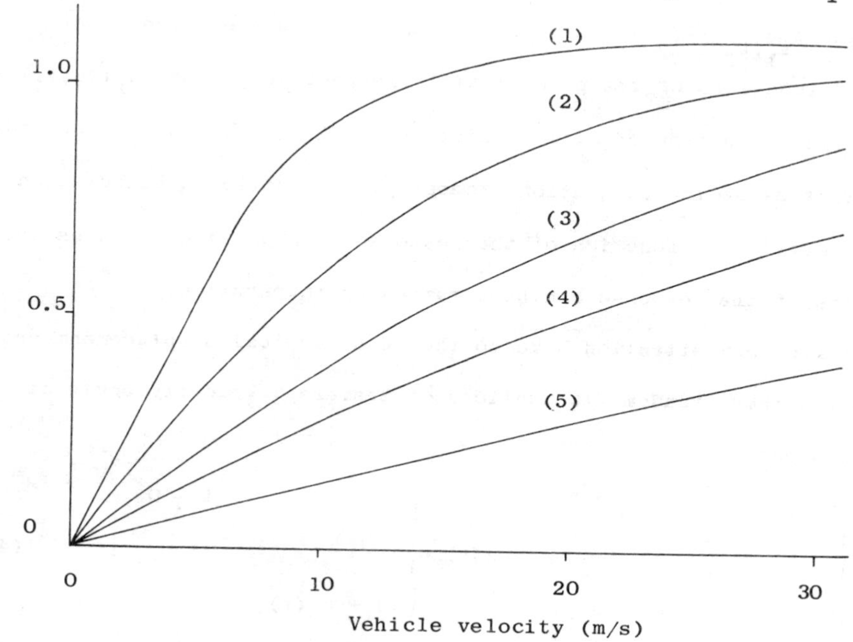

(1) a = 4, α = 1
(2) a = 4, α = 0.5
(3) a = 4, α = 0.2
(4) a = 1, α = 0.2
(5) a = 1, α = 0.1

α: correlation parameter   a: acceleration in m/s

Figure 3 Mean square of relative response W versus instantaneous vehicle velocity

$$R_y(t_1, t_2) = A^2 \langle \sin qx\,(t_1) \sin qx\,(t_2)\rangle$$

$$= \frac{1}{2} A^2 \left[\text{Re}\, \phi_{t_1,t_2}(q, -q) - \text{Re}\, \phi_{t_1,t_2}(q, q)\right]$$

$$= \frac{1}{2} A^2 \Big\{ \cos q_1(t_1 - t_2) e^{-\frac{q^2}{2}\left[\sigma_x^2(t_1) + \sigma_x^2(t_2) - 2 K_x(t_1, t_2\right]}$$

$$+ \cos q_1(t_1 + t_2)\, e^{-\frac{q^2}{2}\left[\sigma_x^2(t_1) + \sigma_x^2(t_2) + 2 K_x(t_1, t_2)\right]} \Big\} \qquad (48)$$

where $\phi_t(\lambda)$ and $\phi_{t_1,t_2}(\lambda_1, \lambda_2)$ are one-dimensional and two-dimensional characteristic functions of the process x(t), respectively, and $K_x(t_1,t_2)$ is the autocovariance function of x(t). Relations (47), (48) with (45) provide the characteristics of the excitation process y(t) required in calculation of the mean and correlation function of the response. In particular cases the characteristics of the response can be obtained by integration.

Let us restrict our attention here to the mean vertical displacement and to the situation where random fluctuations of traversal velocity begin at t = 0. In this case

$$v(t) = \begin{cases} u, & t \leq 0 ; \\ u + v_1(t), & t > 0 ; \end{cases} \qquad x(t) = \begin{cases} ut, & t \leq 0 \\ ut + x_1(t), & t > 0 \end{cases} \qquad (49)$$

where processes $v_1(t)$ and $x_1(t)$ are defined only for positive values of t. The mean of the excitation is

$$\langle y(t)\rangle = \begin{cases} A \sin q_1 t & t \leq 0 \\ A\, e^{\frac{-q^2\sigma_x^2(t)}{2}} \sin q_1\, t, & t > 0 \end{cases} \qquad (50)$$

The mean of the response is expressed by the following integral

$$\langle z(t)\rangle = \int_{-\infty}^{t} p(t - \tau)\, \langle y(t)\rangle\, d\tau \qquad (51)$$

where $\langle y(\tau)\rangle$ is given by (50).

Taking, for illustration, the exponential form of the correlation function of random velocity $R_{v_1}(\tau) = \sigma^2_{v_1} \exp(-\alpha|\tau|)$ we obtain

$$\sigma^2_{x_1}(t) = \frac{2\sigma_{v_1}}{\alpha^2}(\alpha t + e^{-\alpha t} - 1) \qquad (52)$$

The results from calculations of (51) using (50) and (52) are shown in Figure 4.

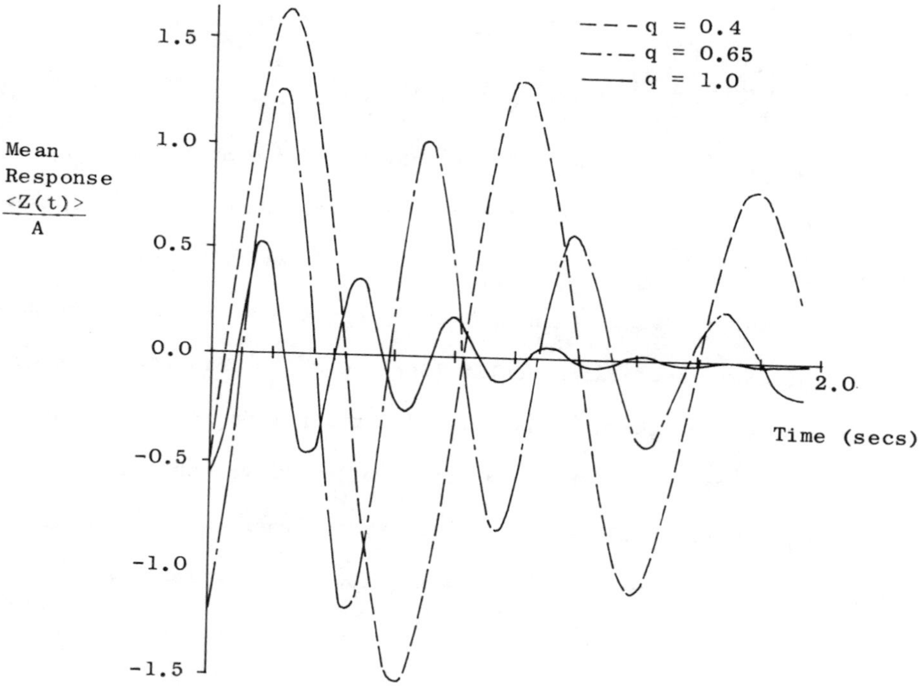

Figure 4 Effect of random velocity variations on mean response (sinusoidal profile, d = 0.4, u = 20 m/s, σ_ν = 2 m/s)

Random Profile: Spatial Description

For full understanding of the problem of the response of a mechanical system travelling with variable velocity it is important to consider the situation where both the profile and traversal velocity are random. Such a problem has been investigated systematically in the paper [10].

In this case, when the profile is described by a random function of horizontal distance x and there is no deterministic relation between x and t, the problem was found to be tractable if x and not t was chosen as the independent variable. The motion described by equation (10) is, therefore (by formal change of variable) expressed by the following equation with random coefficients

$$m\,v^2(x)\,\frac{d^2z}{dx^2} + \left[mv(x)\,\frac{dv(x)}{dx} + c\,v(x)\right]\frac{dz}{dx} + k\,z = c\,v(x)\,\frac{dy}{dx} + ky \qquad (53)$$

where function z(x) describing the vertical displacement is such that z(x) = z(t) for corresponding x and t and v = v(x) is the randomly-varying velocity of the system. For small random fluctuations of the velocity ($v(x) = u + \varepsilon v_1(x)$) and under the assumption that the profile and randomly-varying velocity are stationary and statistically independent the stationary response of the system was investigated by use of the perturbation technique and spectral method. It has been found that the effect of small random velocity fluctuations on response spectral density is of second order in ε. Nevertheless, as analysis of particular practical cases shows, this effect can be considerable.

RESULTS AND DISCUSSION

In this paper we have provided a methodical approach to the study of the non-stationary response generated by variable velocity of a travelling system. Both deterministically and randomly varying velocity was considered. The formulations and solutions of the problems are presented first in general terms to enable analysis of different practical situations. The effect on vertical displacement caused by uniform acceleration of a system traversing the random profile was investigated in detail. Also the detailed analysis of the effect of randomly varying velocity on mean vertical displacement of a system traversing a sinusoidal profile was reported. To give a full picture of the problems associated with variable velocity of travelling systems paper [10] was mentioned in which the analysis of the response to random profile excitation of a system travelling with randomly varying velocity was performed.

In order to obtain quantitative results a tractable, simplified model of a travelling system is taken into consideration. However, to make the results practically relevant, a realistic profile correlation function and typical parameters of a vehicle model were assumed in computations; according to previous experimental investigations of vehicle suspension parameters a damping ratio $\zeta = 0.35$ and natural frequency $\omega_o = 2\pi . f_o$, $f_o = 1.58$ Hz were used. Also typical values of the constant mean velocity were taken.

The results can be summarised as follows:

1. When a linear system traverses a stationary and Gaussian random profile with deterministically variable velocity, the response is Gaussian but non-stationary.

The results plotted in Figure 2 show that the difference between mean-square displacement response of an uniformly accelerating system and one travelling with zero acceleration is small. This result confirms qualitatively the result obtained earlier in [7], for another form of correlation function of a random profile. Figure 2 shows that the effect of uniform acceleration is to decrease the mean-square response only 2 - 3% below the zero-acceleration value for a wide range of profile correlation parameters α. It should be noted that this small difference is associated with the values of system parameter of typical road vehicles (e.g. for a damping ratio of $\zeta = 0.01$ a difference of about 100% occurs). The same assertion holds for relative vertical displacement. Thus, in so far as a single degree of freedom model is appropriate and vertical displacement is concerned, the result obtained (together with that of [7]) indicates a possible simplification of engineering vehicle analysis by neglecting the non-stationary effect generated by uniform acceleration. Figure 3 shows the mean square of relative non-stationary displacement as a function of velocity. It is seen that for typical vehicle velocities and for correlation parameter values greater than 0.3, the mean square of the relative motion is of the same order as that of the road roughness.

2. When a linear system traverses a sinusoidal profile (or more generally deterministically described profile) with stationary and Gaussian randomly varying traversal velocity, the response is non-Gaussian and non-stationary. The mean and correlation function of this response can be determined by using the characteristic functions of the integral of random velocity. Figure 4 shows that the effect of random variations of velocity starting at $t = 0$ is to reduce the mean vertical displacement to zero over a period of time which depends on the wavelengths of the profile ($s = \frac{2\pi}{q}$); the shorter this wavelength is then the shorter the transient time required.

3. When a linear system traverses a stationary and Gaussian random profile with a stationary and Gaussian random traversal velocity and these two randomnesses are assumed to be statistically independent, then the response of a system (travelling from the 'infinite' past) is stationary but non-Gaussian. It turns out that the spatial description of such a problem with use of horizontal distance x as the independent variable makes the analysis possible and leads to interesting results. This requires, however, the

specification of the statistics of the random velocity in terms of horizontal distance.

ACKNOWLEDGEMENT

The authors wish to express their gratitude to Professor J.D. Robson for his continued interest in this work and for valuable discussions.

REFERENCES

1 J.D. Robson, C.J. Dodds, D.B. Macvean, and V.R. Paling, Random Vibrations Udine, Springer-Verlag (1971).

2 Y.K. Lin, Probablistic Theory of Structural Dynamics McGraw-Hill (1967).

3 K. Sobczyk, Methods of Statistical Dynamics, Polish Scientific Publ. (in Polish). (1973).

4 C.J. Dodds and J.D. Robson, The description of road surface roughness. Jour. Sound Vibr. 31 (2) 175-183 (1973).

5 J.B. Roberts, On the harmonic analysis of evolutionary random vibrations. Jour. Sound Vibr. 2(3), 336-352 (1965).

6 J.K. Hammond, On the response of single and multi-degree of freedom systems to non-stationary random excitation. Jour. Sound Vibr. 7(3), 393-416 (1968).

7 Y.J. Virchis and J.D. Robson, Response of an accelerating vehicle to random road undulation. Jour. Sound Vibr. 18(3), 423-427 (1971).

8 I.G. Parkhilovski, Investigation of the probabilistic characteristics of the surface of various types of roads. Automobilnaya Promysh, 18-22 (1968).

9 R.V. Rotenberg, Vehicle Suspension - Vibrations and Smoothness of Ride, Mashinostrogenye, Moscow (in Russian) (1972).

10 K. Sobczyk, D.B. Macvean and J.D. Robson, Response to profile-imposed excitation with randomly varying traversal velocity. (submitted to Jour. Sound Vibr.) (1976).

11 C.W. Clenshaw and A.R. Curtis, A method for numerical integration on an automatic computer. Numerishe Math. 2, 197-205 (1960).

Dr. K. Sobczyk, Institute of Fundamental Technological Research, Polish Academy of Sciences, Warsaw, Poland.

Dr. D.B. Macvean, Department of Mechanical Engineering, University of Glasgow, Scotland.

DISCUSSION

It was suggested (Shinozuka) that (i) road roughness in the lateral direction should be included in the model and (ii) that problems of this type should be treated by digital simulation since this permits the inclusion of more realistic conditions and also provides the actual time history of motion which might be required for example in fatigue studies. The author replied that an aim of the presentation had been to point out that even for a simple model the variable velocity of the vehicle results in non-stationary problems of a type for which a methodical treatment had been offered resulting in integrals (requiring numerical evaluation) that are simpler than those obtained hitherto. The analysis also indicated when it might be necessary to carry out a more detailed study, e.g. in the case of randomly fluctuating velocity the approximations to order ε^2 resulted in departures from Gaussianity, whilst approximations to order ε did not. It was added (Robson) that the order of magnitude of the results had been useful since (in so far as mean square displacement responses are concerned) the use of constant velocity theories had been justified because there was little difference in the results for accelerating and nonaccelerating vehicles.

It was noted (Vanmarke) that the practical situation of a vehicle at constant velocity moving over a road surface having a sudden change in character (smooth to rough) is a cause of passenger discomfort and is another case of nonstationary response.

Finally attention was drawn (Nigam) to some work on a similarly posed problem that had been done at Kanpur and reported in [12] in relation to aircraft taxiing and takeoff and landing. Both time and frequency domain methods had been employed, the latter approach being used when the problem is posed in space co-ordinates since the situation is then one of a variable filter excited by a homogeous (stationary) process. It was noted too that in their experience comparisons of mean square accelerations had shown greater departures between stationary and nonstationary situations than had mean square displacement.

Additional references [13], [14] were submitted by F. Kozin [Ed.].

REFERENCES

12 D. Yadav, Response of Moving Vehicles to Ground Induced Excitation. Ph.D Thesis, Indian Institute of Technology, Kanpur (1976).

13 F. Kozin, J.L. Bogdanoff and L.J. Cote, Introduction to a statistical theory of land locomotion I - IV. Jour. of Terramechanics Vol. 2 (1965), Vol. 3 (1966).

14 F. Kozin, J.L. Bogdanoff and L.J. Cote, An Atlas of Off-Road Ground Surface, Power Spectral Densities and Data Acquisition Techniques. Tech. Report 9387 LL109, Land Locomotion Laboratories. U.S. Army Tank Automotive Centre. (Sept. 1966).

L FRÝBA

Response of bridges to moving random loads

INTRODUCTION

The random stresses in bridges are caused by irregular composition and speed of trains or vehicles in traffic flow, by random track irregularities, random motions of vehicles, etc.

In the stochastic approach, the bridge is assumed to be a mechanical system (e.g. the simple beam) of given properties. Then, going from the knowledge of the statistical characteristics of the load and/or of track irregularities, the appropriate statistical characteristics of the deflections and stresses of the bridge may be analysed.

Therefore, two basic cases of dynamic effects on bridges have been investigated. The first case idealizes the random stresses in short span bridges, while the second case assumes the long span bridges.

The bridge is usually idealized by a Bernoulli-Euler beam with viscous damping which is described by the following differential equation:

$$EJ \frac{\partial^4 v(x,t)}{\partial x^4} + \mu \frac{\partial^2 v(x,t)}{\partial t^2} + 2\mu\omega_b \frac{\partial v(x,t)}{\partial t} = p(x,t) \qquad (1)$$

where

$v(x,t)$ - vertical deflection of the beam at point x and time t

EJ - constant bending stiffness of the beam

μ - constant mass per unit length of the beam

ω_b - coefficient of the viscous damping of the beam

$p(x,t)$ - external load per unit length of the beam

BEAM UNDER MOVING RANDOM FORCE

In the first case, a concentrated force P(t) is assumed, see Figure 1. The force is random variable in time and it moves with a constant speed c along the beam from left to its right hand side. Therefore, the load p(x,t) in the equation (1) is

$$p(x,t) = \delta(x - ct)\, P(t) \tag{2}$$

Here $\delta(x)$ denotes the Dirac delta function.

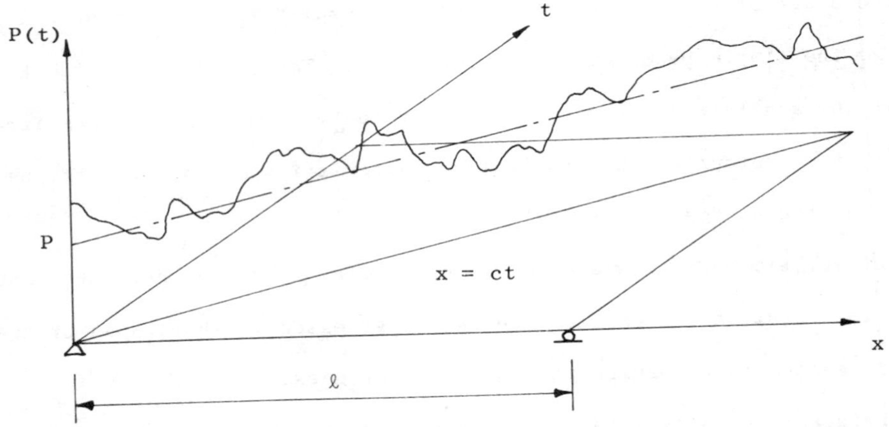

Figure 1 Beam under moving random force P(t)

The force P(t) has a constant mean value P and a zero mean random value $\overset{o}{P}(t)$

$$P(t) = P + \overset{o}{P}(t) \tag{3}$$

$$E\left[P(t)\right] = P \tag{4}$$

where $E[\,]$ represents the linear operator of the mean value.

The convariance of the force P(t) is

$$C_{PP}(t_1,t_2) = E\left[\overset{o}{P}(t_1)\,\overset{o}{P}(t_2)\right] \tag{5}$$

and the convariance of the load p(x,t) is calculated from the equations (2) to (4) with respect to the definition of the covariance

$$C_{pp}(x_1,x_2,t_1,t_2) = E\left[\overset{o}{p}(x_1,t_1)\ \overset{o}{p}(x_2,t_2)\right]$$

$$= \delta(x_1-ct_1)\ \delta(x_2-ct_2)\ C_{PP}(t_1,t_2) \qquad (6)$$

where $\overset{o}{p}(x,t)$ is the zero mean value of p(x,t).

Going from the knowledge of the statistical characteristics of the force, the appropriate characteristics of the response of the beam may be calculated. The mean value of the deflection $E[v(x,t)]$ is given in Reference [1] as the response of the beam to a moving constant force P.

Among the statistical characteristics of the second order the variance σ^2 is the most important and for the deflection of a simple beam it takes the following form (in accordance to the procedure of the correlation method applied to the given problem, see Reference [2]):

$$\sigma_v^2(x,t) = \sum_{j=1}^{\infty} \frac{v_j^2(x)}{M_j^2\ \omega_j'^2} \int_0^t\int_0^t e^{-\omega_b(t-\tau_1)} \sin\omega_j'(t-\tau_1)$$

$$\cdot\ e^{-\omega_b(t-\tau_2)} \sin\omega_j'(t-\tau_2)\ \sin\frac{j\pi c\tau_1}{l}\ \sin\frac{j\pi c\tau_2}{l}$$

$$\cdot\ C_{PP}(\tau_1,\tau_2)\ d\tau_1\ d\tau_2 \qquad (7)$$

where

$v_j(x) = \sin j\pi x/l$ - normal modes of a simple beam

$M_j = \int_0^l \mu\ v_j^2(x)\ dx$ - generalized mass of the beam

l - length (span) of the beam

$\omega_j = 2\pi f_j$ - circular natural frequency of the undamped beam

$\omega_j'^2 = \omega_j^2 - \omega_b^2$ - circular natural frequency of the damped beam

f_j - natural frequency of the beam

Mean value and the variance (7) of the deflection of the beam are functions of the co-ordinate x and of the time t and, therefore, the response of the beam is nonstationary for any random behaviour of the moving force P(t).

The numerical results have been arranged in the form

$$V_v(z) = \frac{\sigma_v(1/2, t)}{v_o} = V_p \ y(z) \tag{8}$$

where

$V_v(z)$ - coefficient of variation of the deflection at the beam span midpoint

$z = ct/l$ - dimensionless position co-ordinate of the force

$\sigma_v(x, t)$ - standard deviation of the deflection of the beam

$v_o = Pl^3/(48EJ)$ - the deflection at the middle of the beam span under a load P at the same point

V_P - coefficient of variation of the force P(t)

y(z) - response of the beam to a moving random force P(t) necessary for the calculation of the coefficient of variation (8)

The function y(z) was calculated for several basic types of covariances of the force P(t) : white noise, constant covariance, cosine wave, exponential, and exponential cosine, see Reference 2 .

For wide-band spectrum of the random force (white noise) where the mutual relations of the covariance C_P and spectral density S_P are

$$C_P(\tau) = S_P \ \delta(\tau) \tag{9}$$

$$S_P(\omega) = S_P \tag{10}$$

the equation (8) gives:

$$V_P = (S_P \ \omega_1)^{1/2}/P \tag{11}$$

$$y(z) = \sum_{j=1}^{\infty} \frac{96\, v_j(x)}{\pi^4 (j^4-\beta^2)} \left[\!\left[\frac{a(j^4-\beta^2)^{1/2}}{16} \right.\right.$$

$$\cdot \left\{ \frac{a+b}{(a+b)^2 + d^2} \left[\sin 2bz + e^{-2dz} \sin 2az \right.\right.$$

$$\left. + \frac{d}{a+b} (\cos 2bz - e^{-2dz} \cos 2az) \right]$$

$$+ \frac{a-b}{(a-b)^2 + d^2} \left[- \sin 2bz + e^{-2dz} \sin 2az \right.$$

$$\left. + \frac{d}{a-b} (\cos 2bz - e^{-2dz} \cos 2az) \right]$$

$$- \frac{2d}{b^2 + d^2} \left\{ \left(\frac{b}{d}\right) \sin 2bz + \cos 2bz - e^{-2dz} \right\}$$

$$- \frac{2d}{a^2 + d^2} \left[1 - e^{-2dz} \left(\cos 2az - \frac{a}{d} \sin 2az\right) \right]$$

$$\left.\left.\left. + \frac{2}{d} (1 - e^{-2dz}) \right\} \right]\!\right]^{1/2} \tag{12}$$

Here

$$\alpha = c/(2f_1 l) \tag{13}$$

- dimensionless speed parameter

$$\beta = \omega_b/\omega_1 \tag{14}$$

- dimensionless damping parameter

$$a = \pi\,(j^4 - \beta^2)^{1/2}/\alpha$$

$$b = j\pi$$

$$d = \pi\beta/\alpha$$

The maximum values of function y(z) from the equation (12) are presented in Figure 2 as functions of the speed parameter α for various damping β.

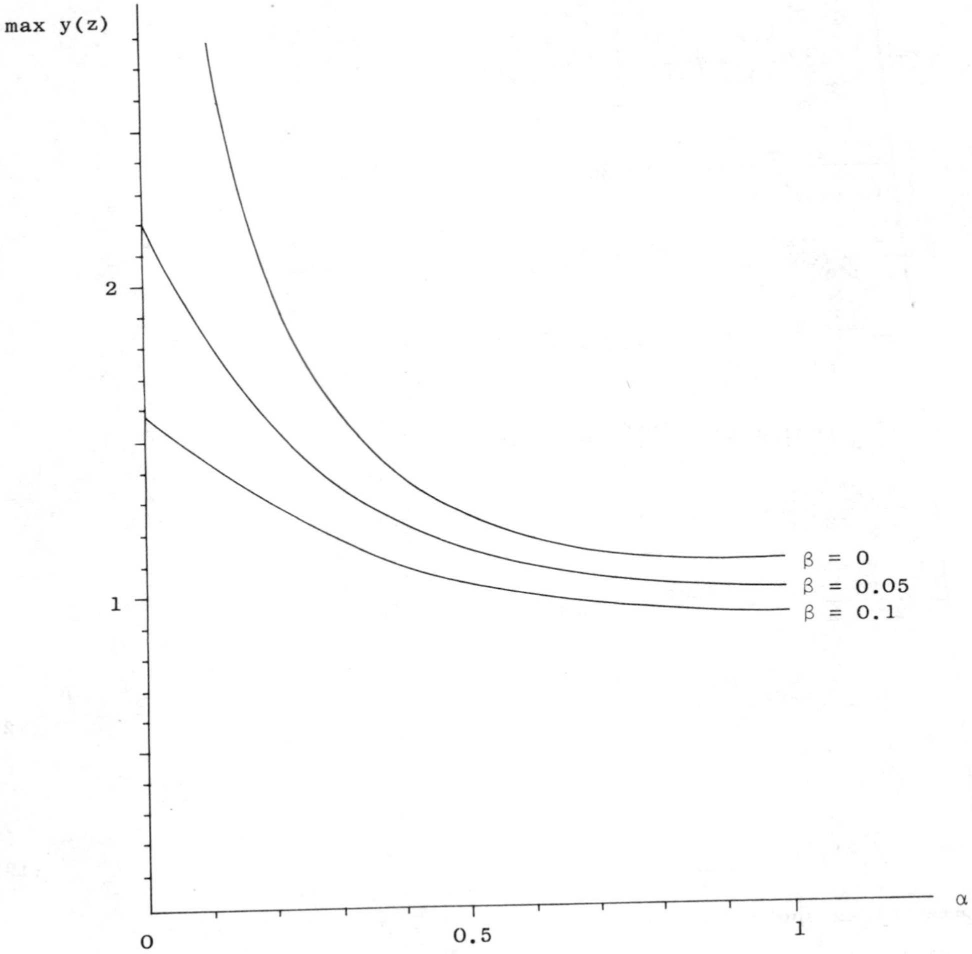

Figure 2 Maxima of y(z) as functions of speed parameter α for various damping β. Wide-band spectrum (white noise) of the moving random force

BEAM UNDER MOVING CONTINUOUS RANDOM LOAD

In the second case, an infinite strip of load p(x,t) is assumed, see Figure 3. The load moves along the beam with a constant speed c, it possesses a constant mean value p; however, it is random variable in length co-ordinate with respect to the moving co-ordinate system s = t - x/c. Moreover, the

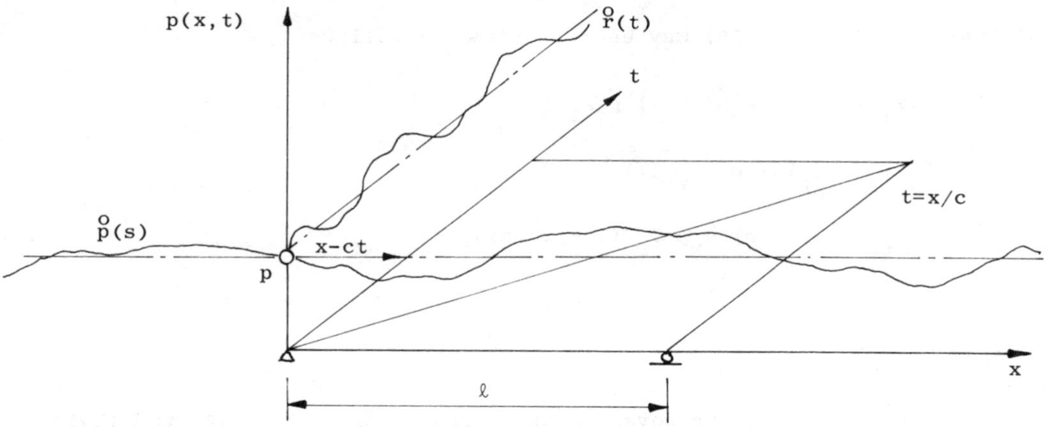

Figure 3 Beam under moving random strip of load p(x,t)

load is random variable in time t. It is supposed that both stochastic processes are stationary and independent and that the random components are small in comparison to the mean value of the load p.

Under these assumptions the load p(x,t) is defined as follows:

$$p(x,t) = \left[p + \varepsilon \overset{o}{p}(s)\right] \cdot \left[1 + \varepsilon \overset{o}{r}(t)\right] \tag{15}$$

where

$p = E\left[p(x,t)\right]$ - mean value of the load

$$\overset{o}{p}(x,t) = \varepsilon \overset{o}{p}(s) + \varepsilon p\overset{o}{r}(t) + \varepsilon^2 \overset{o}{p}(s) \overset{o}{r}(t) \tag{16}$$

- zero mean random value of the load,

$\overset{o}{p}(s)$ - zero mean value of the stationary process which is random variable with respect to the moving co-ordinate $s = t - x/c$

$\overset{o}{r}(t)$ - zero mean value of the stationary process which is random variable in time t

$\varepsilon << 1$

Neglecting small values of the third and fourth order of ε, the covariance of the load (15) and (16) may be simplified as follows:

$$C_{pp}(x_1,x_2,\tau) = E\left[\overset{o}{p}(x_1,t)\ \overset{o}{p}(x_2,t+\tau)\right]$$

$$= C_{pp}(\tau+\tau_o) + p^2\ C_{rr}(\tau)$$

$$+ p\ C_{pr}(\tau + x_1/c) + p\ C_{rp}(\tau - x_2/c) \qquad (17)$$

where

$\tau_o = (x_1 - x_2)/c$

C_{pp}, C_{rr}, C_{pr}, C_{rp} are the covariances of functions p(s), r(t), and their cross-covariances, respectively.

The spectral density of the load p(x,t) is:

$$S_{pp}(x_1,x_2,\omega) = e^{i\omega\tau_o}\ S_{pp}(\omega) + p^2\ S_{rr}(\omega)$$

$$+ p\ e^{i\omega x_1/c}\ S_{pr}(\omega) + p\ e^{-\omega x_2/c}\ S_{rp}(\omega) \qquad (18)$$

where the spectral densities S_{pp}, S_{rr}, S_{pr}, S_{rp} have similar notations as the appropriate covariances mentioned above.

The mean value of the deflection of the beam is derived from equations (1) and (15) in Reference [1] including the effect of the mass of the moving load. The variance may be calculated either from covariance (17)

$$\sigma_v^2(x) = \sum_{j=1}^{\infty} \frac{v_j^2(x)}{M_j^2} \int_{-\infty}^{\infty}\int_{-\infty}^{\infty}\int_{0}^{1}\int_{0}^{1} v_j\ (\xi_1)\ v_j\ (\xi_2)$$

$$.\ h_j(\tau_1)\ h_j(\tau_2)\ C_{pp}(\xi_1,\ \xi_2,\ \tau_1 - \tau_2)$$

$$.\ d\ \xi_1\ d\ \xi_2\ d\ \tau_1\ d\ \tau_2 \qquad (19)$$

or from spectral density (18)

$$\sigma_v^2(x) = \frac{1}{2\pi} \sum_{j=1}^{\infty} \frac{v_j^2(x)}{M_j^2} \int_{-\infty}^{\infty} \Bigg[S_{pp}(\omega) |H_j(\omega)|^2 \cdot \left| \int_0^1 v_j(\xi)\, e^{-i\omega\xi/c}\, d\xi \right|^2 + p^2 S_{rr}(\omega) \cdot \left| H_j(\omega) \right|^2 \left(\int_0^1 v_j(\xi)\, d\xi \right)^2$$

$$+ p\, S_{pr}(\omega) \left| H_j(\omega) \right|^2 \int_0^1 \int_0^1 v_j(\xi_1)\, v_j(\xi_2) \cdot e^{i\omega\xi_1/c}\, d\xi_1 d\xi_2 + p\, S_{rp}(\omega) \left| H_j(\omega) \right|^2 \cdot \int_0^1 \int_0^1 v_j(\xi_1)\, v_j(\xi_2)\, e^{-i\omega\xi_2/c}\, d\xi_1 d\xi_2 \Bigg] d\omega \qquad (20)$$

There was used the notation $h_j(t)$ for the impulse function and $H_j(\omega)$ for its frequency response.

It may be observed from the equations (19) and (20) that the response of the beam is a stationary process assuming a stationary input and an infinite strip of load.

The following form was given to the coefficient of variation of the deflection at the mid-span point of the beam:

$$V_v = \sigma_v(1/2)/v_o = (V_p^2 y_p^2 + V_r^2 y_r^2 + V_{pr}^2 y_{pr}^2 + V_{rp}^2 y_{rp}^2)^{1/2} \qquad (21)$$

where

$v_o = 5pl^4/(384EJ)$ - deflection at the middle of the simple beam under the continuous load p

V_p, V_r, V_{pr}, V_{rp} are the coefficients of variation of the load

y_p, y_r, y_{pr}, y_{rp} are the responses of the beam.

The numerical values of quantities V_i and y_i in (21) were calculated for four basic types of covariances of the load : white noise, constant covariance, cosine wave, and exponential covariance.

E.g. for white noise of the load, it holds true:

$$V_p = \frac{(S_p\ \omega_1)^{1/2}}{p} \tag{22}$$

$$y_p = \left[\frac{\pi\alpha}{8\{[1 - \alpha^2(1 + 2m)]^2 + 4\alpha^2\beta^2\}} \right]^{1/2} \tag{23}$$

$$V_r = (S_r\ \omega_1)^{1/2} \tag{24}$$

$$y_r = \frac{1}{2\ \beta^{1/2}\ (1 - \alpha^2 m)^{1/2}} \tag{25}$$

$$y_{pr} = y_{rp} = 0$$

where

S_p, S_r are constant spectral densities of the load $\overset{o}{p}$ and $\overset{o}{r}$, respectively

$$m = p/(\mu\ g) \tag{26}$$

is the dimensionless parameter covering the approximate effect of the mass of the moving load (g is the gravitational constant).

The values of y_p from (23) are plotted as functions of the dimensionless speed parameter α (13) for several values of dimensionless damping β (14) and mass m (26) in Figure 4.

PREDICTED LIFE OF BRIDGES

The stochastic approach to the stress analysis of bridges has been applied to the prediction of life of railway bridges, see Reference [3].

The stress-time history in a selected point of a bridge is calculated or measured and then classified using the 'rain-flow' counting method. The

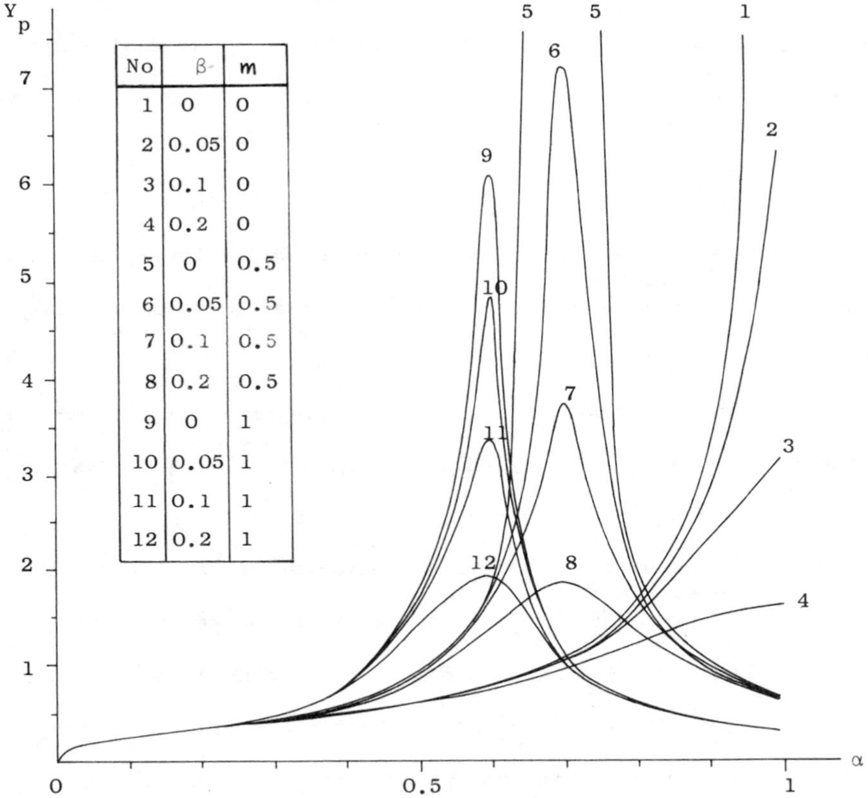

No	β	m
1	0	0
2	0.05	0
3	0.1	0
4	0.2	0
5	0	0.5
6	0.05	0.5
7	0.1	0.5
8	0.2	0.5
9	0	1
10	0.05	1
11	0.1	1
12	0.2	1

Figure 4 Response of the beam y_p as function of speed parameter α for various damping β and mass of the load . Load is random variable in the length coordinate, white noise spectrum.

Palmgren-Miner theory of cumulative damage enables to estimate the life of the bridge.

Large-scale calculations were carried out for several typical trains of seven European railway administrations. The trains are running at various speeds over 10 different spans (2, 3, 4, 5, 7, 10, 15, 20, 30 and 50 m) of typical railway bridges. The calculations used four forms of the Wöhler fatigue lines, see Reference [3].

A sample of these calculations in Table 1 gives the numbers of trains to failure for six typical trains, for five spans and for four forms of Wöhler line.

CONCLUSIONS

The case solved in the second section is typical of problems involving short-span bridges and longitudinal beams, both structures that are usually subjected to loads produced by one vehicle axle only.

The case solved in the third section is illustrative of other extreme - large-span bridges that are usually loaded with a number of axles. That is why their stresses depend on the traffic flow.

Going from the knowledge of the statistical characteristics of the load the present paper enables to calculate the stochastic response of a beam.

The response of a beam to a moving random force is a nonstationary process with a variable variance. On the other hand, a stationary process occurs if the beam is subjected to a moving stationary and infinite strip of load.

The statistical characteristics of the beam depend on the assumed type of the covariance of the load, on the speed of the movement and on the mass of the load. The increasing damping of the beam causes a diminishing of the variance for all investigated types of covariances.

It has appeared that the number of trains to failure is increasing with increasing span. The assumed type of the train (total weight, number of axles and speed) influences the predicted life of the bridge. The differences have appeared among the assumed types of Wöhler fatigue lines with respect to the number of trains to failure.

REFERENCES

1 L. Frýba, Vibration of solids and structures under moving loads. Academia Prague, Noordhoff International Publishing Groningen (1972).

2 L. Frýba, Nonstationary response of a Begam to a moving random force. Journal of Sound and Vibration, 46 (1976), No. 3 (pp. 323-338).

3 Report of the Office for Research and Experiments of the International Union of Railways, ORE D 128/RP 5. Bending moment spectra and predicted lives of railway bridges. Utrecht (1976) - manuscript.

Doc. Ing. Ladislav Frýba, Dr. Sc., Railway Research Institute, Prague, Czechoslovakia.

Span (in metres)	Wöhler line Number	Type of train 1	2	3	4	5	6
		Number of trains to failure (in millions)					
2	1	0.256	0.181	0.0295	0.0139	0.0185	0.0275
	2	0.266	0.187	0.0300	0.0139	0.0185	0.0278
	3	0.311	0.233	0.0328	0.0152	0.0203	0.0308
	4	0.497	0.392	0.0429	0.0204	0.0272	0.0418
4	1	1.12	0.782	0.144	0.0916	0.0954	0.135
	2	1.25	0.960	0.148	0.0959	0.0955	0.141
	3	1.68	1.25	0.197	0.130	0.132	0.188
	4	5.62	4.27	0.551	0.409	0.402	0.542
10	1	2.24	2.62	0.447	0.342	0.373	0.444
	2	2.41	3.56	0.503	0.410	0.424	0.505
	3	3.57	4.89	0.706	0.576	0.602	0.699
	4	16.9	31.1	2.94	3.13	2.89	2.80
20	1	5.23	2.93	1.32	2.33	2.20	1.20
	2	6.89	3.65	1.55	2.60	2.55	1.31
	3	9.74	4.66	2.09	2.82	2.79	1.82
	4	70.1	14.3	7.75	3.80	3.93	6.28
50	1	20.7	8.22	1.58	1.59	2.32	1.65
	2	27.5	8.22	1.69	1.59	2.32	1.79
	3	38.7	11.9	1.88	1.52	2.44	1.95
	4	341	42.9	2.44	1.52	2.77	2.47

TABLE 1

Predicted Lives of Bridges

G COUPRY

Mean number of loads and acceleration in roll of an airplane flying in turbulence

INTRODUCTION

For many years, the continuous turbulence approach is used, in parallel with discrete gust patterns, to describe the behaviour of an aircraft flying in atmospheric turbulence. This approach is based on models of the spectral density of turbulence, on the calculation of the aircraft transfer function to turbulence, and the derivation of the usual numbers $\bar{A}$ and N_o, which define respectively the energy of the aircraft reyponse and the mean number of zero crossings of any output. Here lies some inconsistencies, due to the fact that, for the turbulence spectra used, Rice's integrals, that give N_o, are infinite. To cope with this difficulty, most engineers choose an arbitrary cut-off frequency, which limits the domain of integration.

This trouble disappears once isotropy of turbulence and unsteadiness of aerodynamic loads are taken into account. Even in the simple case of a rigid aircraft, with the single degree of freedom of plunge, Rice's integral converges and gives definite results. The work presented here follows and completes a previous work of I. Kaynes [1] of R.A.E., which takes also into account these two effects.

In its first part, the paper will remind the reader how the transverse coherence function of turbulence can be computed, for the usual spectra, and presented in closed form. In its second part, the calculation of $\bar{A}$ and N_o will be given, for a rigid aircraft with one degree of freedom, both for loads and accelerations in roll, assuming that the strip theory is valid and that the lifting line theory can be used.

In the third part, approximate simple formulas will be derived, and used to compare the mean number of loads, as predicted, with measurements achieved on Caravelle.

TRANSVERSE COHERENCE FUNCTION OF TURBULENCE

Setting of the Problem

When computing the response of an aircraft to turbulence, one of two assumptions is usually made (Figure 1). According to the first one, the aircraft

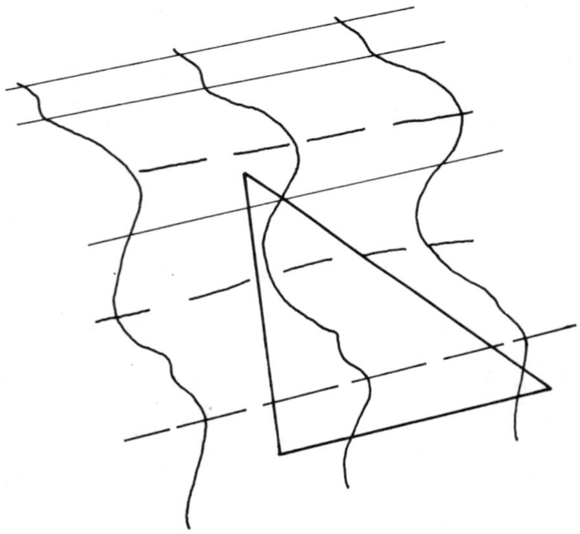

FIG. 1a Spanwise uniform turbulence

is assumed to fly through cylindrical random waves, and then Rice's integral that give N_o do not converge; according to the second one, the turbulent field is isotropic, i.e. no longer uniform in span. This section will remind the reader how the isotropic effect can be taken into account,

As a matter of fact, it can be concluded from the large number of spectra collected all around the world that atmospheric turbulence is not far from

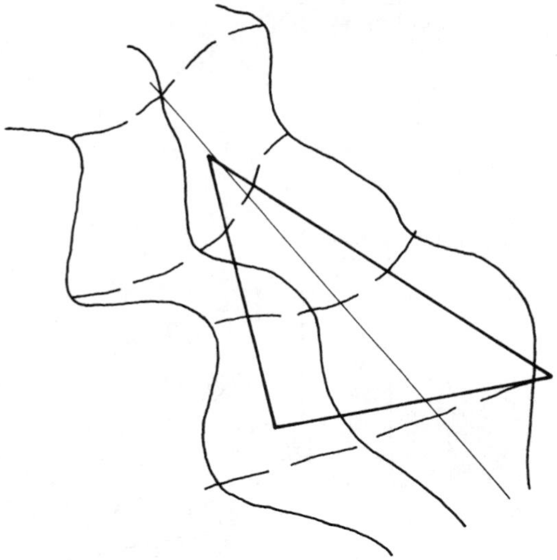

FIG. 1b Isotropic turbulence

being isotropic, and is fairly well represented, in most cases, by Bullen's spectral model.

Derivation of the Coherence Function

We will assume that the turbulent field is isotropic, and, using Taylor's hypothesis, we will express the cross-correlation of the vertical component of the turbulent velocity between two points M(x,y) and M'(m',y') of a lifting surface (Figure 2) flying at mean speed V_o by:

$$R_W(M,M',\tau) = g(([V_o\tau - (x - x')]^2 + (y - y')^2)^{\frac{1}{2}}) \tag{1}$$

where g(r) is the transverse correlation function of the turbulent field. Fourier transform yields the cross power spectrum

$$S_W(M,M',\omega) = G(\omega)\ C\ (\omega; \frac{|y-y'|}{V_o})\ \exp\ (-i\omega\frac{x-x'}{V_o}) \tag{2}$$

where G(ω) is Bullen's transverse spectrum:

$$G(\omega) = \frac{L\sigma w^2}{\pi V_o} \quad \frac{1 + 2\,\omega^2 K^2 (P+1)}{(1 + \omega^2 K^2)^{P+3/2}} \tag{3}$$

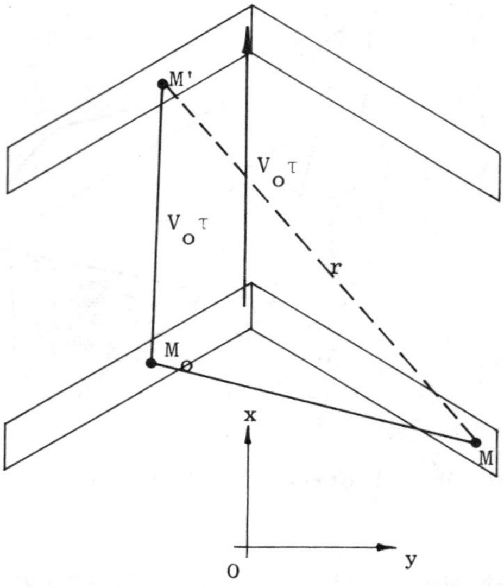

FIG. 2 $\qquad r = \sqrt{(V_o\tau-(x-x'))^2+(y-y')^2}$

This spectrum agrees reasonably well with most of the experimental results. It corresponds to Van Karman's spectrum for p = 1/3 and to Dryden's model for p = 1/2. In Equation (3), K has the dimension of a time, related to the scale of turbulence L by

$$K = \frac{L\Gamma\ (P)}{V_o\ \sqrt{\pi}\ \Gamma\ (P + \frac{1}{2})}$$

In Equation (2), where one has written

$$\eta = \frac{|y - y'|}{V_o}$$

$C(\omega;\eta)$ is the spanwise coherence function which is expressed as:

$$C(\omega;\eta) = 1 - \frac{|\eta|}{G(\omega)} \int_0^{+\infty} G(\sqrt{\omega^2 + \nu^2})\ J_1\ (|\eta|\nu)\ d\nu$$

This last function is related with the spanwise lack of coherence of the turbulent field. $C(\omega;\eta)$ can be calculated in closed form with the help of Nielsen's integral. Assuming that the wavelengths of interest fall in the inertial subrange, one finds:

$$C(\omega;\eta) = \frac{2}{\Gamma(P+\frac{1}{2})} \left(\frac{\omega\eta}{2}\right)^{P+\frac{1}{2}} K_{P+\frac{1}{2}}\ (\omega\eta)$$

This function depends only on the product:

$$\xi = \omega\eta$$

For instance, for Van Karman's model:

$$C(\xi) = \frac{2}{\Gamma(5/6)} \left(\frac{\xi}{2}\right)^{5/6} K_{5/6}\ (\xi)$$

and, for the Dryden's model which will be used in the next sections:

$$C(\xi) = \xi\ K_1\ (\xi)$$

This function is plotted in Figure 3.

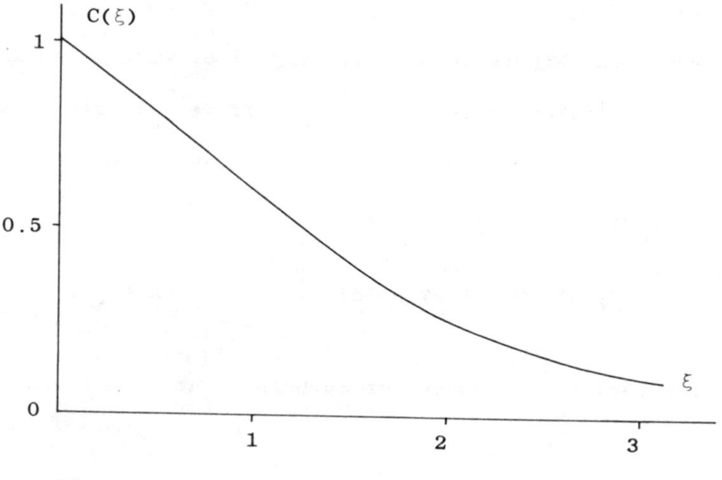

FIG. 3 Function $C(\xi)$ for Dryden's model

NUMBER OF LOADS AND ACCELERATION IN ROLL

Setting of the Problem

Calculation of the transfer function of a flexible aircraft can obviously be made, even if the turbulent field is assumed to be isotropic. The main lines of this derivation have been presented in a recent article [2]. Now, we will work out a much simpler problem, in which we shall replace the actual aircraft by a lifting line, perfectly rigid, that has the only degree of freedom of plunge. This simplification relies on the feeling, shared by all passengers flying in turbulence, that only the vertical motion can be sensed in the vicinity of the c.g.

For this simple model, we shall compute the numbers $\bar{A}$ and N_o related with the vertical acceleration and the acceleration in roll, when the plane is flying through an isotropic turbulence. The lift and the roll moment will be computed using the strip theory and the Sears function $\tilde{s}(\omega)$ which defines the unsteady forces induced by a sinusoidal air motion.

Statistics of Loads

The aircraft is replaced (Figure 2) by a lifting line, with sweep angle φ and spanwise dimension 2b, flying at mean speed V_o. If we call $s(y,\alpha)$ the impulse function associated to the Sears function $\tilde{s}(\omega)$, we can express the lift of the aircraft, at time t, by:

$$P(t) = \int_{-b}^{+b} \int_{-\infty}^{+\infty} s(y,\alpha) \quad W(y,t-\alpha) \, d\alpha \, dy$$

where $W(y,t)$ is the vertical component of turbulence at time t and location y. At time $t+\tau$, the lift will be:

$$P(t+\tau) = \int_{-b}^{+b} \int_{-\infty}^{+\infty} s(y', \alpha')\, W\, (y', t+\tau - \alpha')\, d\alpha'\, dy'$$

and the lift correlation:

$$R_p(\tau) = \int_{-b}^{+b} \int_{-b}^{+b} s(y, -\tau) * s(y',\tau) * R_W(y,y',\tau)\, dy\, dy'$$

where * stands for time convolution.

After Fourier transform, and assuming the chord 2c is constant, one obtains the spectral density of the lift:

$$\phi_P\, (\omega) = |\tilde{s}(\omega)|^2 \int_{-b}^{+b} \int_{-b}^{+b} S_W\, (y,\, y',\omega)\, dy\, dy'$$

and, using equation (2)

$$\phi_P\, (\omega) = G(\omega)\, |\tilde{s}(\omega)|^2 \int_{-b}^{+b} \int_{-b}^{+b} C(\omega\, \frac{|y-y'|}{V_o})$$

$$\exp(- \frac{i\omega\lambda}{V_o}\, (|y| - |y'|))\, dy\, dy'$$

where λ stands for tgφ.

Assuming that the transfer function between load and lift is not very different from 1/M in the range of frequency concerned, one can express the p.s.d. of vertical acceleration as:

$$\phi_{\ddot{z}}\, (\omega) = \frac{16\, \pi^2\, \rho^2\, V_o^{\,2}\, b^2\, c^2}{M^2}\, G(\omega)\, |\sigma(\frac{c\omega}{V_o})|^2\, h(\frac{b\omega}{V_o})$$

where ρ is the density of atmosphere, M the mass of the aircraft, σ (x) the reduced Sears function:

$$\tilde{s}(\omega) = 2\pi\, \rho\, V_o\, C\, \tilde{\sigma}\, (\frac{c\omega}{V_o})$$

and:

$$h(x) = \frac{1}{4x^2} \int_{-x}^{+x} \int_{-x}^{+x} C(|\eta - \eta'|) \exp(-i\lambda(|\eta| - |\eta'|)\, d\eta\, d\eta'$$

The calculation of $\bar{A}$ then follows from the definition:

$$\bar{A}_{\ddot{y}_2} = \frac{\sigma_{\ddot{y}_2}}{\sigma_W}$$

where:

$$\sigma_W^2 = \int_0^{+\infty} G(\omega)\, d\omega \; ; \qquad \sigma_{\ddot{y}_2}^2 = \int_0^{+\infty} \phi_{\ddot{y}_2}(\omega)\, d\omega$$

One obtains:

$$\bar{A}_{\ddot{y}_2}^2 = \frac{8\,\pi\,\rho^2\,V_o^2\,b^2\,c^2\,L}{M^2}\, f(\xi,\delta)$$

For the Dryden's model:

$$f(\xi,\delta) = \int_0^{+\infty} \frac{1 + 3\,\xi^2\,x^2}{(1 + \xi^2\,x^2)^2}\, \tilde{\sigma}^2(\delta x)\, h(x)\, dx \tag{4}$$

with:

$$\xi = \frac{L}{b} \; ; \quad \delta = \frac{c}{b}$$

The mean number of zero crossings is given by Rice's formula:

$$N_{o\ddot{y}_2} = \frac{1}{2\,\pi\,\sigma_{\ddot{y}_2}} \left[\int_0^{+\infty} \omega^2\, \phi_{\ddot{y}_2}(\omega)\, d\omega \right]^{\frac{1}{2}}$$

and can be expressed as:

$$N_{o\ddot{y}_2} = \frac{V_o}{4\,\pi\,b} \sqrt{\frac{g(\xi,\delta)}{f(\xi,\delta)}} \tag{5}$$

where

$$g(\xi,\delta) = \int_0^{+\infty} x^2\, \frac{1 + 3\,\xi^2\,x^2}{(1 + \xi^2\,x^2)^2}\, |\tilde{\sigma}(\delta x)|^2\, h(x)\, dx \tag{6}$$

Statistics of Acceleration in Roll

Following exactly the same line as in the last subsection, we can derive the power spectral density of the roll acceleration as

$$\emptyset''_{\psi}(\omega) = \frac{16\ \pi^2\ \rho^2\ V_o^2\ b^4\ c^2}{I_x^2}\ G(\omega)\left|\tilde{\sigma}\left(\frac{c\omega}{V_o}\right)\right|^2\ \tilde{h}\left(\frac{b\omega}{V_o}\right)$$

where I_x is the inertia around the longitudinal axis and:

$$\tilde{h}\ (x) = \frac{1}{4x^2}\int_{-x}^{+x}\int_{-x}^{+x} \eta\ \eta'\ C(|\eta - \eta'|)\ \exp(-i\lambda(|\eta| - |\eta'|))\ d\eta\ d\eta'$$

One then deduces the corresponding $\bar{A}$ and N_o:

$$\left.\begin{aligned} \bar{A}^2_{\ddot{\psi}} &= \frac{4\ \pi\ \rho^2\ b^3\ c^2\ L\ V_o^2}{I_x^2}\ \tilde{f}\ (\xi,\delta) \\ N_{o\ddot{\psi}} &= \frac{V_o}{4\pi b}\sqrt{\frac{\tilde{g}(\xi,\delta)}{\tilde{f}(\xi,\delta)}} \end{aligned}\right\} \qquad (7)$$

where:

$$\left.\begin{aligned} \tilde{f}(\xi,\delta) &= \int_0^{+\infty} \frac{1 + 3\ \xi^2\ x^2}{(1 + \xi^2\ x^2)^2}\ |\tilde{\sigma}\ (\delta x)|^2\ \tilde{h}\ (x)\ dx \\ \tilde{g}(\xi,\delta) &= \int_0^{+\infty} x^2\ \frac{1 + 3\ \xi^2\ x^2}{(1 + \xi^2\ x^2)^2}\ |\tilde{\sigma}\ (\delta x)|^2\ \tilde{h}\ (x)\ dx \end{aligned}\right\} \qquad (8)$$

APPROXIMATE FORMULAS AND COMPARISON WITH FLIGHT RESULTS

Approximation for Loads

As can be seen on Figure 4, the quantity h(x) remains nearly equal to unity in the range of frequency in which the turbulence spectrum has most of its energy. It follows that:

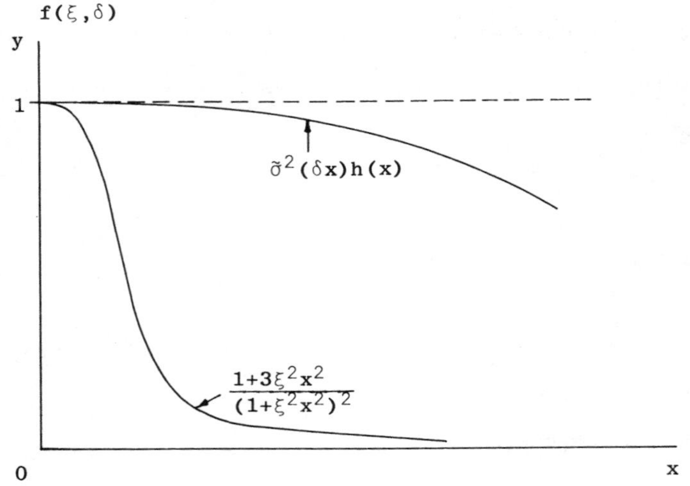

$$f(\xi,\delta) = \int_0^{+\infty} \frac{1+3\xi^2x^2}{(1+\xi^2x^2)^2}\tilde{\sigma}^2(\delta,x)h(x)\,dx \simeq \int_0^{+\infty} \frac{1+3\xi^2x^2}{(1+\xi^2x^2)^2}\,dx = \frac{\pi}{\xi} = \frac{b}{L}$$

$$\overline{A}^2 \simeq \frac{8\pi^2p^2v_o^2b^3c^2}{M^2}$$

Figure 4 Variation of Function $f(\xi\delta)$ with x.

$$f(\xi,\delta) = \int_0^{+\infty} \frac{1+3\,\xi^2\,x^2}{(1+\xi^2\,x^2)^2}\,\tilde{\sigma}\,(\ x)\ h(x)\ dx \approx \int_0^{+\infty} \frac{1+3\,\xi^2\,x^2}{(1+\xi^2\,x^2)^2}\,dx$$

$$= \frac{\pi}{\xi} = \pi\,\frac{b}{L} \qquad (9)$$

As can be seen on Figure 5, for all practical values of the scale L of turbulence

$$x^2 \quad \frac{1+3\,\xi^2\,x^2}{(1+\xi^2\,x^2)^2} \approx \frac{3}{\xi^2}$$

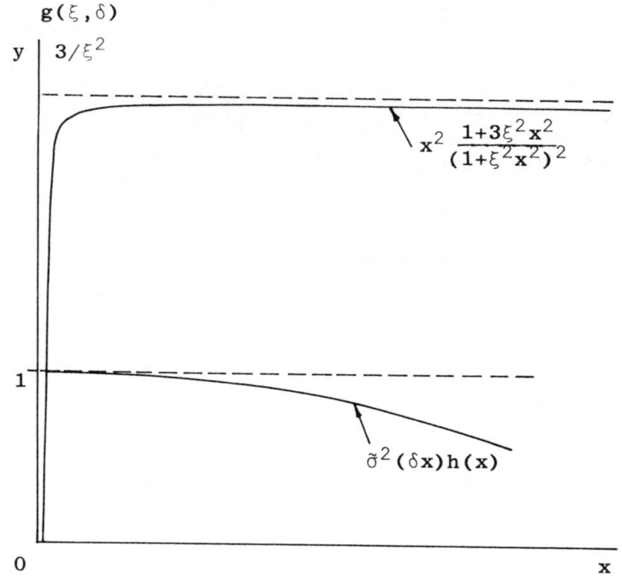

$$g(\xi,\delta) = \int_0^{+\infty} x^2 \frac{1+3\xi^2x^2}{(1+\xi^2x^2)^2} \tilde{\sigma}^2(\delta x)h(x)\,dx \simeq \frac{3}{\xi^2}\int_0^{+\infty} \tilde{\sigma}^2(\delta x)h(x)\,dx = \frac{3b^2}{L^2}p(\delta)$$

$$N_o = \frac{v_o}{2\pi\sqrt{bL}}\sqrt{\frac{3}{\pi}p(\delta)}$$

Figure 5 Variation of Function g(ξδ) with x

Consequently:

$$g(\xi,\delta) = \int_o^{+\infty} x^2 \frac{1 + 3\xi^2 x^2}{(1 + \xi^2 x^2)^2} \tilde{\sigma}^2 (\delta x)\ h(x)\ du \approx \frac{3}{\xi^2} \qquad (10)$$

$$\int_o^{+\infty} \tilde{\sigma}^2 (\delta x)\ h(x)\ dx = \frac{3b^2}{L^2} p\ (\delta)$$

where:

$$p(\delta) = \int_o^{+\infty} \tilde{\sigma}\ (\delta x)\ h(x)\ dx$$

$p(\delta)$ is fairly well approximated, for all values of δ and φ of interest, by the simple function:

$$p(\delta) = 0.606\ (1 - 0.3414\varphi^2)\ \delta^{-\frac{1}{2}} \tag{11}$$

where φ is expressed in radians (Figure 6).

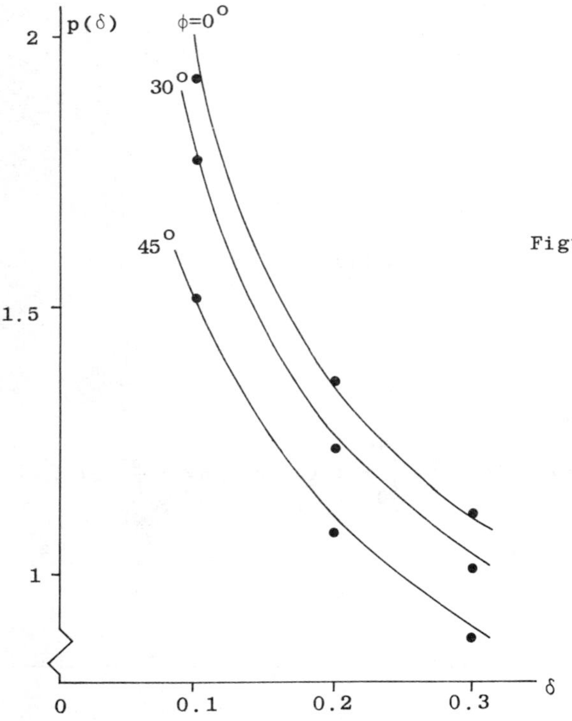

Figure 6 —p(δ) exact
• Approximate formula

Using (9), (10) and (11) in formula (5), one obtains:

$$N_{o_{\frac{1}{2}}} = 0.122\ (bc)^{-\frac{1}{4}}\ L^{-\frac{1}{2}}\ V_0\ (1 - 0.1707\varphi^2) \tag{12}$$

and the mean number of crossing per unit length:

$$\tilde{N}_{o_{\frac{1}{2}}} = 0.122\ L^{-\frac{1}{2}}\ (bc)^{-\frac{1}{4}}\ (1 - 0.1707\varphi^2) \tag{13}$$

Approximations for Acceleration in Roll

Following the same line as in the last subsection, we will derive an approximate formula for $N_{o_{\ddot{\psi}}}$. As most of the energy of the integrals $\tilde{f}\ (\xi,\delta)$ and $\tilde{g}(\xi,\delta)$ come from the inertial subrange, we will write:

$$f(\xi,\delta) \approx \frac{3}{\xi^2}\, \varphi(\delta)\ ;\quad \tilde{g}(\xi,\delta) \approx \frac{3}{\xi^2}\, \tau(\delta) \tag{14}$$

with:

$$\varphi\,(\delta) = \int_0^{+\infty} \tilde{\sigma}^2\ (\delta x)\ \tilde{h}(x)\ dx;\quad \tau(\delta) = \int_0^{+\infty} \tilde{\sigma}^2\ (x)\ \frac{\tilde{h}(x)}{x^2}\, dx \tag{15}$$

The function:

$$u(\delta) = \frac{\varphi(\delta)}{h(\delta)}$$

is fairly well approximated, for a straight wing (Figure 7), by the simple function:

$$U(\delta) = 1.301\ \delta^{-\frac{1}{2}} \tag{16}$$

which yields the approximate formula:

$$N_{o_{\ddot{\psi}}} = 0.108\ b^{-\frac{3}{4}}\ c^{-\frac{1}{4}}\ V_o$$

and the mean number of zero-crossings per unit-length

$$\tilde{N}_{o_{\ddot{\psi}}} = 0.108\ b^{-\frac{3}{4}}\ c^{-\frac{1}{4}} \tag{17}$$

Comparison with Flight Measurements

Formula (13) gives the mean number of zero-crossings of load per unit length, for an airplane that flies in a patch of turbulence. For the purpose of comparison with the mean number of loads measured in flight, it is necessary to take into account the proportion P_e of time spent in turbulence; formula (13) then becomes:

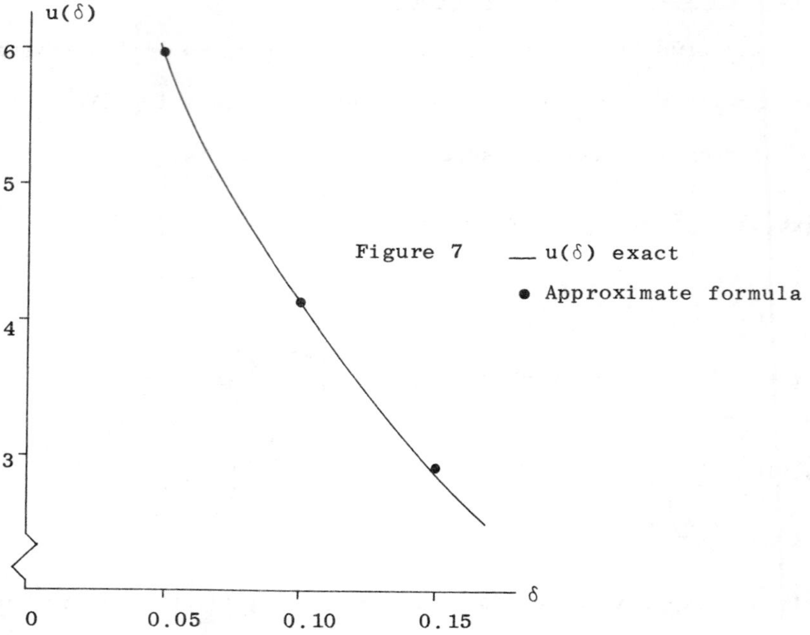

$$N_{o_2''} = 0.122\ L^{-\frac{1}{2}}\ (bc)^{-\frac{1}{4}}\ P_e(1-0.1707\,\varphi^2) \tag{18}$$

The parameter P_e, as proposed by NASA, is given in Figure 8 as a function of altitude z.

Comparison between prediction by formula (18) and flight measurement obtained by Caravelle has been made for two altitudes

$$\begin{cases} z = 5000 \text{ ft}; & P_e = 0.30 \\ z = 30000 \text{ ft}; & P_e = 0.030 \end{cases}$$

According to a general agreement, the corresponding choices of scales of turbulence have been 300 m and 1000 m.

The result of the comparison is given in Figure 9, which presents the cumulative frequency distribution of loads for each of these altitudes. On this chart, the circles are corresponding to the theoretical prediction, and show a good agreement with the flight measurements.

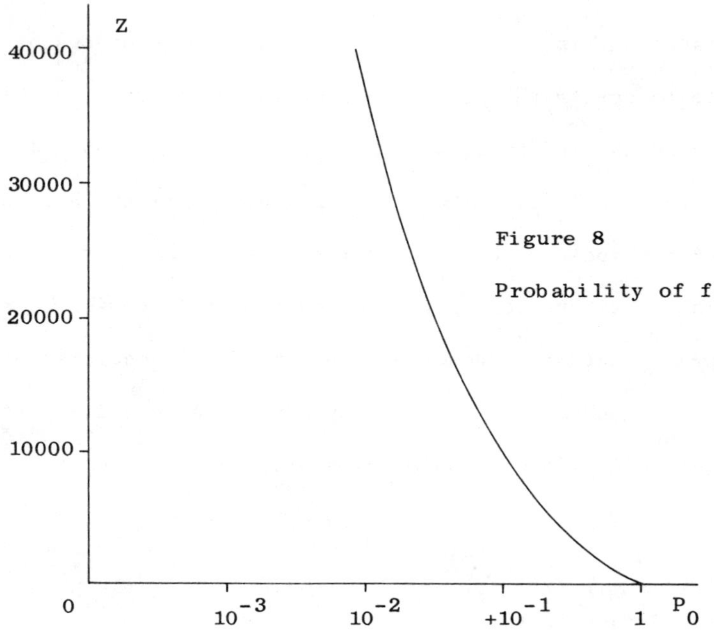

Figure 8

Probability of flying in turbulence

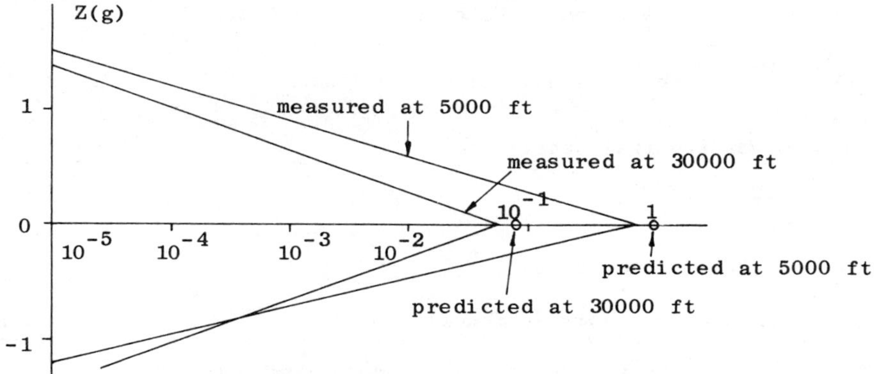

Figure 9 Comparison of measured and predicted number of loads per kilometer.

CONCLUSION

Introducing the effects of the isotropy of turbulence and unsteadiness of loads makes it possible to derive the mean number of loads and roll acceleration without any arbitrary assumption the Rice's integral calculation. To be complete, it is necessary to calculate the transfer function of the aircraft, including the main flexible modes. However, at the first stage of the design, the very simple formulas that have been presented here make it possible to predict with some accuracy, the mean number of loads and roll accelerations that the aircraft will meet during its life. They provide also a first information of the fatigue life, for a given mission profile.

NOMENCLATURE

V_o	Mean speed of the aircraft
W	Vertical component of turbulence
g(r)	Transverse correlation function of turbulence
$R_W(M,M',\tau)$	Cross correlation of W between points M and M'
$S_W(M,M',\omega)$	Cross power of W between points M and M'
$G(\omega)$	Power spectral density of W
σ_W^2	Variance of W
L	Scale of turbulence
$C(\omega;\eta)$	Transverse coherence function of W
$K_1(x)$	Modified Bessel Function of second kind and order 1
P(t)	Lift as a function of time t
$R_p(\tau)$	Lift correlation
$\phi_P(\omega)$	p.s.d. of lift
$\phi_Z''(\omega)$	p.s.d. of vertical acceleration

$\tilde{s}(\omega)$	Sears function
M	Mass of the airplane
b	half-span
c	half-cord
$N_{oZ''}$	Mean number of zero crossings of Z" per second
$\tilde{N}_{oZ''}$	Mean number of zero crossings of Z" per unit length
$N_{o\psi''}$	Mean number of zero crossings of ψ'' per second
$\tilde{N}_{o\psi''}$	Mean number of zero crossings of ψ'' per unit length
$\ddot{\psi}$	roll acceleration
ρ	air density
ξ	L/b
δ	c/b

REFERENCES

1 I.W. Kaynes, Aircraft centre of gravity response to two dimensional spectra of turbulence, ARC R & M 3665 (1971).

Dr. G. Coupry, 29 avenue de ladivision Leclerc, Chatillon sous Bagneux (Hauts de Seine) 92320 Chatillon, France.

DISCUSSION

Dr. Kozin asked how Rice's results were used to predict fatigue life.

Dr. Coupry replied that it is usual to use Miner's cumulative damage hypothesis but new techniques are being developed to predict the crack propagation (rather than crack initiation) phase. Professor Lin commented on the reason for the unbounded form of the zero crossing. It has been suggested that this is due to the assumption of uniform spanwise distribution of turbulence, but it could be due to the approximation of the Sears function where the unboundedness comes from the infinity behaviour of the transfer function. Dr. Coupry replied that he thought that there were two reasons. For boundedness the integral needs to vary as $\frac{1}{\omega^2}$. The isotropic assumption gives a decrease as $\frac{1}{\omega}$ and also the Sears assumption gives a decrease as $\frac{1}{\omega}$. Thus when both assumptions are used there is a decrease proportional to $\frac{1}{\omega^2}$.

There has also been a lot of discussion about the loads. We hope that there are a finite number of loads and also that there are a finite number of zeros of the turbulence field itself. If Rice's integral is used with the Von Karman model even for turbulence itself Rice's integral goes to infinity. The reason is that the models used here are valid for the inertial sub range, (i.e. at low frequencies); the next sub range is the viscous range but here the spectral density decreases as ω^{-7} which makes the integral converge. However this viscous sub range is of no interest to the aeroplane because it starts at wavelengths less than 10 mm. The effect of averaging along the spanwise direction does not change the results until wavelengths greater than about 10 m.

Professor Shinozuka thought that for a flexible aircraft it would be necessary to use a more extensive form of transfer function. Then when the bending moment is evaluated at the wing root, say, it is necessary to use the

Rice formula for the complete transfer function. This may not converge fast enough. Dr. Coupry was not sure on this point but said that the problem was worse than that, for even using the isotropic turbulence theory and the lifting surface theory, which takes into account the unsteadiness of the aerodynamic forces, and the real shape of the aeroplane the comparison between the flight results and the predictions on such aeroplanes as the Boeing 747, European Airbus, Concorde, IL62 is really poor. There must be some other effect causing the order of magnitude difference but we don't know what it is.

In reply to a question (Grossmayer) on the nature of turbulence as indicated by the aircraft measurements, Dr. Coupry said that at medium to high altitudes it is reasonable to assume isotropy. But close to the ground and in jet streams this is not true. On the question (Kozin) of the possibility of large aircraft changing the turbulence field Dr. Coupry said that the only effects were the wakes left by such aircraft which persisted in the atmosphere for a relatively long time. Professor Lin made the point that maybe the precise model of the turbulence field would not have a significant effect. In random response to jet noise for example providing that the transfer function is represented accurately reasonable results can be obtained by assuming white noise excitation. Dr. Coupry replied that the exact form of Sears function had been used. The approximation is after that stage when long wavelengths are considered.

Y K LIN, S MAEKAWA, H NIJIM and L MAESTRELLO

Response of periodic beam to supersonic boundary-layer pressure fluctuations

INTRODUCTION

In recent years there has been considerable interest in free and forced vibrations of periodic structures; i.e., structures composed of identical subunits. The research activities are motivated primarily by applications in the aerospace industry since an aircraft fuselage is a good example of a periodic structure, and the immediate objectives vary from panel flutter to acoustic fatigue, and to noise transmission into the cabin. Although an aircraft fuselage is a very complicated multi-panel system its dynamic behaviour is similar to that of a beam on evenly spaced elastic supports. The one-dimensional beam problem, however, is more suitable for fundamental studies since basic concepts can be developed without the burden of mathematical details. Thus, the analysis of the present paper will be restricted to one spatial coordinate.

At the first sight the problem of a periodic beam may not appear more difficult than that of any other structure if one accepts the linearity assumption and uses a normal-mode formulation. In practice, however, the normal modes of a periodic beam of many spans cannot be calculated accurately due to close clustering of natural frequencies in frequency bands. The futility of the normal mode approach in dealing with a large number of spans has led to two alternatives: the wave propagation approach [1,2,3] and the transfer matrix approach [4,5,6]. The two alternative methods are closely related however. The so-called free wave propagation constants in the first method are the natural logarithms of the eigenvalues of the basic transfer matrix in the second method [7]. The computational simplicity in both methods is obtained by utilizing the fact that the entire system is composed of identical sub-units in the formulation.

In general the spatial periodicity is destroyed under the excitation of a random forcing field. However, it is still possible to formulate the problem on a point load basis; i.e., considering a single point load first, and then the results are superimposed by numerical integration to obtain the structural response for a distributed random load. With only one span loaded by a single point load, the unloaded spans still form spatially periodic groups (only one periodic group if the loaded span is an end span); and, with some necessary adjustments, the analysis of periodic structure still applies [4]. However, the computation can become very unwieldy since each integration over a spatial coorindate results in a double integration when computating the spectral density or the correlation of the response.

One type of random excitation which does not destroy the spatial periodicity of the system is that which is convected as a frozen-pattern at a given velocity. Known as Taylor's hypothesis this is a frequently made assumption in the analysis of airplane response to atmospheric turbulence. Under such an assumption the computation of response spectral density becomes quite simple [3, 5]. Unfortunately, significant decays in the correlation have been found in experimental measurements of boundary-layer turbulences. Thus, calculations based on frozen-pattern models are just crude estimates, at least for structural response to boundary-layer turbulence, which stand to be improved when a better method becomes available.

In this paper use will be made of a scheme [8] in which a decaying turbulence is treated as superposition of frozen-pattern components, thus allowing the structural response to be similarly superposed and the advantage of the frozen-pattern analysis maximally utilized. The fundamental solution required for the construction of the total response is one corresponding to the excitation of a frozen-pattern sinusoid. To obtain this fundamental solution the formulation will follow Mead's wave-propagation method [3], but will take into account the effect of free stream velocity on the same side of the turbulence excitation and the effect of a cavity on the opposite side of the excitation. As a numerical example, the spectral density of the structural response will be computed and the results will be compared with experimental measurements.

SUPERPOSITION SCHEME

Measured frequency cross-spectra of random boundary-layer turbulent pressures have the general form of

$$\bar{\Phi}_p(\xi,\omega) = \bar{\Phi}_p(0,\omega)\ \psi(\xi)\ \exp(-i\omega\xi/U_c) \tag{1}$$

where ξ is the spatial separation, U_c is the convection velocity, and $\psi(\xi)$ is an even non-negative definite function of ξ, which has an absolute maximum equal to one at $\xi = 0$, and approaches to zero at large absolute values of ξ. On the right hand side of Equation (1), the symbol $\bar{\Phi}_p(0,\omega)$ stands, of course, for the usual spectrum.

It can be shown that if Taylor's hypothesis of frozen-pattern turbulence were valid the function $\psi(\xi)$ would reduce to a constant (equal to one). Therefore, $\psi(\xi)$ represents the decay in statistical correlation as the turbulence is convected down-stream. Indeed, the decay in a boundary-layer turbulence is very significant.

In seeking an easy way to calculate structural response to random excitation of this type, Lin and Maekawa [8] have proposed a scheme in which a general decaying turbulence is constructed by superposition of infinitely many frozen-pattern ones. Briefly, a general turbulent pressure is assumed to be expressible in a Stieltjes integral

$$p(x,t) = \int_{-\infty}^{\infty} \hat{p}(x - ut)\, dG(u) \tag{2}$$

where both $\hat{p}$ and G are random function. The physical implication of Equation (2) is obvious; namely, each component pressure $\hat{p}$ is a frozen-pattern, convected at a velocity u. Each component pressure can again be constructed from frozen-pattern sinusoids. Thus,

$$p(x,t) = \iint_{-\infty}^{\infty} e^{i(ukt - kx)}\, dF(k, u)\, dG(u) \tag{3}$$

It is clear from Equation (3) that the structural response can also be constructed from a fundamental solution obtained by taking the excitation to be just a frozen-pattern sinusoid. Let $H(x,k)\exp(i\omega t)$ be the steady-state solution for

$$\mathcal{L}\{H(x,k)\exp(i\omega t)\} = \exp[i(\omega t - kx)] \tag{4}$$

where $\mathcal{L}$ represents, symbolically, a linear operator in x and t pertaining to the linear problem at hand. Then one basic ingredient required for the construction of structural response is $H(x,k)$ (to be called the wave-number response function in the sequel). Of course, $H(x,k)$ must satisfy all the necessary boundary conditions.

We shall assume that the cross-correlation function of the boundary-layer pressure $E[p(x_1,t_1)\, p(x_2,t_2)]$, where E denotes an assemble average, is dependent only on $\xi = x_1 - x_2$ and $\tau = t_1 - t_2$. Then it can be shown that the theoretical cross-spectrum of p must have the form of [8]

$$\Phi_p(\xi,\omega) = \int_{-\infty}^{\infty} \frac{1}{|u|} e^{-i\omega\xi/u}\, S_p(\omega/u, u)\, du \tag{5}$$

If one equates Equations (1) and (5) one finds a formula to compute S_p as follows:

$$S_p(\omega/u,\ u) = |\omega/u|\ \bar{\Phi}_p(0,\omega)\Psi(\omega/u - \omega/U_c) \tag{6}$$

where Ψ is the Fourier transform of ψ; i.e.,

$$\Psi(v) = \frac{1}{2}\int_{-\infty}^{\infty} \psi(\xi)\ e^{i\xi v}\ d\xi$$

In terms of the H and Ψ functions the expression for the cross-spectrum of the structural deflection response is quite simple; i.e.,

$$\Phi_w(x_1,x_2;\omega) = \bar{\Phi}_p(0,\omega)\int_{-\infty}^{\infty} H(x_1,k)\ H^*(x_2,k)\ \Psi(k - \omega/U_c)dk \tag{7}$$

where an asterisk denotes the complex conjugate. When $x_1 = x_2$ the cross-spectrum reduces, of course, to the spectrum. For derivation of Equation (7) the reader is referred to Reference [8].

In experimental works accelerometers are often used to measure the structural response. To compare with such experimental results the deflection spectrum, Equation (7), must be multiplied by a factor ω^4 to convert into the acceleration spectrum.

WAVE-NUMBER RESPONSE FUNCTION

The structural model chosen for the present study is an infinitely long uniform beam on evenly spaced elastic supports as shown in Figure 1. The beam is backed on the lower side by a space of depth d which is filled with an initially quiescent fluid of density ρ_2 and sound speed a_2. On the upper side the beam is exposed to the excitation of a supersonic boundary-layer turbulent pressure p. The fluid on the upper side of the beam which 'carries' the turbulence has a free stream velocity U_∞, density ρ_1 and sound speed a_1.

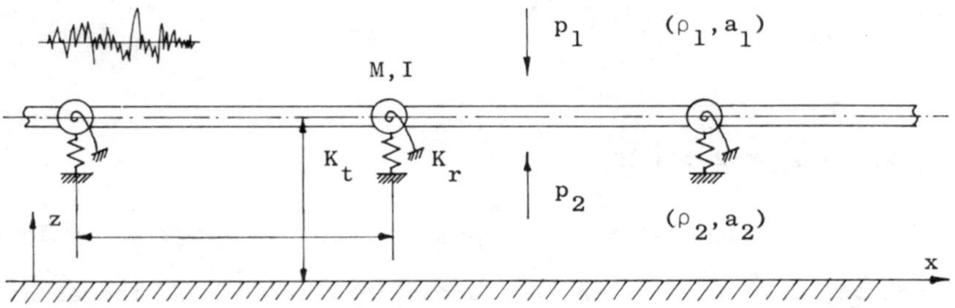

FIGURE 1. An Infinite beam under the Excitation of Boundary-Layer Turbulence.

As the beam responds to the excitation its motion will generate additional pressures in the fluid media on the upper and the lower sides. Denoting such pressures by p_1 and p_2, respectively, the governing equation of the beam motion not directly over an elastic support is given by

$$B \frac{\partial^4 w}{\partial x^4} + \frac{m\partial^2 w}{\partial t^2} = p + (p_1 - p_2)_{z = d} \tag{8}$$

where B is the bending rigidity and m is the mass per unit length of the beam.

For the purpose of determining the wave-number response function H(x,k) the turbulent pressure p should be replaced by $\exp[i(\omega t - kx)]$ and the structural response w by $H(x,k) \exp(i\omega t)$. Thus Equation (8) becomes

$$(BH^{(4)} - m\omega^2 H)e^{i\omega t} = e^{i(\omega t - kx)} + (p_1 - p_2)_{z=d} \tag{9}$$

The forcing function $\exp\,[i(\omega t - kx)]$ gives every span the same excitation but with a phase-lag $\mu_0 = k\ell$ from one span to the next. In this sense μ_0 may be considered as the imposed phase-lag of the excitation. To satisfy the spatial periodicity in the structural response Mead [1] suggested the following series form:

$$H(x,k) = \sum_{n=-\infty}^{\infty} A_n \exp(-i\mu_n x/\ell) \tag{10}$$

where

$$\mu_n = 2n\pi + \mu_0 = 2n\pi + k\ell \tag{11}$$

Without the elastic supports the wave-number response function would be just the one term [8] associated with the forcing phase-lag μ_0. The elastic supports give rise to multiple reflections, thereby admitting other μ values.

For the induced additional pressure p_1 we shall make the usual approximation that it can be calculated without regard to the presence of the turbulence. Then p_1 is governed by the equation

$$\left(\frac{\partial}{\partial t} + U_\infty \frac{\partial}{\partial x}\right)^2 p_1 - a_1^2\left(\frac{\partial^2}{\partial x^2} + \frac{\partial^2}{\partial z^2}\right) p_1 = 0 \tag{12}$$

and subject to the conditions that p_1 can propagate only in the region $z > d$, and that

$$\left(\frac{\partial p_1}{\partial z}\right)_{z=d} = -\rho_1 \left(i\omega + U_\infty \frac{\partial}{\partial x}\right)^2 H\, e^{i\omega t} \tag{13}$$

The solution for p_1 corresponding to each component in the wave-number response function $A_n \exp(-i\mu_n x/\ell)$ is known [9]. The total p_1 can then be obtained by superposition. This pressure, when evaluated at z=d, is given by

$$p_1 = i\,\rho_1 a_1 \sum_{n=-\infty}^{\infty} A_n \frac{(u_n - U_\infty)^2}{(\ell/\mu_n)\left[(u_n - U_\infty)^2 - a_1^2\right]^{\frac{1}{2}}}$$

$$e^{-i\mu_n x/\ell + i\omega t} \tag{14}$$

where $u_n = \omega\ell/\mu_n$.

Some comments about Equation (14) are in order:

1. u_n is the speed at which the component structural motion $A_n \exp[i(\omega t - \mu_n x/\ell)]$ is propagated along the beam.

2. A component structural motion generates no pressure in the adjacent fluid medium if it is propagated at the same velocity as that of the fluid medium (the case of $u_n = U_\infty$).

3. Theoretically, the generated pressure attains an infinite amplitude when the propagation velocity of the structural motion relative to the medium is equal to the speed of sound (the case of $|u_n - U_\infty| = a_1$, the shock wave effect).

4. When this relative velocity is less than the speed of sound; i.e., $|u_n - U_\infty| < a_1$, the generated pressure should provide additional inertia for the structural motion (the apparent mass effect); therefore, a positive imaginary value should be given to the square-root $\left[(u_n - U_\infty)^2 - a_1 2\right]^{\frac{1}{2}}$ in the calculation.

The induced pressure p_2 in the fluid medium $0 \leq z < d$ is governed by

$$\frac{\partial^2 p_2}{\partial t^2} - a_2^2 \left(\frac{\partial^2}{\partial x^2} + \frac{\partial^2}{\partial z^2}\right) p_2 = 0 \tag{15}$$

and subject to the conditions

$$\frac{\partial p_2}{\partial z} = 0, \quad \text{at } z = 0 \tag{16}$$

and

$$\frac{\partial p_2}{\partial z} = \rho_2 \omega^2 H e^{i\omega t}, \quad \text{at } z = d \tag{17}$$

The solution for p_2, when evaluated at $z = d$, is given by

$$p_2 = - \rho_2 \omega^2 \sum_{n=-\infty}^{\infty} A_n \frac{\cot \gamma_n d}{\gamma_n} e^{-i(\mu_n x/\ell - \omega t)} \tag{18}$$

where

$$\gamma_n^2 = (\omega/a_2)^2 - (\mu_n/\ell)^2 \tag{19}$$

At low frequencies γ_n may become imaginary in which case we should take

$$\frac{\cot\gamma_n d}{\gamma_n} = -\frac{\coth|\gamma_n d|}{|\gamma_n|}, \text{ if } \gamma_n \text{ is imaginary} \tag{20}$$

since we expect that the fluid-filled confined space should provide added stiffness for the beam motion when the sound wave lengths in the fluid are much greater than d.

Equations (10), (14) and (18) can now be substituted into Equation (9) to obtain

$$\sum_{n=-\infty}^{\infty} A_n \, \phi(n) \, \exp(-i\mu_n x/\ell) = \exp\,(-i\mu_0 x/\ell) \tag{21}$$

where

$$\phi(n) = B(\mu_n/\ell)^4 - m\omega^2 - i\rho_1 a_1$$

$$\frac{(u_n - U_\infty)^2}{(\ell/\mu_n)\left[(u_n - U_\infty)^2 - a_1^2\right]^{\frac{1}{2}}} - \rho_2\omega^2 \frac{\cot\gamma_n d}{\gamma_n} \tag{22}$$

To determine the amplitudes A_n we, again, follow Mead's procedure and calculate the virtual work done by the external forces acting on and internal forces in the structure through a virtual displacement.

$$\delta A_m \, e^{i\mu_m x/\ell - i\omega t}$$

Excluding the elastic supports, the virtual work done within one span of the beam is

$$\delta W_b = \delta A_m \left\{ \sum_{n=-\infty}^{\infty} A_n \, \phi(n) \int_0^\ell e^{-i\mu_n x/\ell} \, e^{i\mu_m x/\ell} \, dx \right.$$

$$\left. - \int_0^\ell e^{-i\mu_0 x/\ell} \, e^{i\mu_m x/\ell} \, dx \right\} \tag{23}$$

The elastic supports are characterized by a translational spring constant K_t, a translational inertia M, a torsional spring constant K_r, and a torsional inertia I. Thus the virtual work contributed by each elastic spring is

$$\delta W_e = \delta W_t + \delta W_r$$

$$= (K_t - M\omega^2)\ \delta A_m \sum_{n=-\infty}^{\infty} A_n + (K_r - I\omega^2)\ (\mu_m/\ell)\ \delta A_m \sum_{n=-\infty}^{\infty} (\mu_n/\ell)\ A_n \tag{24}$$

Since the structural motion is spatially periodic the virtual work done throughout the entire structure is proportional to that of a periodic unit. Therefore, the principle of virtual work can be stated for a periodic unit as follows:

$$\delta W_b + \delta W_t + \delta W_r = 0 \tag{25}$$

which leads to the simultaneous algebraic equations:

$$A_m \phi(m) + \left[(K_t - M\omega^2)/\ell\right] \sum_{n=-\infty}^{\infty} A_n + \left[(K_r - I\omega^2)/\ell\right] (\mu_m/\ell) \sum_{n=-\infty}^{\infty} (\mu_n/\ell) A_n$$

$$= \begin{cases} 1, & \text{if } m=0 \\ 0, & \text{if } m\neq 0 \end{cases} \tag{26}$$

In actual computations the number of simultaneous equations must be truncated. One can, for example, solve a system of 2N + 1 equations corresponding to $-N \leq m \leq N$. The choice of N must be such that the values of A_m computed do not change appreciably by further increase of the number of equations and that the truncated version of the wave-number response, Equation (10), is sufficiently accurate. These convergence criteria must be programmed in the machine computation.

Equations (26) are derived for finite $(K_t - M\omega^2)$ and finite $(K_r - I\omega^2)$. These equations cannot be reduced to those for supports rigid in translation or rotation. For example, if the supports are rigid in translation then K_t

becomes infinitely large, but the summation of all the A_n must be zero since the deflection at each support is zero, and the product of these two becomes indefinite. To deal with this case substitute [2]

$$A_0 = - \sum_{\substack{n=-\infty \\ n\neq 0}}^{\infty} A_n$$

into Equation (10) to obtain

$$H(x,k) = \sum_{\substack{n=-\infty \\ n\neq 0}}^{\infty} A_n \left(e^{-i\mu_n x/\ell} - e^{-i\mu_0 x/\ell}\right) \tag{27}$$

Correspondingly, virtual displacements are chosen in the form of

$$\delta A_m \left(e^{i\mu_m x/\ell} - e^{i\mu_0 x/\ell}\right) e^{-i\omega t}$$

Then, instead of Equation (26), one obtains from a similar derivation:

$$A_m \phi(m) + \phi(0) \sum_{\substack{n=-\infty \\ n\neq 0}}^{\infty} A_n + \frac{K_r - I\omega^2}{\ell} \left(\frac{\mu_m - \mu_0}{\ell}\right)$$

$$\sum_{\substack{n=-\infty \\ n\neq 0}}^{\infty} \left(\frac{\mu_n - \mu_0}{\ell}\right) A_n = -1, \qquad m \neq 0 \tag{28}$$

It is interesting to note that if the supports are rigid in translation but without constraints in rotation (the case of hinge supports) then the equations for A_m can be de-coupled. For such a case, Equation (28) reduces to

$$A_m \phi(m) + \phi(0) \sum_{\substack{n=-\infty \\ n\neq 0}}^{\infty} A_n = -1, \; m\neq 0 \tag{29}$$

Equation (29) shows that the product A_m (m) is independent of m; i.e.,

$$A_1\phi(1) = A_2\phi(2) = \ldots$$

Thus, substituting

$$A_n = A_m\phi(m) / \phi(n) \tag{30}$$

in Equation (29), one obtains an equation involving only one unknown:

$$A_m\phi(m) \ \phi(0) \sum_{n=-\infty}^{\infty} 1/\phi(n) = -1 \tag{31}$$

which is solved readily to give

$$A_m = -\left\{ \phi(m) \ \phi(0) \sum_{\substack{n=-\infty \\ n\neq 0}}^{\infty} 1/\phi(n)\right\}^{-1} \tag{32}$$

Uncoupled solutions such as Equation (32) are not restricted to the case of hinge supports. In fact, if one of the two infinite sums in Equation (26) can be dropped, then a substitution of the type of Equation (30) is possible which is the key for reducing to only one A_m in each equation. This is the case when either $K_t - M\omega^2 = 0$ or $K_r - I\omega^2 = 0$; i.e., when the elastic supports offer either no translational constraint or no rotational constraint. However, if the rotational constraint is infinite, one again cannot obtain a reduced equation from Equation (26) which is valid only for finite support constraints. To derive such a reduced equation one must use the zero slope condition at the supports

$$\sum_{n=-\infty}^{\infty} \mu_n \ A_n = 0$$

The remaining procedure is very similar to that leading to Equation (29).

A NUMERICAL EXAMPLE

To illustrate the application of the present theory, the spectral densities of the acceleration response at a mid-span location (i.e., $x_1 = x_2 = \ell/2$) have been computed based on the following physical data:

Properties of the beam

B (bending rigidity) = $3.935(10^4)$ N-m^2

ℓ (span-length) = 0.508 m

m (mass per unit length) = 9.746 kg/m

Properties of the surrounding fluid media

$\rho_1 = \rho_2 = \rho$ (density) = 0.11015 kg/m^3

$a_1 = a_2 = a$ (speed of sound) = 261.6 m/sec

U_∞ (free stream velocity on upper side of beam) = 575.6 m/sec

U_c (convection velocity of the turbulence) = 0.75 U_∞

d (cavity depth = 0.1178 m

Properties of the turbulent pressure

$\psi(\xi)$ = decay factor $\simeq \exp\left(-\frac{|\xi|}{\alpha\delta}\right)$

$\bar{\Phi}_p(0,\omega)$ = spectral density = $\frac{\delta}{2U_\infty} \sum_{n=1}^{4} A_n \, e^{-K_n} \, (\omega\delta/U_\infty)$

with

δ (boundary-layer thickness) = 0.279 m, $\alpha = 3$

$A_1 = 4.4 \times 10^{-2}$ $\qquad$ $K_1 = 5.78 \times 10^{-2}$

$A_2 = 7.5 \times 10^{-2}$ $\qquad$ $K_2 = 2.43 \times 10^{-1}$

$A_3 = -9.3 \times 10^{-2}$ $\qquad$ $K_3 = 1.12$

$A_4 = -2.5 \times 10^{-2}$ $\qquad$ $K_4 = 11.57$

The above physical data were taken from a recent experiment on a multi-panel system [10]. The structural specimen used in this experiment was actually a two-dimensional panel array as shown in Figure 2. Therefore, some data have

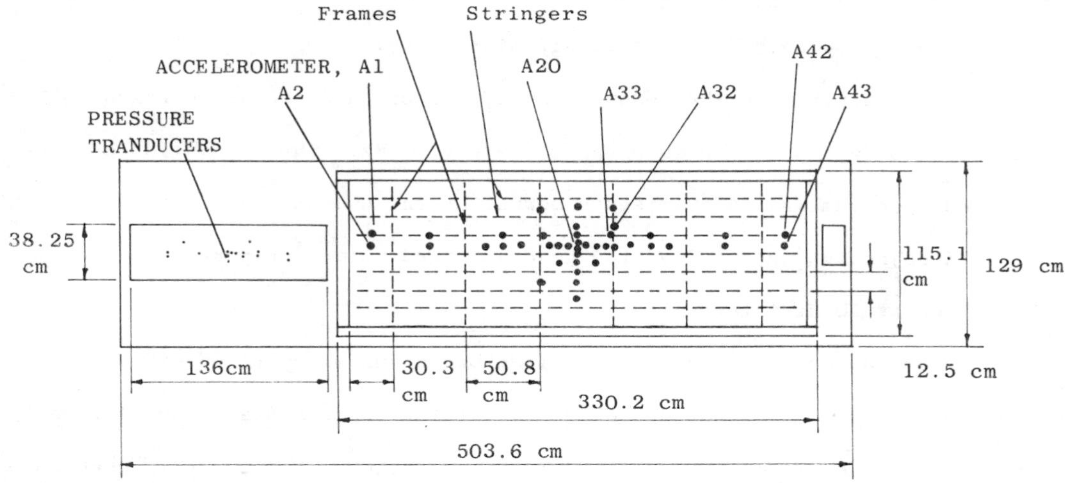

FIGURE 2. The Experimental set-up

been converted to their one-dimensional equivalents. For example, the bending rigidity B of the beam was the average value for the skin and the reinforcing stringers over a unit width, and the specific mass m was obtained similarly. Although the structural specimen had only seven spans and the two end spans were somewhat shorter, it was felt that the theory of an infinite periodic beam on evenly spaced supports should give a reasonable result for the response spectrum at the centre of the middle span where accelerometer A20 was located (referred to Figure 2), and where the effects of the end spans were least important. The translational constraints provided by the supporting frames were sufficiently strong to justify taking the translational spring constant K_t of the supports to be infinite (i.e. the deflections at the supports were assumed to be zero). For the rotational constraints we selected K_r = 60 N-m/rad and I = $3.3(10^{-4})$ kg-m^2. These are the one-dimensional equivalents of the torsional constraints of the frames if the torsional mode of each frame is a half-sine curve.

In order to minimize the computer cost, the computation was carried out only to 3000 Hz and at the intervals of every 50 Hz. The resolution of the computed spectrum was compromised somewhat by the use of coarse intervals, but our main objective was to find the general trend which could be revealed by the values at 50 Hz intervals.

In Figure 3 the computed spectrum is shown along with the experimental spectrum. It should be noted, however, that the experimental results were obtained using a filter of 1 Hz bandwidth; therefore, the curve is much more rugged than the theoretical one. Furthermore, experimentally obtained signals may contain noise other than the structural response. Although the theoretical and the experimental curves show the same general trend, the former is lower than the latter throughout the entire frequency range investigated. This is to be expected since the theoretical curve represents the average between the panel response and the stringer response, whereas the experimental curve shows the panel response alone.

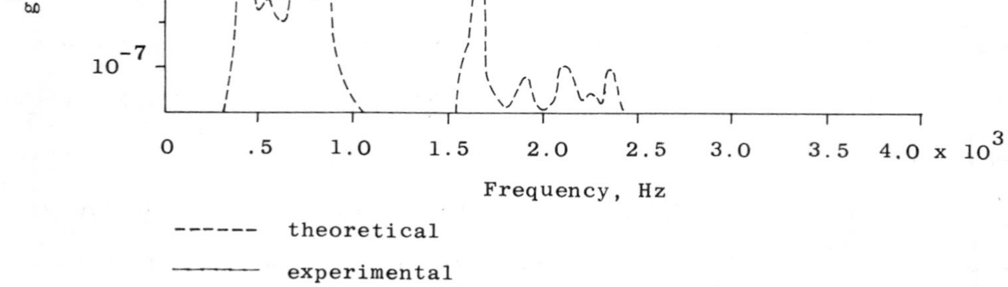

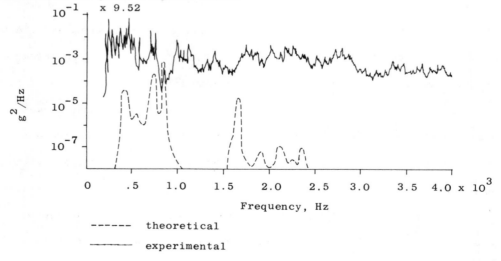

FIGURE 3. Acceleration Spectra of Structural Response

To assess the effects of the rotational constraints at the supports, additional computations were carried out for the case of $K_r = I = 0$, while keeping the other physical constants unchanged. It was found that, except for some shifting of the locations of the peaks and their magnitudes, the general shape of the structural acceleration spectrum remained about the same. In general, the response level was somewhat higher in the region of 200-1000 Hz but it was lower in the region of 1500-2500 Hz.

CONCLUDING REMARKS

As can be seen from Equation (7), the application of the present theory requires the knowledge of the wave-number response function $H(x,k)$, and spectral density of the forcing fields $\bar{\Phi}_p(0,\omega)$ and the decay factor $\psi(\xi)$. The last two are measurable experimentally. Equation (7) is valid for any one-dimensional structure; the essential task in each case is the determination of the H function. Usually, interactions between the structure and adjacent fluids (called fluid loading by some authors) are added sources of complication.

The case of fluid-loaded infinite periodic beam, considered in this paper, is one of the very few where a mathematically exact solution for H can be obtained. If the beam is finite in length then the method of transfer matrix [4, 5, 6] may be more preferable; however, the effect of fluid loading cannot be accounted for exactly (in the mathematical sense) at the present time.

Further extensions to the two dimensional case of panel systems are obvious. If only one row of panels is considered, and if the two parallel edges of the panel row are assumed to be simply supported, then separation of spatial variables is possible in expressing the structural response. This is the well-known Levy's type solution for plate problems. With small modifications, the solution for the one-dimensional beam case can be changed to suit such a panel row problem. When more than one row of panels are included in the structural model the separation of spatial variables in the structural motion is no longer mathematically exact, but a separable form can still be used as an approximation. Although new concepts are not required in treating such two-dimensional problems, the machine computation time can become extremely excessive and burdensome to small research budgets.

REFERENCES

1 D.J. Mead, Free wave propagation in periodically supported, infinite beams. J. Sound Vib. 11, 181 (1970).

2 D.J. Mead and K.K. Pujara, Space-harmonic analysis of periodically supported beams: response to convected random loading. J. Sound Vib. 14, 525 (1971).

3 D.J. Mead, Vibration response and wave propagation in periodic structures. J. Eng. for Industry, Trans. ASME 93(B) 783 (1971).

4 Y.K. Lin and T.J. McDaniel, Dynamics of beam-type periodic structures. J. Eng. for Industry, Trans. ASME 91(B), 1133 (1969).

5 R. Vaicaitis and Y.K. Lin, Response of finite periodic beam to turbulence boundary-layer pressure excitation. AIAA J 10, 1020 (1972).

6 Y.K. Lin, Random vibration of periodic and almost periodic structures. Mechanics Today, Chapter III, ed. S. Nemat-Nasser, Pergamon Press, 93 (1976).

7 G. Sen Gupta, On the relation between the propagation constant and the transfer matrix used in the analysis of periodically stiffened structures. J. Sound Vib. 11, 483 (1970).

8. J.K. Lin and S. Maekawa, A new look at decomposition of turbulence forcing field and the structural response. To be published (1976).

9 P.M. Morse and K.U. Ingard, Theoretical Acoustics, 707 McGraw-Hill Book Company, New York (1968).

10 L. Maestrello, J.H. Monteith, J.C. Manning and D.L. Smith, Measured response of a complex structure to supersonic turbulent boundary layers. AIAA 14th Aerospace Science Meeting, Washington, D.C. (1976).

Professor Y.K. Lin, Department of Aeronautical and Astronautical Engineering, University of Illinois at Urbana-Champaign, 101 Transportation Building, Urbana, Illinois 61801.

Dr. L. Maestrello, NASA Langly Research Center, Hampton, Virginia, U.S.A.

Mr. S. Maekawa, Department of Aeronautical and Astronautical Engineering, University of Illinois at Urbana-Champaign, 101 Transportation Building, Urbana, Illinois 61801.

Mr. N. Nijim, Department of Aeronautical and Astronautical Engineering, University of Illinois at Urbana-Champaign, 101 Transportation Building, Urbana, Illinois 61801.

DISCUSSION

Questioned (Elishakoff) on the results of letting all the spring stiffnesses go to zero, Professor Lin replied that this reduced to the case of an infinite unsupported beam. In this case there is no phase lag but the wave takes on the same form as the forcing function. Dr. Mead welcomed the experimental results because the theory is only directly relevant to infinite structures. It would be valuable to find out how many periods are required in a practical structure to make the application of the infinite structure results valid. He mentioned one experiment in which a periodic structure was subjected to a travelling acoustic field. Professor Lin replied that from a mathematical point of view the difficulty with finite structures is not only on the structural modelling side but also on the fluid side. It is not possible to solve the field equation because of finite length of fluid. Mr. Uhlrich mentioned some recent measurements at Southampton on an eight element finite structure. By averaging the responses at three or four bays he had obtained very good agreement between measured and theoretical propagation constants for both logitudinal and flexural waves.

S NARAYANAN and N C NIGAM

Optimum structural design of sheet-stringer panels subjected to jet noise excitation

INTRODUCTION

Rows of rectangular sheet-stringer panels supported on frames are typical of wings, fuselage and control surfaces of flight vehicle structures. These panels are subjected to random pressure fields generated by jet noise and boundary layer turbulence. This paper deals with the minimum weight design of sheet-stringer panels subjected to jet noise excitation. The pressure field due to jet noise is treated as a stochastic process and the optimum design problem is formulated within the reliability frame-work.

The limitations of the conventional design procedures based on factor of safety are well-known. The reliability based design provides a more realistic treatment of uncertainty inherent in design problems. A comprehensive review of the significant developments in structural reliability is available in reference [1]. Most of the work in this area has been confined to static problems. The extension of reliability based design approach to dynamic problems is recent [2, 3, 4, 5]. Nigam [6] and Narayanan [7] have extended the concepts of reliability based design to the domain of structural optimization in the random vibration environment. In this paper the problem of the minimum weight design of sheet-stringer panels excited by jet noise is formulated with constraints on frequency, stesses, fatigue life and side constraints on design variables. The optimization problem is reduced to a non-linear programming problem and solved using the sequential unconstrained minimization technique of Fiacco and McCormick [8] incorporating the gradient method of Fletcher and Powell [9]. The paper provides an algorithm for automated optimum design of sheet-stringer panels. Results indicate that optimization leads to significant saving in weight and optimum design is sensitive to the constraint on fatigue life.

OPTIMIZATION PROBLEM FORMULATION

The structural optimization problem in random vibration environment can be stated as,

$$\text{Find } \bar{D}^* \text{ to, minimize } W(\bar{D}) \tag{1}$$

subject to,

$$P\left[\bigcup_{\substack{h=1\\ 0\le t\le T}}^{n} \{s_{hk}(\bar{X}(\bar{D},t) \geq r_{hk}\}\right] \leq p_k, \quad k=1,\ldots m_1 \tag{2}$$

$$g_j(\bar{D}) \leq \alpha_j, \quad j = m_1+1,\ldots m \tag{3}$$

$$\omega_\ell^L \leq \omega_\ell(\bar{D}) \leq \omega_\ell^U, \quad \ell = 1,2,\ldots p \tag{4}$$

where, $W(\bar{D})$ is the objective function, $\bar{D}$ is the vector of design variables, s_{hk} is a function of the random dynamic response $\bar{X}(\bar{D},t)$, r_{hk} is the deterministic or random limit on the function s_{hk}, p_k is the specified probability of failure in the kth failure mode, α_j is the constraint on the deterministic function $g_j(\bar{D})$, ω_ℓ^L and ω_ℓ^U are respectively the specified lower and upper bounds on the ℓth natural frequency of the system $\omega_\ell(\bar{D})$. $P[.]$ denotes the probability of the events inside the brackets and U represents the union of events.

In the constraints expressed by the inequalities (2), the probability of failure is defined over a period of time $[0,T]$. This adds a dimension of difficulty to the optimization problem. To reduce the problem to a standard nonlinear programming problem the probability of the union of events specifying failure must be evaluated. This can be done if acceptable upper bounds for the probabilities expressed by the left hand side of the inequalities are established.

Let

$$E_{hk} = \left[\underset{0 \leq t \leq T}{s_{hk}} (\bar{X}(\bar{D},t)) \geq r_{hk} \right] \tag{5}$$

and $$P\left[E_{hk}\right] \leq q_{hk}(\bar{D},T) \tag{6}$$

Under the above assumption inequalities (2) can be replaced by

$$P\left[\bigcup_{h=1}^{n} E_{hk} \right] \leq p_k, \quad k = 1, \ldots m_1 \tag{7}$$

Bounds on either side of the probability $P\left[\bigcup_{h=1}^{n} E_{hk} \right]$ can be established considering the nature of the individual events E_{hk}. When the events E_{hk} are independent and when q_{hk}'s are small, a close upper bound for $P\left[\bigcup_{h=1}^{n} E_{hk} \right]$ can be obtained as,

$$P\left[\bigcup_{h=1}^{n} E_{hk} \right] \leq \sum_{h=1}^{n} q_{hk}(\bar{D},T) \tag{8}$$

If the events E_{hk} are fully correlated,

$$P\left[\bigcup_{h=1}^{n} E_{hk} \right] \leq \max_{h} \{q_{hk}(\bar{D},T)\} \tag{9}$$

It can also be shown that,

$$P\left[\bigcup_{h=1}^{n} E_{hk} \right] \leq q_{1k} + \sum_{h=2}^{n} a_h q_{hk} \tag{10}$$

where, a_h is defined in terms of the survival events

$$\bar{E}_{hk} = \left[\underset{0 \leq t \leq T}{s_{hk}} (\bar{X}(\bar{D},t)) < r_{hk} \right].$$

$$a_h = P\left[\bar{E}_{1k} \cap \bar{E}_{2k} \cap \ldots \cap \bar{E}_{h-1,k} \middle| E_{hk}\right], \quad h = 2,3,\ldots n \tag{11}$$

The values of a_h depend on the ordering of the component events in the particular failure mode. Evaluation of (11) would involve multiple integrations even if the needed joint distributions were available. A useful set of upper bounds for the conditional probability a_h can be obtained when all except one of the survival events are eliminated. From equation (11)

$$a_h \leq P\left[\bar{E}_{\ell k} | E_{hk}\right] = a_{h\ell} \ , \ \ell = 1,\dots,h-1 \tag{12}$$

The closest upper bound to a_h is obtained by selecting the ℓ which minimizes the set of upper bounds. Choose,

$$a_h^* = \min_{\ell=1}^{h-1} (a_{h\ell}) \tag{13}$$

Vanmarcke [10] has given an approximate result for the probabilities $a_{h\ell}$ in terms of the correlation coefficients of the individual events.

$$a_{h\ell} \simeq 1 - \frac{P\left[S_{hk}(\bar{X}(\bar{D},t)) \geq \max\left(r_{hk},\ r_{hk} - \mu_{hk}^M + \dfrac{\sigma_{hk}^M \mu_{\ell k}^M}{\sigma_{\ell k}^M |\rho_{h\ell}^M|}\right)\right]}{P\left[S_{hk}\ (\bar{X}(\bar{D},t)) \geq r_{hk}\right]} \tag{14}$$

The quantities in equation (14) are defined in terms of the statistics of the safety margin $M_{ik} = r_{ik} - s_{ik}$. μ_{ik}^M is the mean of M_{ik} and σ_{ik}^M is the standard deviation of M_{ik} and ρ_{ij}^M is the correlation coefficient between the random variables M_i and M_j.

From equations (13) and (14) a close upper bound a_h^* can be determined. On substituting in equation (10)

$$P\left[\bigcup_{h=1}^{n} E_{hk}\right] \leq \sum_{h=1}^{n} a_h^* \, q_{hk} \tag{15}$$

where $a_1^* = 1$, $a_2^* = a_2 = P\left[\bar{E}_{1k} | E_{2k}\right]$,

$a_3^* = \min\{P\left[\bar{E}_{1k} | E_{3k}\right],\ P\left[\bar{E}_{2k} | E_{3k}\right]\}$ and so on.

Thus inequalities (2) can be replaced by any one of the following forms:

$$\max_{h=1}^{n} \{q_{hk}(\bar{D},T)\} \leq p_k, \quad k = 1,\ldots m_1 \tag{16}$$

or

$$\sum_{h=1}^{n} a_h^*(\bar{D})\, q_{hk}(\bar{D},T) \leq p_k, \quad k = 1,\ldots m_1 \tag{17}$$

or

$$\sum_{h=1}^{n} q_{hk}(\bar{D},T) \leq p_k, \quad k = 1,\ldots m_1 \tag{18}$$

The structural optimization problem in random vibration environment can now be stated in the standard nonlinear programming format as, Find $\overline{D^*}$, to minimize $W(\bar{D})$ (1)

Subject to:

$$\max_{h=1}^{n} \{q_{hk}(\bar{D},T)\} \leq p_k, \quad k = 1,\ldots m_1 \tag{16}$$

or

$$\sum_{h=1}^{n} a_h^*(\bar{D})q_{hk}(\bar{D},T) \leq p_k, \quad k = 1,\ldots m_1 \tag{17}$$

or

$$\sum_{h=1}^{n} q_{hk}(\bar{D},T) \leq p_k, \quad k = 1,\ldots m_1 \tag{18}$$

$$g_j(\bar{D}) \leq \alpha_j, \quad j = m_1+1,\ldots m \tag{3}$$

$$\omega_\ell^L \leq \omega_\ell(\bar{D}) \leq \omega_\ell^U, \quad \ell = 1,\ldots p \tag{4}$$

In structural systems 'failure' can be specified in terms of (i) first excursion above a given level, (ii) fraction of time spent above a given level, and (iii) cumulative damage, as in fatigue. In the above formulation, first two modes of failure are specified by inequalities (2) and reduced to one of the inequalities (16) or (17) or (18). The fatigue failure is expressed in terms of the expected rate of fatigue damage determined on the basis of Palmgren-Miner cumulative damage hypothesis and specified through inequalities (3). In the class of problems, where the response can be treated as stationary and the duration of interest $[0,T]$ is large, it is possible to assume time independence of response statistics to simplify probability estimates.

SHEET-STRINGER PANELS - RESPONSE ANALYSIS

Consider the sheet-stringer panel shown in Figure 1. Instead of the multi-panel system if only a single panel between two stringers is considered, it

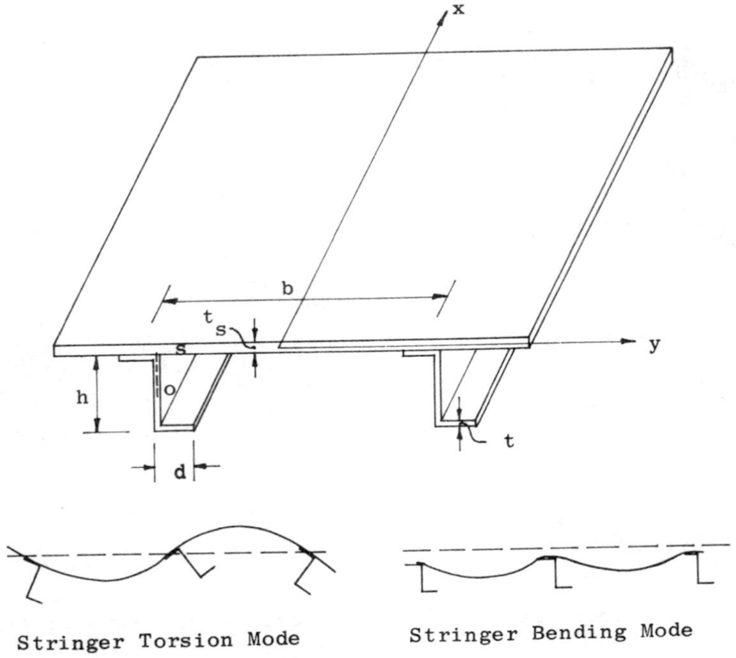

Figure 1 Sheet-Stringer combination and normal modes corresponding to bounding frequencies

has been noted by Lin [12] that the natural frequencies of such a system closely equal the limiting frequencies of the multipanel system. The lowest frequency corresponds to the stringer torsion mode and the highest frequency corresponds to the stringer bending mode in each frequency band. The interaction of intermediate modes has been considered by Lin [13] by way of a statistical argument and he achieved good agreement between measured and computed stresses. Clarkson [14] has enumerated certain general principles for the design of fatigue resistant structures. These principles specify conditions on the frequency which have to be satisfied by adjusting the parameters of the panel, such as stringer spacing, their torsional and bending stiffnesses. The conditions on the frequencies are specified to reduce the response levels by keeping away from frequencies where the noise level is predominant and to make a reasonably simplified analysis based on the bounding frequencies only. In the formulation of the optimization problem for the sheet-stringer panels the response analysis is made considering only the contribution from the two bounding frequencies in the first frequency band.

Considering the single panel between a pair of stringers and the two frames $x = 0$ and $x = a$, the governing differential equation for the plate motion is

$$D\left[\frac{\partial^4 w}{\partial x^4} + \frac{2\partial^4 w}{\partial x^2 \partial y^2} + \frac{\partial^4 w}{\partial y^4}\right] + c\dot{w} + m\ddot{w} = p(x,y,t) \tag{19}$$

where w is the transverse displacement of the plate, $D = Et_s^3/12(1-q^2)$ is the bending rigidity of the plate, m is the mass per unit area of the plate, c is the damping coefficient per unit area of the plate, $p(x,y,t)$ is the random pressure field.

It is assumed that the sheet is simply supported at the frames. The damping is assumed to be viscous and small. Under these assumptions, the normal mode approach can be used for analysing the plate vibrations. The transverse deflection in terms of normal modes and generalized coordinates is expressed as:

$$w(x,y,t) = \sum_{j=1}^{\infty} \sum_{k=1}^{\infty} \phi_j(x)\ \eta_k(y)\ q_{jk}(t) \tag{20}$$

As the plate is simply supported at frames the mode shapes along the x direction can be assumed to be sinusoidal.

$$\phi_j(x) = \sin \frac{j\pi x}{a} \tag{21}$$

In the sequel it is assumed that only the first mode in the x direction makes a significant contribution to the response. This is reasonable as the jet noise is assumed to be fully correlated along the x direction. In view of this the joint acceptance in the x direction for even number modes vanish and progressively decrease in magnitude for the odd number modes as shown in the work of Bozich [15].

The governing differential equation in generalized coordinates can be expressed as

$$M_j\ddot{q}_j(t) + C_j\dot{q}_j(t) + K_jq_j = p_j(t) \tag{22}$$

where

$$w(x,y,t) = \sum_{j=1}^{\infty} f_j(x,y)\ q_j(t) \text{ and}$$

and

$$f_j(x,y) = \sin \frac{\pi x}{a}\ \eta_j(y),$$

M_j, K_j, C_j and p_j are the generalized mass, stiffness, damping and force respectively.

Following Lin [12], the frequency equation and the mode shape of the plate for the stringer torsion mode and the stringer bending mode are given by

$$\cosh \frac{K_1 b}{2} \left(K_2 \; \beta \sin \frac{K_2 b}{2} + 2DK_2^2 \cos \frac{K_2 b}{2}\right) +$$

$$\cos \frac{K_2 b}{2} \left(\beta K_1 \sinh \frac{K_1 b}{2} + 2DK_1^2 \cosh \frac{K_1 b}{2}\right) = 0 \qquad (23)$$

$$\eta_t(y) = \cosh K_1 y - \alpha \cosh K_2 y \qquad (24)$$

$$K_1' \sinh \frac{K_1' b}{2} \left(\gamma \cos \frac{K_2'}{2} - 2DK_2'^3 \sin \frac{K_2' b}{2}\right) +$$

$$K_2' \sin \frac{K_2' b}{2} \left(\gamma \cosh \frac{K_1' b}{2} - 2DK_1'^3 \sinh \frac{K_1' b}{2}\right) = 0 \qquad (25)$$

$$\eta_b(y) = \cosh K_1' y + \alpha' \cos K_2' y \qquad (26)$$

The various quantities in equations (23) - (26) are defined in the Appendix.

Jet Noise Pressure Field

The cross power spectral density of the jet noise pressure field is assumed to be of the form

$$\Phi_{pp}(x_1,y_1;x_2,y_2;\omega) = \Phi_p(\omega) \; e(x_1,y_1;x_2,y_2;\omega) \qquad (27)$$

where $\Phi_p(\omega)$ is the power spectral density of the normalizing homogeneous pressure field and $e(x_1,y_1;x_2,y_2;\omega)$ is in the form of a correlation coefficient. The jet noise is assumed to be a weakly stationary Gaussian random process. It is assumed to be fully correlated in the x direction and homogeneous in the y direction. The power spectral density of the jet noise is taken to be

$$\Phi_{pp}(x_1,y_1;x_2,y_2;\omega) = \Phi_p(\omega) \cos \left\{\frac{\omega \cos\theta}{v} (y_1-y_2)\right\} \qquad (28)$$

where ω is the jet noise frequency, v is the speed of sound in air, θ the angle of predominant direction of the jet noise with the y axis. The form of this function is the same as used by Wallace [16].

It can be shown that the power spectral density of the transverse displacement is

$$\Phi_{WW}(x_1,y_1;x_2,y_2;\omega) = \sum_{j=1}^{\infty} \sum_{k=1}^{\infty} \frac{f_j(x_1,y_1) f_k(x_2,y_2)}{Z_j(\omega)\ Z_k^*(\omega)} \quad I_{jk}(\omega) \tag{29}$$

where

$$Z_j(\omega) = \frac{1}{-M_j\omega^2 + iC_j\omega + K_j}$$

is the complex impedance function in the jth mode, asterisk denoting the complex conjugate, and

$$I_{jk}(\omega) = \Phi_p(\omega) \int_A \int_A f_j(x_1,y_1)\ f_k(x_2,y_2)$$

$$\times\ e(x_1,y_1;x_2,y_2;\omega)\ dA_1\ dA_2 \tag{30}$$

The power spectral densities of the stresses at different locations can be obtained as

$$\Phi_{s_x s_x}(x_1,y_1;x_2,y_2;\omega) =$$

$$\sum_{j=1}^{\infty} \sum_{k=1}^{\infty} \frac{g_{jx}(x_1,y_1)\ g_{kx}(x_2,y_2)}{Z_j(\omega)\ \ Z_k^*(\omega)} \quad I_{jk}(\omega) \tag{31}$$

$$\Phi_{s_y s_y}(x_1,y_1;x_2,y_2;\omega) =$$

$$\sum_{j=1}^{\infty} \sum_{k=1}^{\infty} \frac{g_{jy}(x_1,y_1)\ g_{ky}(x_2,y_2)}{Z_j(\omega)\ \ Z_k^*(\omega)} \quad I_{jk}(\omega) \tag{32}$$

$$\Phi_{s_x s_y}(x_1, y_1; x_2, y_2; \omega) = \tag{33}$$

$$\sum_{j=1}^{\infty} \sum_{k=1}^{\infty} \frac{g_{jx}(x_1, y_1)\; g_{ky}(x_2, y_2)}{Z_j(\omega)\; Z_k^*(\omega)}\; I_{jk}(\omega)$$

where

$$g_{jx} = - \frac{Et_s}{1-\gamma^2} \left\{ \frac{\partial^2 f_j}{\partial x^2} + \gamma \frac{\partial^2 f_j}{\partial y^2} \right\}$$

and

$$g_{jy} = - \frac{Et_s}{1-\gamma^2} \left\{ \frac{\partial^2 f_j}{\partial y^2} + \gamma \frac{\partial^2 f_j}{\partial x^2} \right\}$$

From the power spectral densities, the variance and the correlation coefficients between stresses at different locations can be obtained by integration

$$\sigma_{sx}^2(x,y) = \int_{-\infty}^{\infty} \Phi_{s_x s_x}(x,y;x,y;\omega)\; d\omega \tag{34}$$

$$\sigma_{sy}^2(x,y) = \int_{-\infty}^{\infty} \Phi_{s_x s_y}(x,y;x,y;\omega)\; d\omega \tag{35}$$

$$\rho_{sx\,sx}(x_1,y_1;x_2,y_2) = \frac{\int_{-\infty}^{\infty} \Phi_{sxsx}(x_1,y_1;x_2,y_2;\omega)\,d\omega}{\sigma_{sx}(x_1,y_1)\;\sigma_{sx}(x_2,y_2)} \tag{36}$$

$$\rho_{sysy}(x_1,y_1;x_2,y_2) = \frac{\int_{-\infty}^{\infty} \Phi_{sysy}(x_1,y_1;x_2,y_2;\omega)\,d\omega}{\sigma_{sy}(x_1,y_1)\;\sigma_{sy}(x_2,y_2)} \tag{37}$$

$$\rho_{sxsy}(x_1,y_1;x_2,y_2) = \frac{\int_{-\infty}^{\infty} \Phi_{sxsy}(x_1,y_1;x_2,y_2;\omega)\,d\omega}{\sigma_{sx}(x_1,y_1)\;\sigma_{sy}(x_2,y_2)} \tag{38}$$

MINIMUM WEIGHT DESIGN OF SHEET-STRINGER PANELS-PROBLEM FORMULATION

Objective Function

Since the distance 'a' between the frames is specified in the design and since the density of the material is constant the objective function is taken to be the area of cross-section of the sheet-stringer panel. The design variables in the problem are the stringer spacing b, the web height of the stiffener h, the flange width of the stiffener d, the thickness of the plate t_s and the thickness of the stringer t. Thus each possible design is completely defined by the components of the vector,

$$\bar{D}^T = \{d_1, d_2, d_3, d_4, d_5\}^T = \{b, h, d, t_z, t\}^T \tag{39}$$

The objective function is,

$$W(\bar{D}) = bt_s + t(h + 2d) \tag{40}$$

Frequency Constraint

In the stringer torsion mode adjacent panels vibrate, out of phase while in the stringer bending mode they vibrate in phase (Figure 1). In the frequency range of 400-1000 Hz, the jet noise pressure correlation is high and positive up to about two feet. If the stringer pitch is of the order of six inches or so at least four panels will have pressures in phase. Thus the stringer torsion mode will not be excited appreciably by the jet noise as compared to the stringer bending mode. To reduce the response, use can therefore be made of the pressure spectrum to ensure that the frequency of the sheet corresponding to the stringer bending mode is well above the frequency of maximum sound energy. This condition can be satisfied if the lower bounding frequency itself is above the frequency of maximum sound energy.

This requirement takes the form of the frequency constraint

$$\omega_t(\bar{D}) \geq \omega^L \tag{41}$$

where $\omega_t(\bar{D})$ is the frequency of the plate vibrations corresponding to the stringer torsion mode. To ensure that the stringer torsion frequency does not in any case exceed the stringer bending frequency the second frequency constraint is expressed as

$$\omega_b(\bar{D}) \geq \omega_t(\bar{D}) \tag{42}$$

where $\omega_b(\bar{D})$ is the frequency of the plate vibrations corresponding to the stringer bending mode.

Fatigue damage constraint

As already noted the critical stresses are identified as $s_x(a/2,\ b/2)$, $s_y(a/2,\ b/2)$ and $s_y(a/2,0)$. These stresses are Gaussian as the excitation is Gaussian. The expected rate of fatigue damage is calculated on the basis of these stresses using the Palmgren-Miner cumulative damage rule. For the crossing rate of the stress cycles an average frequency is used. The fatigue damage constraint is expressed as,

$$\frac{1}{2\pi\beta}\left(\frac{\omega_t + \omega_b}{2}\right)\{\sqrt{2}\ \sigma_{sx}(a/2,b/2)\}^{\alpha}\Gamma(1 + \alpha/2) \leq D_1 \tag{43}$$

$$\frac{1}{2\pi\beta}\left(\frac{\omega_t + \omega_b}{2}\right)\{\sqrt{2}\ \sigma_{sy}(a/2,b/2)\}^{\alpha}\Gamma(1 + \alpha/2) \leq D_1 \tag{44}$$

$$\frac{1}{2\pi\beta}\left(\frac{\omega_t + \omega_b}{2}\right)\{\sqrt{2}\ \sigma_{sy}(a/2,0)\}^{\alpha}\Gamma(1 + \alpha/2) \leq D_1 \tag{45}$$

where α,β are the constants in the fatigue law $NS^{\alpha} = \beta$, D_1 is the specified damage level and $\Gamma(.)$ Gamma function.

Stress Constraint

Exceedance of the critical stresses above the yield stress of the material is considered as failure. The probability of failure constraint based on the stress response can be expressed considering the individual stresses at the critical locations or as an overall probability of failure constraint. The stress constraints can be expressed in one of the following three ways.

(i) Failure based on the individual stresses:

$$\text{erfc}\ \{S_y/\ [\sqrt{2}\ \sigma_{sx}(a/2,b/2)]\} \leq p_1 \tag{46}$$

$$\text{erfc}\ \{S_y/\ [\sqrt{2}\ \sigma_{sy}(a/2,b/2)]\} \leq p_1 \tag{47}$$

$$\text{erfc}\ \{S_y/\ [\sqrt{2}\ \sigma_{sy}(a/2,\ 0)]\} \leq p_1 \tag{48}$$

(ii) Failure based on the upper bound given by inequality (18),

$$\text{erfc}\ \{S_y/\ [\sqrt{2}\ \sigma_{sx}(a/2,b/2)] + \text{erfc}\ \{S_y/\ [\sqrt{2}\ \sigma_{sy}(a/2,b/2)]\}$$

$$+ \text{erfc}\ \{S_y/\ [\sqrt{2}\ \sigma_{sy}(a/2,\ 0)]\} \leq p_2 \tag{49}$$

(iii) Failure based on inequality (17),

$$\text{erfc}\ \{S_y/\ [\sqrt{2}\ \sigma_{sy}(a/2,\ 0)]\} + C_1\ \text{erfc}\ \{S_y/\ [\sqrt{2}\ \sigma_{sy}(a/2,b/2)]\}$$

$$+ C_2\ \{\text{erfc}\ S_y/\ [\sqrt{2}\ \sigma_{sx}(a/2,b/2)]\} \leq p_2 \tag{50}$$

where,

$$C_1 = 1 - \text{erfc}(a_1)/\text{erfc}\{S_y/\ [\sqrt{2}\sigma_{sy}(a/2,b/2)]\}$$

$$a_1 = \max\ \{ \frac{S_y}{\sqrt{2}\ \sigma_{sy}(a/2,b/2)},$$

$$\frac{S_y \sigma_{sy}(a/2,b/2)}{\sqrt{2}\ \sigma^2_{sy}(a/2,0)\ |\rho_{sysy}(a/2,b/2;a/2,0)|}\}$$

$$C_2 = \min\ \{1 - \frac{\text{erfc}(a_2)}{\text{erfc}\{S_y/\ [\sqrt{2}\sigma_{sx}(a/2,b/2)]\}},$$

$$1 - \frac{\text{erfc}\ (a_3)}{\text{erfc}\{S_y/[\sqrt{2}\ \sigma_{sx}(a/2,b/2)]\}}\ \}$$

$$a_2 = \max\{\ \frac{S_y}{\sqrt{2}\ \sigma_{sx}(a/2,b/2)}\ ,$$

$$\frac{S_y \sigma_{sx}(a/2,b/2)}{\sqrt{2}\ \sigma^2_{sy}(a_2,0)\ |\rho_{sxsy}\ (a/2,b/2;\ a/2,0)|}\ \}$$

$$a_3 = \max\ \{\ \frac{S_y}{\sqrt{2}\ \sigma_{sx}(a/2,b/2)}\ ,$$

$$\frac{S_y\ \sigma_{sx}(a/2,b/2)}{\sqrt{2}\ \sigma^2_{sy}(a/2,b/2)\ |\rho_{sxsy}(a/2,b/2;a/2,b/2)|}\ \}$$

In the above constraints, erfc(.) denotes the complementary error function and S_y is the yield stress of the material which is assumed to be deterministic.

Other Constraints

It has been assumed that the mode shape, in the x direction is sinusoidal. This assumption is valid if the frame spacing 'a' is greater than one and a half times the stringer spacing [12]. This constraint is expressed as,

$$a \geq 1.5\ b \tag{51}$$

Side constraints on the thickness of sheet and stringer material specify limits on them in the usual ranges of thickness of aluminium sheets

$$0.02 \leq t_s \leq 0.2 \tag{52}$$

$$0.02 \leq t \leq 0.2 \tag{53}$$

In the problem formulation it should be ensured that there is no tendency for the stringer spacing to reduce indefinitely in the optimization process. This constraint is expressed as,

$$b \geq 2h \tag{54}$$

Side constraints on the minimum dimensions of the web height and the flange width of the stringer are also specified.

$$h \geq 0.5 \tag{55}$$

$$d \geq 0.5 \tag{56}$$

The non-negativity restrictions of the design variables are also included in the constraints.

$$d_i \geq 0, \qquad i = 1,\ldots 5 \tag{57}$$

<u>Problem statement</u>: The minimum weight design problem for the sheet-stringer panels subjected to jet noise can be stated as,

$$\text{Find } \overline{D^*}, \quad \text{to minimize } W(\bar{D}) = bt_s + t(h + 2d)$$

Subject to

$$g_1(\bar{D}) = 1 - \omega^L/\omega_t(\bar{D}) \geq 0$$

$$g_2(\bar{D}) = 1 - \omega_t(\bar{D})/\omega_b(\bar{D}) \geq 0$$

$$g_3(\bar{D}) = 1 - \frac{1}{2\pi\beta D_1}\left(\frac{\omega_t+\omega_b}{2}\right)\{\sqrt{2}\ \sigma_{sx}(a/2,b/2)\}^{\alpha}\ \Gamma(1 + \tfrac{\alpha}{2}) \geq 0$$

$$g_4(\bar{D}) = 1 - \frac{1}{2\pi\beta D_1}\left(\frac{\omega_t+\omega_b}{2}\right)\{\sqrt{2}\ \sigma_{sy}(a/2,b/2)\}^{\alpha}\ \Gamma(1 + \alpha/2) \geq 0$$

$$g_5(\bar{D}) = 1 - \frac{1}{2\pi\beta D_1}\left(\frac{\omega_t+\omega_b}{2}\right)\{\sqrt{2}\ \sigma_{sy}(a/2,0)\}^{\alpha}\ \Gamma(1 + \alpha/2) \geq 0$$

$$g_6(\bar{D}) = 1\text{-erfc}\{S_y/\left[\sqrt{2}\ \sigma_{sx}(a/2,b/2)\right]\}/p_1 \geq 0$$

$$g_7(\bar{D}) = 1\text{-erfc}\{S_y/\left[\sqrt{2}\ \sigma_{sy}(a/2,b/2)\right]\}/p_1 \geq 0$$

$$g_8(\bar{D}) = 1\text{-erfc}\{S_y/\left[\sqrt{2}\ \sigma_{sy}(a/2,0)\right]\}/p_1 \geq 0$$

$$g_9(\bar{D}) = 1\text{-}1.5b/a \geq 0$$

$$g_{10}(\bar{D}) = 1-t_s/0.2 \geq 0$$

$$g_{11}(\bar{D}) = 1-t/0.2 \geq 0$$

$$g_{12}(\bar{D}) = 1-0.02/t_s \geq 0$$

$$g_{13}(\bar{D}) = 1-0.02/t \geq 0$$

$$g_{14}(\bar{D}) = 1-2h/b \geq 0$$

$$g_{15}(\bar{D}) = 1-0.5/h \geq 0$$

$$g_{16}(\bar{D}) = 1-0.5/d \geq 0$$

$$d_i \geq 0, \quad i = 1.,,,5$$

The above formulation is modified if the stress constraint is expressed in the form of inequality (49) or (50). In that case constraints g_6, g_7, g_8 are replaced by a single constraint of the form

$$g_6(\bar{D}) = 1-\{\mathrm{erfc}(S_y/[\sqrt{2}\,\sigma_{sx}(a/2,b/2)]) + \mathrm{erfc}(S_y/[\sqrt{2}\,\sigma_{sy}(a/2,b/2)]) + \mathrm{erfc}(S_y/[\sqrt{2}\,\sigma_{sy}(a/2,0)])\}/p_2 \geq 0$$

or

$$g_6(\bar{D}) = 1-\{\mathrm{erfc}(S_y/[\sqrt{2}\,\sigma_{sy}(a/2,0)]) + C_1\,\mathrm{erfc}(S_y/[\sqrt{2}\,\sigma_{sy}(a/2,b/2)]) + C_2\,\mathrm{erfc}(S_y/[\sqrt{2}\,\sigma_{sx}(a/2,b/2)])/p_2 \geq 0.$$

In such a case, constraints g_9 to g_{16} are numbered consecutively from g_7 to g_{14}.

SOLUTION PROCEDURE

The structural optimization problem is solved by the sequential unconstrained minimization technique of Fiacco and McCormick [8] using the penalty function (the P-function) of the form

$$P(\bar{D},r) = W(\bar{D}) + r \sum_{j=1}^{n} \frac{1}{g_j(\bar{D})} \qquad (58)$$

where n is the total number of inequality constraints of the form $g_i(\bar{D}) \geq 0$. r is the penalty parameter greater than zero. From the optimization problem statement it may be noted that constraints g_1 to g_{16} have been normalized to have values between zero and one.

The Fletcher and Powell [9] gradient algorithm was used for the unconstrained minimization incorporating a golden section rule for the one dimensional search. The penalty parameter r was reduced to one tenth of its original value at the end of each unconstrained minimization. An extrapolation scheme was adopted to locate improved starting points at the end of two P-function minimizations. Details of the convergence criteria and other particulars are given in [7].

RESULTS

The optimization problem is solved for two sets of design data, the difference in them being in the levels of the white noise power spectral density of the jet noise pressures $\Phi_p(\omega)$ and the limiting value specified on the frequency corresponding to the stringer torsion mode. These are

(i) $\Phi_p(\omega) = 0.41\times10^{-4}\ (\text{lbs/in}^2)^2/(\text{rad/sec})$,

$\omega^L = 1884$ rad/sec.

(ii) $\Phi_p(\omega) = 0.7\times10^{-5}\ (\text{lbs/in}^2)^2/(\text{rad/sec})$,

$\omega^L = 942$ rad/sec.

The rest of the design specifications are the same in both the cases. These are given below:

$a = 20.0$ in, $\zeta = 0.02$, $S_y = 20000$ lbs/in^2,

$\alpha = 6.0, \quad \beta = 6.4 \times 10^{31}, \quad v = 1100$ ft/sec,

$\theta = \pi/4$ radians, $\rho = 0.1$ lbs/in^3.

The results for the two sets are given in Tables 1 and 2 respectively. Tables 1a, 1b and 1c give the optimum designs for a constant fatigue damage level $D_1 = 5 \times 10^{-6}$ and failure probabilities $p_2 = 10^{-3}$, $p_2 = 10^{-4}$ and $p_2 = 10^{-5}$ respectively. Table 1d gives the optimum design for $D_1 = 1 \times 10^{-6}$ and $p_2 = 10^{-5}$. In all these four cases, the failure probability constraint is expressed as in inequality (52). The results of the optimization cycles in these cases are plotted in Figure 2. Table 1e gives the optimum design when $D_1 = 5 \times 10^{-6}$ and

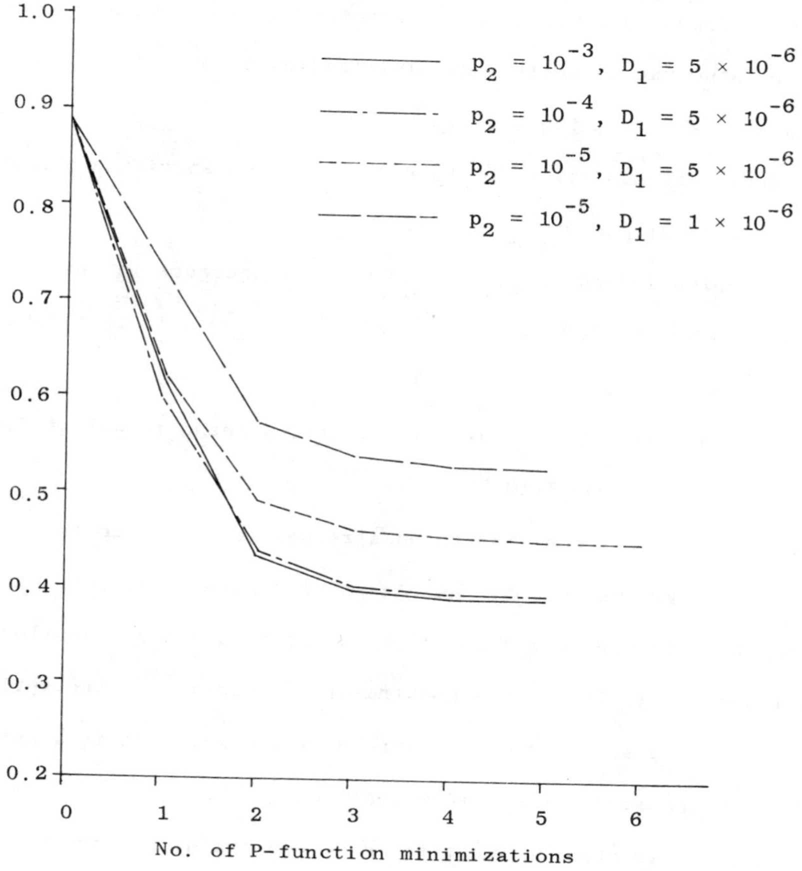

Figure 2 Sequence of unconstrained optima (probability of failure constraint as in inequality 52).

$p_1 = 10^{-4}$. In this case the failure probability constraints are expressed as in inequalities (46), (47) and (48). Table 2a gives the optimum design when the failure probability is expressed as in inequalities (46), (47) and (48), while Table 2b gives the optimum design when the failure probability is expressed as in inequality (50).

SUMMARY AND CONCLUSIONS

The minimum weight design problem for sheet-stringer panels subjected to jet noise excitation has been formulated within reliability frame-work. The problem has been reduced to a nonlinear programming problem and solved on a digital computer using the unconstrained minimization technique. Following conclusions can be drawn from the results:

1. The optimum design is sensitive to the constraints on specified probability of failure and fatigue damage level.

   Comparison of Tables 1b and 1c shows 14.2 percent increase in weight for a decrease in probability of failure from 10^{-4} to 10^{-5}. In 1a stress constraints are not active.

   Comparison of Tables 1c and 1d shows 17 percent increase in weight for a change in damage level from 5×10^{-6} to 1×10^{-6}.

2. In the above formulation, constraints on stresses were imposed in three different ways (equations 46-50). Results indicate that the optimum design is insensitive to the manner in which the constraints are applied. This is obviously so if the stress constraint is inactive at the optimum. In other cases, it is so, because the stress at one location is significantly more than the stresses at other locations.

3. From the results it is observed that the side constraint on the stringer spacing is active in all cases. This is to be expected for a single panel

as the reduction in stringer spacing reduces the area of sheet and the size of stringers. For a structural system, such as a fuselage, minimum weight of a panel does not always imply minimum weight of the total system. It is meaningful, therefore, to consider the area of a single panel per unit length of the stringer spacing as the objective function.

$$W(\bar{D}) = t_s + t(h + 2d)/b \tag{59}$$

Results for the above objective function and second set of design data are given in Table 3. Comparison with Table 2 shows that the formulation based on (59) leads to a reduction in weight of 15.4%. Objective function (59) is recommended for further work.

APPENDIX

$$K_1 = ((\omega_t^2 m/D)^{1/2} + \pi^2/a^2)^{1/2}$$

$$K_2 = ((\omega_t^2 m/D)^{1/2} - \pi^2/a^2)^{1/2}$$

$$\beta = E\, C_{WS}(\pi^4/a^4) + GC(\pi^2/a^2) - \rho I_s \omega^2$$

$$\alpha = \cosh\, (K_1 b_2)/\cos\, K_2 b/2$$

$$K_1' = ((\omega_b^2 m/D)^{1/2} + \pi^2/a^2)^{1/2}$$

$$K_2' = ((\omega_b^2 m/D)^{1/2} - \pi^2/a^2)^{1/2}$$

$$\gamma = EI(\pi^4/a^4) - \rho A \omega^2$$

$$\alpha' = K_1' \sinh(K_1' b/2)$$

where,

A - Area of cross-section of stringer

E - Young's modulus of stringer material

G - Shear modulus of stringer material

C - Torsion constant of stringer cross-section

C_{WS} - Warping constant of stringer cross-section with respect to point s (Figure 1).

I_s - Polar moment of inertia about horizontal axis through O

a - Frame spacing

ρ - Mass density of stringer material

ν - Poisson ratio

REFERENCES

1 A.H.S. Ang, Chairman, Structural Safety, A Literature Review, Proc. ASCE, JSD, 98, 845 (1972).

2 M. Shinozuka, Safety against dynamic forces. Proc. ASCE, JSD, 100, 1821 (1974).

3 J.N. Yang and E. Heer, Reliability of randomly excited structures. AIAA J. 9, 1262 (1971).

4 J.N. Yang and M. Shinozuka, On the first excursion probability in stationary narrow band random vibration. Trans. ASME, J. App. Mech. 38, 1017 (1971).

5 R.L. Racicot, Random vibration analysis, application to wind loaded structures. Case Western Res. Univ. Ph.D Thesis (1969).

6 N.C. Nigam, Structural optimization in random vibration environment. AIAA J. 10, 551 (1972).

7 S. Narayanan, Structural optimization in random vibration environment. Ph.D Thesis, Indian Institute of Technology, Kanpur (1975).

8 A.V. Fiacco and G.P. McCormick, The sequential unconstrained minimization technique for nonlinear programming. A Primal-Dual Method. Management Science, 10, 360 (1964).

9 R. Fletcher and M.J.D. Powell, A rapidly convergent descent method for minimization. Computer J. 6, 163 (1963).

10 E.H. Vanmarcke, Matrix formulation of reliability analysis and reliability based design. J. Computers and Structures, 3 (1973).

11 B.L. Clarkson, Stresses in skin panels subjected to random acoustic loading. Aeronautical J. Roy. Aero. Soc. 72, 1000 (1968).

12 Y.K. Lin, Free vibration of continuous skin-stringer panels. Trans. ASME J. App. Mech. 27, 669 (1960).

13 Y.K. Lin, Stresses in continuous skin-stiffener panels under random loading. J. Aero. Sp. Sc. 29, 67 (1962).

14 B.L. Clarkson, Design of fatigue resistant structures. Noise and Acoustic Fatigue in Aeronautics, E.J. Richards and D.J. Mead (Ed.) John Wiley & Sons, 354 (1968).

15 D.J. Bozich, Spatial correlation in acoustic structural coupling. J. Ac. Soc. of America, 36, 52 (1964).

16 C.E. Wallace, Stress response and fatigue life of acoustically excited sandwich panels. Acoustical fatigue in Aero-space Structures, Trapp and Fomey (Ed.), Syracuse Univ. Press (1965).

Professor S. Narayanan, Department of Applied Mechanics, Indian Institute of Technology, Madras, India.

Professor N.C. Nigam, Department of Aeronautical Engineering, Indian Institute of Technology Kanpur, Kanpur, India.

		r	d_1	d_2	d_3	d_4	d_5	$W(\bar{D})$	$W(\bar{D}) + r \Sigma \frac{1}{g_j(\bar{D})}$	$\frac{W(\bar{D})}{b}$
(a)	INITIAL	1.384×10^{-2}	5.0	2.4	0.6	0.12	0.08	0.888	2.367	0.176
	OPTIMAL	1.384×10^{-5}	3.621	1.801	0.516	0.061	0.0597	0.390	0.395	0.108
(b)	INITIAL	1.384×10^{-2}	5.0	2.4	0.6	0.12	0.08	0.888	2.367	0.176
	OPTIMAL	1.384×10^{-6}	3.536	1.765	0.516	0.0604	0.064	0.393	0.394	0.111
(c)	INITIAL	1.384×10^{-2}	5.0	2.4	0.6	0.12	0.08	0.888	2.367	0.176
	OPTIMAL	1.384×10^{-8}	3.706	1.853	0.523	0.0676	0.069	0.451	0.451	0.121
(d)	INITIAL	1.384×10^{-2}	5.0	2.4	0.6	0.12	0.08	0.888	2.367	0.176
	OPTIMAL	1.384×10^{-5}	4.051	2.018	0.522	0.0779	0.0694	0.528	0.551	0.130
(e)	INITIAL	1.51×10^{-2}	5.0	2.4	0.6	0.12	0.08	0.888	2.367	0.176
	OPTIMAL	1.51×10^{-6}	3.546	1.770	0.518	0.0606	0.0634	0.393	0.395	0.110

(a) $p_2 = 10^{-3}$, $D_1 = 5\times10^{-6}$, Active constraints g_5, g_{12}

(b) $p_2 = 10^{-4}$, $D_1 = 5\times10^{-6}$, Active constraints g_5, g_6, g_{12}

(c) $p_2 = 10^{-5}$, $D_1 = 5\times10^{-6}$, Active constraints g_6, g_{12}

(d) $p_2 = 10^{-5}$, $D_1 = 1\times10^{-6}$, Active constraints g_5, g_{12}

(e) $p_1 = 10^{-4}$, $D_1 = 5\times10^{-6}$, Active constraints g_8, g_{14}

TABLE 1. OPTIMUM DESIGN FOR $\Phi_p(\omega) = 0.41\times10^{-4}$ $(lbs/in^2)^2/(rad/sec)$, $\omega^L = 1884$ rad/sec.

		r	d_1	d_2	d_3	d_4	d_5	$W(\bar{D})$	$W(\bar{D}) + r\,\Sigma \frac{1}{g_j(\bar{D})}$	$\frac{W(\bar{D})}{b}$
(a)	INITIAL	1.703×10^{-2}	8.0	2.0	1.0	0.08	0.08	0.96	1.923	0.12
	OPTIMAL	1.703×10^{-7}	2.897	1.447	0.501	0.035	0.0361	0.190	0.191	0.0655
(b)	INITIAL	1.649×10^{-2}	8.0	2.0	1.0	0.08	0.08	0.96	1.923	0.12
	OPTIMAL	1.649×10^{-7}	3.044	1.521	0.501	0.0361	0.0321	0.191	0.191	0.0655

(a) $p_1 = 10^{-3}$, $D_1 = 1\times10^{-6}$, Active constraints g_5, g_{14}, g_{16}

(b) $p_2 = 10^{-3}$, $D_1 = 1\times10^{-6}$, Active constraints g_5, g_{12}, g_{14}

Note: g_{12} in cases 1a, 1b, 1c and 1d and g_{14} in cases 1e, 2a and 2b refer to geometric constraints on stringer spacing.

TABLE 2. OPTIMUM DESIGN FOR $\Phi_p(\omega) = 0.7\times10^{-5}$ $(lbs/in^2)^2/(rad/sec)$, $\omega^L = 942$ rad/sec.

		r	d_1	d_2	d_3	d_4	d_5	$W(\bar{D})$	$W(\bar{D}) + r \Sigma \frac{1}{g_j(\bar{D})}$
(a)	INITIAL	2.061×10^{-3}	8.0	2.0	1.0	0.08	0.08	0.12	0.24
	OPTIMAL	2.061×10^{-9}	5.954	1.702	1.06	0.0403	0.023	0.05505	0.05506
(b)	INITIAL	2.044×10^{-3}	8.0	2.0	1.0	0.08	0.08	0.12	0.2403
	OPTIMAL	2.044×10^{-9}	6.70	1.822	1.276	0.042	0.02	0.05518	0.05518

(a) $p_2 = 10^{-3}$, $D_1 = 1.0\times10^{-6}$, Active constraints g_5

(b) $p_2 = 10^{-3}$, $D_1 = 1.0\times10^{-6}$, Active constraints g_5, g_{13}

TABLE 3. OPTIMUM DESIGN FOR $\Phi_p(\omega) = 0.7\times10^{-5}$ $(\text{lbs/sec}^2)^2/(\text{rad/sec})$, $\omega^L = 942$ rad/sec.

$W(\bar{D}) = t_s + (2d+h)/b$

DISCUSSION

When asked about error analysis (Bendat) the author stated that confidence limits, standard errors etc. had not been obtained.

It was suggested (Clarkson) that rather than restrict considerations to a certain basic design one might consider permitting, for example, a variation in thickness of the sheet. The author agreed that such methods were being introduced within the framework of control theory but had not been considered in their (the authors) work.

In reply to comments (Lin) that it is difficult to avoid the main response 'hump' occurring between 50 Hz and 1000 Hz for sheet-stringer panels the author noted that it so happened that in the example the frequency constraint never became active and agreed that the band was so wide as to be not particularly meaningful. Indeed if one wanted to stay above 1000 Hz the design may become too heavy and, if so, such a constraint would not be useful.

When asked (Shinozuka) if the minimum was sharp, the author replied that different starting points did not result in convergence to exactly the same optimum design values but were close.

E H VANMARCKE

Method of spectral moments to estimate structural damping

INTRODUCTION

This note briefly describes a method introduced by the writer [11-13], now referred to as "the method of moments" [7], for determining damping values of structures from field measurements of their response to random excitation. To evaluate the equivalent damping of structures that may be modelled as a single-degree-of-freedom system, the procedure requires computation of the first three moments (i.e., area, first and second moment) of the estimated response power spectral density in a frequency band which includes the funda-mental frequency. No prior smoothing of the spectrum is necessary to obtain relatively stable estimates of the damping. New information is provided herein on the reliability of the resulting damping estimates. The same method may be used to estimate the properties of higher modes in multi-degree-of-freedom systems by successively isolating portions of the power spectrum containing higher mode peaks.

CURRENT METHODS

Much work has been done in the field of identification of linear dynamic systems. Attention is restricted here to methods for estimating the critical damping ratio(s) of large structures such as tall buildings or offshore towers from measurements of their response to wind or wave excitation. Available procedures operate either in the time or in the frequency domain. Auto-correlation function techniques are discussed by Cherry and Brady (1965) and have been applied by Davenport (1970). Essentially, the damping is estimated from the decay rate of the auto-correlation function. A widely used spectral method is to determine the damping ratio from the half-power bandwidth, i.e., the absolute value of the difference between the frequencies at which the spectral density is equal to one-half times its maximum value. When bandwidths are narrow, the accuracy of this measurement is often not satisfactory. It is difficult to estimate single spectral ordinates with high confidence. It is well known in classical spectral analysis [1,3,9,10] that the goals of high resolution and small statistical uncertainty are in basic conflict. Gersch et al. [6] proposed a procedure based on maximum likelihood, and more recently, a least squares estimation procedure [7].

BACKGROUND FOR THE METHOD

Spectral Moments and Related Parameters

The most important characteristic of a zero mean stationary random motion X(t) is undoubtedly its mean square value or average total power. The one-sided spectral density function $G(\omega)$, $\omega \geq 0$, normalized with respect to the mean square value, indicates how the power is distributed over all frequencies. It is argued here that a first-order characterization of the frequency content of X(t) requires the knowledge of two spectral parameters, ω_2 and δ, which depend on the first few moments of $G(\omega)$. Define

$$\lambda_i = \int_0^\infty \omega^i G_x(\omega)\,d\omega \qquad i = 0,\ 1,\ 2$$
$$\omega_1 = \lambda_1/\lambda_0 \qquad \omega_2 = (\lambda_2/\lambda_0)^{\frac{1}{2}} \tag{1}$$

λ_0 is the area under $G_x(\omega)$; ω_1 may be interpreted as the distance from the frequency origin to the centre of "spectral mass" $G(\omega)$; and ω_2 as the "radius of gyration" about the frequency origin. The spectral parameter δ is defined as follows [11]:

$$\delta = \left[1 - \frac{\lambda_1^2}{\lambda_0\lambda_2}\right]^{\frac{1}{2}} \qquad (2)$$

It is a unitless measure of the variability in frequency content, i.e. of the bandwidth or the dispersion of G(ω) about its centroid. It is easy to show that

$$\delta = \frac{(\omega_2^2 - \omega_1^2)^{\frac{1}{2}}}{\omega_2} = \frac{\omega_s}{\omega_2}$$

where ω_s may be interpreted as the "centroidal radius of gyration" of $G_x(\omega)$. From Schwarz' inequality, $0 \leq \lambda_1^2/\lambda_0\lambda_2 \leq 1$, and hence, δ is always between zero and one, being relatively small for narrow-band processes and relatively large for wide-band processes. A pure sinusoid and a wide-band process may have a common value of λ_0 and ω_2, but the value of δ will always be different; that of a sinusoid is zero since there is no variability in its frequency content. The time domain interpretation of the spectral parameters just defined is considered elsewhere [12,14].

Relations between Spectral Moments and Structural Damping

It is straightforward to compute the spectral parameters for any specified spectral density function. Consider the stationary response of a simple linear viscoelastic system to ideal white noise excitation: ω_2 equals the natural frequency of the oscillator and δ is a function of the damping ratio only. The s.d.f. is (see for example, [4]):

$$G(\omega) = G_0 \, |H(\omega)|^2 = \frac{G_0}{(\omega_n^2 - \omega^2)^2 + 4\zeta^2\omega_n^2\omega^2} , \quad 0 \leq \omega \leq \infty \qquad (3)$$

where H(ω) is the system transfer function, ω_n is the natural frequency and ζ is the damping ratio. The parameter $\delta^2 = 1 - \lambda_1^2/\lambda_0\lambda_2$ is

$$\delta^2 = 1 - \frac{1}{1-\zeta^2}\left\{1 - \frac{1}{\pi}\tan^{-1}\left[\frac{2\zeta\sqrt{1-\zeta^2}}{1-2\zeta^2}\right]^2\right\} = \frac{4\zeta}{\pi}(1-1.1\zeta+\ldots) \simeq \frac{4\zeta}{\pi} \tag{4}$$

The approximation $\delta^2 \simeq 4\zeta/\pi$ is very satisfactory when the damping is light (say, $\zeta < 0.10$).

It is interesting to compare these values of δ^2 with those obtained if the white noise excitation bandwidth is limited to an interval which includes the oscillator's fundamental frequency. A convenient choice is $\omega_a = r\omega_n$ and $\omega_b = \omega_n/r$, in which r must be chosen between 0 and 1; for r=0, the complete moments are obtained. The partial spectral moments of $G(\omega)$ are defined as follows:

$$\lambda_{i,r} = \int_{\omega_a}^{\omega_b} \omega^i\, G(\omega)\,d\omega = \int_{\omega_n r}^{\omega_n/r} \omega^i\, G(\omega)\,d\omega \tag{5}$$

By Taylor series expansion of the first three partial spectral moments, the bandwidth parameter based on partial spectral moments, δ_r, can be approximated [11,13] . The result, for relatively light damping values ($\zeta < 0.1$), is

$$\delta_r^2 = 1 - \frac{\lambda_{1,r}^2}{\lambda_{0,r}\,\lambda_{2,r}} \simeq \left[\frac{4\zeta}{\pi} \cdot \frac{1-r}{1+r}\right] \tag{6}$$

The parameter δ_r decreases from its maximum value at r=0 to the limit $\delta_r=0$ which is approached when $r\to 1$.

THE ESTIMATION PROCEDURE

The spectral moments and the bandwidth parameter may also be computed from a sample power spectral density $\hat{G}(\omega_k)$ which is proportional to the squared norm of the discrete Fourier transform of the sample record x(t). The discrete Fourier transform is

$$F(\omega_k) = \sum_{n=1}^{N} x(n\Delta t)\, e^{-i\omega_k n\Delta t}$$
$$\omega_k = \frac{2\pi k}{N\Delta t} \quad k = 1,2,\ldots,N \tag{7}$$

Δt is the sampling interval, N is the number of sample points, and $T = N\Delta t$ is the record length. The spacing between ω_k and ω_{k+1} is $\Delta\omega = 2\pi/T$. Multiplying $F(\omega_k)$ by its complex conjugate yields $|F(\omega_k)|^2$ which is proportional to the sample spectrum $\hat{G}(\omega_k)$. The proportionality factor (which is inversely proportional to T) is not needed in the estimation procedure. The estimated spectral characteristics $\hat{\lambda}_i$ and $\hat{\delta}$ based on complete spectral moments of $G_T(\omega)$ are:

$$\hat{\lambda}_i = \sum_{n=1}^{N} \omega_k^i \, \hat{G}(\omega_k) \, \Delta\omega \qquad (8)$$

$$\hat{\delta}^2 = 1 - \hat{\lambda}_1^2/\hat{\lambda}_0\hat{\lambda}_2 \qquad (9)$$

Similarly, the estimates $\hat{\lambda}_{i,r}$ and $\hat{\delta}_r$ are obtained if the spectral moments are computed by integrating over a limited range of frequencies between ω_a and ω_b (as defined earlier). If x(t) is the recorded response of a one-degree system to wide-band excitation, the damping factor ζ is therefore estimated as follows:

$$\hat{\zeta} \simeq \frac{\pi}{4} \hat{\delta}_r^2 \frac{1+r}{1-r} \qquad (10)$$

which reduces to $\hat{\zeta} \simeq (\pi/4)\,\hat{\delta}^2$ if complete moments are computed.

Partial moments lead to more reliable damping estimates for two reasons. First, for real wind or wave excitation, the input spectral density is approximately constant within some frequency interval (ω_a, ω_b) centred around the eigenfrequency of the structural mode under study, but obviously not over the entire frequency range. Secondly, high and low frequency ends of the estimated spectral density are unreliable due to finite record length, digitization of the record ("aliasing"), non-stationary effects, measuring equipment limitation, etc. ("Aliasing" is the effect whereby the part of the spectrum corresponding to frequencies in excess of the Nyquist frequency $\omega_c = \pi/\Delta t$ are "folded" into the frequency range $0 < \omega < \omega_c$ and added to the spectral density in this range).

The transfer function and response spectral density function just studied are applicable when (i) the input is an applied force and the output the system's displacement, or (ii) the input is a base acceleration and the output is the relative displacement. Response spectral densities appropriate for other input-output parameters are quite similar and their spectral moments are directly related to those already given. In fact, it may be shown that for lightly damped systems, Equations (6) and (10) remain valid when the frequency ratio r is relatively close to one.

RELIABILITY OF SPECTRAL MOMENT ESTIMATES

For a given record length T, estimates of spectral moments may be expected to be much more reliable than those of individual spectral ordinates. This follows essentially from the fact that $\hat{\lambda}_i$ is the sum of a large number of statistically independent contributions $\omega_k^i\ \hat{G}(\omega_k)\Delta\omega$. The coefficient of variation (= the ratio of the standard deviation to the mean) of individual spectral ordinates $\hat{G}(\omega_k)$ is approximately one. (It does not decrease as N increases because the frequency interval $\Delta\omega$ is inversely proportional to N). The coefficient of variation of the estimated spectral moment $\hat{\lambda}_i$ is approximately

$$\text{c.o.v. of } \hat{\lambda}_i = \frac{(\text{Var}[\hat{\lambda}_i])^{\frac{1}{2}}}{E[\hat{\lambda}_i]} \simeq \frac{1}{T^{\frac{1}{2}}} \frac{\left[\int_0^\infty \omega^{2i} G^2(\omega)\, d\omega\right]^{\frac{1}{2}}}{\int_0^\infty \omega^i G(\omega)\, d\omega}$$

For the spectral density function given by Equation (3) the result for the moments λ_i, i=0,1,2, is [8]:

$$\text{c.o.v. of } \hat{\lambda}_i \simeq \left[\frac{1}{T}\ \frac{1}{2\omega_n\zeta}\right]^{\frac{1}{2}}$$

Consider, for example, a 30 minute (T = 30x60 sec) record of the motion of a tall building with ω_n= 1 r.p.s. and a damping of about 1%. The computed coefficient of variation is 0.17, i.e., one sixth of c.o.v. of individual spectral ordinates.

There are other sources of error in the estimated spectra, and hence, in the damping estimate. Potentially the most serious distortion arises due to the fact that $\hat{G}(\omega_k)$ is really a convolution of $G(\omega)$ with the so-called sampling function †. As T becomes large, the sampling function becomes an impulse and there will be no distortion. At smaller values of T, the adverse effect of the convolution(of $G(\omega)$ with the sampling function) would be a spreading, flattening of the spectral peak at $\omega = \omega_n$, and hence, overestimating the damping. The condition that one width of the sampling function be less than a fraction of the half-power bandwidth produces an important lower bound on T. For linear one-degree systems, this means that the record length must exceed the value $4\pi(\omega_n\zeta)^{-1}$ (see footnote). For example, for $\omega_n = 1$ sec. and $\zeta = 0.01$, the record length must be at least 1,250 seconds or about 10 minutes. Incidentally, this condition also guarantees that the coefficient of variation of the estimated spectral moments will be less than 0.20.

CONCLUSIONS AND EXTENSIONS

A new method has been presented for determining the equivalent viscous damping of structures from records of their response to random excitation caused by wind and waves. From a Fourier decomposition of the vibration, an estimate of the spectral density function can be determined. For a single-degree-of-

---

† The sampling function has the form $\left[T \frac{\sin^2(\omega T/2)}{(\omega T/2)^2}\right]$. It becomes more peaked as T increases, but has constant area equal to 2π. The width of the central lobe of the sampling function is approximately $2\pi/T$. The condition $2\pi/T < .5\ \omega_n\zeta$ leads to $T > 4\pi(\omega_n\zeta)^{-1}$.

freedom system, the proposed procedure requires the computation of the first three spectral moments of the estimated spectral density in a frequency band which includes the fundamental frequency. The spectral parameter δ^2 is approximately proportional to the damping.

The procedure may be extended to estimate modal damping values of multi-degree-of-freedom systems by successively isolating portions of the estimated spectral density $\hat{G}(\omega)$ which contain individual modal peaks. The dominant mode is isolated first by defining suitable frequency band limits $\omega_a = r\omega_1$ and $\omega_b = \omega_1/r$. From the partial moments, both ζ_1 and $G_o(\omega_1)$ can be estimated. Next, the expected contributions due to the first mode to higher mode spectral peaks of $\hat{G}(\omega)$ are subtracted. The second modal damping can then be estimated from partial moments involving frequency contributions in the neighbourhood of ω_2; and so on. Several cycles of iteration may be necessary to insure that contributions from lower modes to higher mode spectral peaks are correctly accounted for.

The following features of the proposed estimation procedure based on partial spectral moments are notable:

(i) low and high frequency portions of the estimated spectra, which are unreliable due to time interval size, finite record length and non-stationary effects, measuring equipment limitations, etc., can be eliminated.

(ii) the record length needed to obtain reasonably stable estimates of partial spectral moments is relatively small.

(iii) smoothing of the "raw" spectral estimates is unnecessary; estimated spectral moments and parameters based upon them, may be expected to change very little as a result of smoothing.

Finally, the method of spectral moments can be used to estimate the parameters of rather arbitrary spectral density functions. The basic idea is to equate the <u>theoretical</u> spectral moments or the related quantities λ_0, ω_2 and δ (which are expressed in terms of the desired parameters) and the corresponding values <u>estimated</u> from a recorded trace.

REFERENCES

1 R.B. Blackman and J.W. Tuckey The Measurement of Power Spectra from the Point of View of Communication Engineering, New York, Dover Publication, (1959).

2 S. Cherry and A.G. Brady Determination of Structural Analysis of Random Vibrations, Proc. 3rd World Conf. Earthquake Eng. Wellington, New Zealand, Vol.II pp 1950-1967. (1965).

3 S.H. Crandall Measurement of Stationary Random Processes, Random Vibration, Vol. 2. MIT Press, (1963).

4 S.H. Crandall and W.D. Mark Random Vibration in Mechanical Systems. Academic Press, New York. (1963).

5 A.G. Davenport, M. Hogan, and B.J. Vickery An Analysis of Records of Wind Induced Building Motion and Column Strain taken at the John Hancock Center (Chicago), Boundary Layer Wind Tunnel Laboratory, U. Western Ontario, London, Canada, (1970).

6 W. Gersch, N.N. Nielsen and H. Akaike Maximum Likelihood Estimation of Structural Parameters from Random Vibration Data. J. Sound Vib. Vol.31 No.3 pp 295-308 (1973).

7 W. Gersch and D. Foutch Least Squares Estimates of Structural System Parameters Using Covariance Function Data. IEEE Trans. on Automatic Control, Vol. AC-19 No.6 (1974).

8 D.B. Harris Effects of Finite Record Length on a Damping Estimation for Tall Buildings. S.B. Thesis in MIT Dept. of Electrical Eng. and Computer Science. Supervised by E.H. Vanmarcke. (1973).

9 J.M. Jenkins General Considerations in the Analysis of Spectra. Technometrics, 3, pp 133-166. (1961).

10 E. Parzen Mathematical Considerations in the Estimation of Spectra. Technometrics, 3, pp 167-190. (1961).

11 E.H. Vanmarcke Parameters of the Spectral Density Function: Their Significance in the Time and Frequency Domains. Research Report R70-58, MIT Dept. of Civil Eng.(1970).

12 E.H. Vanmarcke Properties of Spectral Moments with Applications to Random Vibration, J.Eng.Mech.Div. ASCE Vol 98 (1972).

13 E.H. Vanmarcke, R. Dobry and G. Madera Estimation of Dynamic Properties of Structures and Soils, in Appl. of Prob. in Soil and Struct. Engr. Hong Kong U. Press pp 640-660 (1971)

14 E.H. Vanmarcke On the Distribution of the First-Passage Time for Normal Stationary Random Processes J. Appl. Mech. (1975).

E.H. Vanmarcke, Department of Civil Engineering, Massachusetts Institute of Technology, Cambridge, Massachusetts 02139, USA.

* Presented orally at the Closing Session.

W SCHIEHLEN

Random vibrations of magnetically levitated vehicles on flexible guideways

The motion of a vehicle on a double track guideway, Figure 1, can be split into the travelling motion, the heave-pitch-roll motion and the lateral yaw motion. By assumption of a symmetric vehicle, equal parallel tracks and vanishing external lateral forces the heave-pitch motion remains as the essential motion. As shown in [1] the state equations are composed by the vehicle, suspension and guideway dynamics as well as by the measurement and control dynamics. The state equation reads as

$$\left.\begin{aligned} \dot{x}(t) &= A(t)\,x(t) + V(t)\,z(t) + W(t)\,r(t) \\ x_{\nu+} &= Jx_{\nu-}\,, \quad \nu = 0,1,2,\ldots, \quad x_{0-} = x(t_0) \end{aligned}\right\} \tag{1}$$

where x(t) is the state vector, z(t) the deterministic disturbance vector and r(t) the stochastic disturbance vector characterized by a white noise vector process with zero mean and intensity matrix Q. Further, the system matrix A(t), the deterministic disturbance matrix V(t) and the stochastic disturbance matrix W(t) are periodic with period T. The matrix J characterized the jumping states due to the simply supported guideway beams.

For the analysis of the steady-state stochastic response of the periodically time-variant system Eqn. (1) reduces to a time-invariant system and the deterministic disturbance vector is omitted. Using the Lyapunov-transformation

$$x(t) = Z(t)\,\bar{x}(t), \qquad Z(t) = Z(t+T) \tag{2}$$

with the periodic matrix Z(t) one obtains from (1)

$$\dot{\bar{x}}(t) = \bar{A}\,\bar{x}(t) + \bar{W}(t)r(t) \tag{3}$$

where $\bar{A}$ is the time-invariant system matrix and W(t) is the periodically time-invariant stochastic disturbance matrix.

The random vibrations of (3) can be found by the covariance analysis method, cf [2]. The steady-state covariance matrix P(t) of the system (3) follows from

$$\overline{P}(t) = \overline{P}(t+T) = \sum_{k=1}^{\infty} (G_k e^{ik\Omega t} + G_k^* e^{-ik\Omega t}) \tag{4}$$

where the complex response matrices G_k are defined by the algebraic Lyapunov equation

$$(\overline{A} - \frac{1}{2} ik\Omega E)\, G_k + G_k\, (\overline{A} - \frac{1}{2} ik\Omega E)^T + Q_k = 0 \tag{5}$$

The corresponding intensity matrices Q_k are found by Fourier series expansion,

$$\overline{W}(t)\, Q\overline{W}^T(t) = \sum_{k=1}^{\infty} (Q_k e^{ik\Omega t} + Q_k^* e^{-ik\Omega t}) \tag{6}$$

However, the covariance matrix P(t) according to (4) is not correct due to the jumping states. It can be shown that the correct covariance matrix $P_J(t)$ is given by

$$\overline{P}_J(t) = \overline{P}_J(t+T) = e^{\overline{A}\tau} (S - \overline{P}_0)\, e^{\overline{A}^T \tau} + \overline{P}(\tau) \tag{7}$$

where $0 \leq \tau \leq T$ and S follows the algebraic Stein matrix equation. For vanishing jumping states, J = E and $S = \overline{P}_0$, results (4) and (7) coincide. The steady-state covariance matrix P_J, given in the original state variables

$$P_J(t) = P_J(t+T) = Z(t)\, \overline{P}_J(t)\, Z^T(t) \tag{8}$$

is likewise periodically time-variant.

REFERENCES

1 K. Popp and W. Schiehlen Dynamics of magnetically levitated vehicles on flexible guideways. Proceedings of IUTAM Symposium on the Dynamics of Vehicles on Roads and Railway Tracks, Delft, August 18-22, 1975, pp 479-503. Amsterdam: Swets & Zeitlinger, (1976).

2 W. Schiehlen Parameterregte Zufallsschwingungen. Z. angew. Math. Mech. Vol. 55, pp T67-T68. (1975).

W. Schiehlen, Technical University of Munich, West Germany.

* Presented orally at the Closing Session.

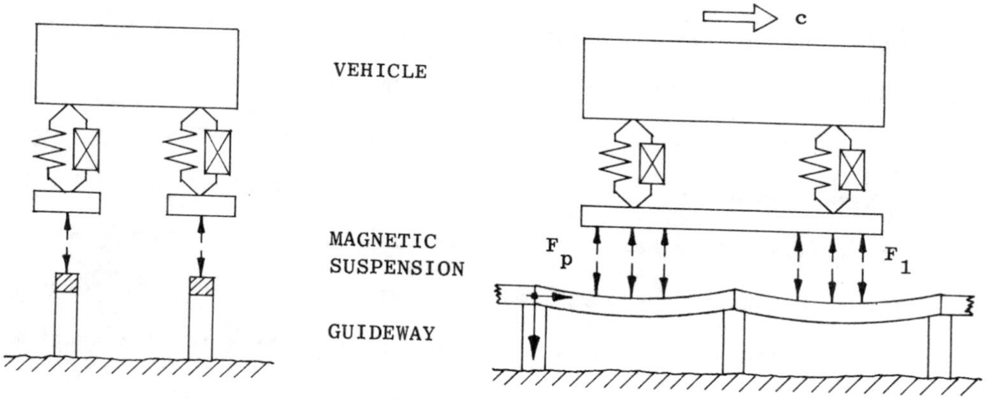

Figure 1 Vehicle-guideway model

J MURZEWSKI

Design problems of systems subject to random loads

INTRODUCTION

Decisions and responsibility are the specific problems of designer. Both depend on requirements. The requirements in civil engineering are defined in the standard specifications. The engineers' practice and judgement contributed to the shape of the specifications. New, inconventional developments are based on probabilistic considerations. A first step in applications of the probabilistic theories is given by introduction of the limit states method of design [1]. Semi-probabilistic design formats of this kind have been adopted since few years in some European and North-American countries. They have been studied also by some international organisations (European Concrete Committee, European Convention for Steel Construction etc.) and they are recommended by the International Standard Organisation [2] for the general use in structural analysis and design.

THE LIMIT STATES

There are two kinds of limit states: ultimate limit states and serviceability limit states. Both are considered from the point of view of a structural engineer. The first limit states are concerned with the collapse danger and the second limit states are concerned with the stiffness conditions.

If the serviceability limit states are considered, characteristic values of mechanical properties (strengths) and characteristic values of actions (loads) will be applied. If ultimate limit states are analysed, design values of strengths and design values of load will be taken into account.

There are no final agreements about definitions of the characteristic and design values, although their probabilistic sense is commonly recognised. Consistent definitions, which shall satisfy the requirements of simple design problems, are given below.

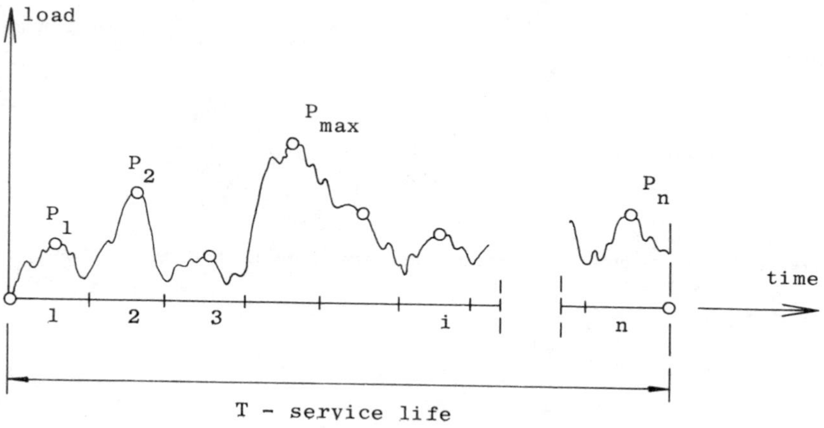

Figure 1

There is supposed that a structural member is subject to a stationary ergodic random loading process P(t) during the service time of the structure, $0 \leqslant t \leqslant T$ (Fig 1). The service life T is divided in n control periods (eg years) in such a way that the load peak P_i of an i-th period, i=1,2...n, is stochastically independent from a load peak P_j in another control period $j \neq i$. The maximum load

$$P_{max} = \max_{i=1}^{n} P_i \tag{1}$$

is characteristic for the structural member under consideration. P_{max} may be evaluated "a posteriori", ie after the service life, as the last value of the statistical series of load peaks

$$F(P_{max}) = P_n, \text{ if } P_1 \leqslant P_2 \leqslant \dots\dots \leqslant P_n. \tag{2}$$

The value P_{max} can be also estimated "a priori", if load peaks are measured during one control period (eg. a year) for a sample of N similar structural members.

$$F(P_k) = 1 - \frac{1}{n+1} \qquad (3)$$

where $F(P_i)$ - the estimated cumulative distribution function.

The characteristic value $P_{max} = P_k$ of a variable load P(t) is the fractile (3) of the distribution of $n > 1$ load peaks. The case n = 1 corresponds to a constant random load,

$$P(t) = \text{const.} \qquad (4)$$

The median is the characteristic value in this case

$$P_k = \check{P} \qquad (5)$$

The fractiles (3) for normal distributions are formulated as follows,

$$P_k = \overline{P} + k_n \, \mu_P \qquad (6)$$

for the Gauss distribution

$$P_k = \check{P} \exp \,(k_n \, \nu_P) \qquad (7)$$

for the log-normal distribution,

where $\overline{P}$, μ_P - mean and standard deviation

$\check{P}$, ν_P - median and logarithmic coefficient of variation.

The characteristic index k_n is defined by an equation

$$\tfrac{1}{2} + \tfrac{1}{2} \operatorname{erf} \frac{k_n}{\sqrt{2}} = 1 - \frac{1}{1+n} \, , \qquad (8)$$

where

$$\operatorname{erf} x = \frac{1}{\sqrt{\pi}} \int_{-x}^{+x} e^{-\xi^2} \, d\xi \text{ - the error function.}$$

If the number n is large ($n \gg 1$), asymptotic distribution functions of extremes may be applied to characterise a random maximum load $P = P_{max}$, $0 < t < T$,

$$F_{max}(P) = [\, F(P) \,]^n \dashrightarrow \begin{cases} \exp\left(-\exp \dfrac{\check{P}_n - P}{U}\right), & (9) \\[2ex] \exp\left(- \sqrt[u]{\dfrac{\check{P}_n}{P}}\right), \quad P > 0 & (10) \end{cases}$$

The central parameters $\check{P}_m$ or $\check{P}_n$ are the other estimates of characteristic values (2), (3),

$$P_k = \check{P}_n \quad \text{or} \quad P_k = \check{P}_n \tag{11}$$

The extremal distribution function (9) of the first kind (or the Gumbel) will be actual, if the Gauss distribution law characterises a single load peak and the extremal distribution function (10) of the second kind (or the Fréchet) corresponds to the log-normal primary distribution $F(P_i)$.

A characteristic value of the carrying capacity of structure has to be defined in a similar way. Its definition is easy for simple, statically determined systems (eg a chain subject to tension). The characteristic strength R_k of such a system may be determined from one realization (eg testing one chain). The value R_k is equal to the minimum strength of its m elements (links)

$$R_k = \min_{i=1}^{m} R_i \tag{12}$$

The characteristic strength R_k of the entire system can be estimated also from a sample of single elements (links) in such a way that the empirical cumulative distribution function is calculated and its fractile is taken

$$F(R_k) = \frac{1}{1+m} \tag{13}$$

We have in particular cases

$$R_k = \overline{R} - k_m \, \mu_R \tag{14}$$

for the Gauss distribution

$$R_k = \check{R} \exp\left(-k_m \, \nu_R\right) \tag{15}$$

for the log-normal distribution,

where $\overline{R}$, μ_R - mean and standard deviation,

$\check{R}$, ν_R - median and logarithmic c.o.v.,

$$\tfrac{1}{2} - \tfrac{1}{2} \operatorname{erf} \frac{k_m}{\sqrt{2}} = \frac{1}{1+m} .$$

If the number of elements of a system is large ($m >> 1$), asymptotic external distribution may be applied:

- the Gumbel distribution function for the Gauss primary distribution function $F(R_i)$,

$$F_{min}(R) = 1 - [1-F(R)]^m \longrightarrow 1 - \exp(- \exp \frac{R - \hat{R}}{W}) \tag{16}$$

- the Weibull distribution function for the log-normal $F(R_i)$,

$$F_{min}(R) \longrightarrow 1 - \exp(-\sqrt[w]{\frac{R}{\breve{R}}}), \quad R > 0 \tag{17}$$

The central parameters $\hat{R}$, $\breve{R}$ are called characteristic extremal values in statistics of extremes. They are the other estimates of the characteristic values (13).

Design values of the structural load, P_d, and strength, R_d, are defined as fractiles of the extremal distributions

$$F_{max}(P_d) = 1 - \omega_P \tag{18}$$

$$F_{min}(R_d) = 1 - \omega_R \tag{19}$$

where ω_P, ω_R - small probabilities of over-loading and under-strength, respectively.

The design values P_d, R_d are defined as follows,

- in the case of the Gumbel extremal distributions

$$P_d = \check{P} + t_\omega U , \qquad R_d = \hat{R} - t_\omega W, \tag{20}$$

- in the cases of the log-extremal distributions (the Frechet or the Weibull)

$$P_d = \check{P} \exp(t_\omega u), \qquad R_d = \breve{R} \exp(-t_\omega w), \tag{21}$$

where

$t_\omega = -\ln[- \ln(1-\omega)]$ - extremal tolerance index,

$\omega = \omega_P$ or ω_R, respectively.

The design values can be determined also from the primary distribution functions

$$[F(P_d)]^n = 1 - \omega_P, \qquad [1 - F(R_d)]^m = \omega_R \tag{22}$$

hence, for the Gauss distribution

$$P_d = \overline{P} + t_{\omega/n}\ \mu_P, \qquad R_d = \overline{R} - t_{\omega/m}\ \mu_R \tag{23}$$

and for the log-normal distribution

$$P_d = \breve{P}\ \exp(+\ t_{\omega/n}\ \nu_P), \qquad R_d = \breve{R}\ \exp(-t_{\omega/m}\ \nu_R), \tag{24}$$

where $t_{\omega/n}$, $t_{\omega/m}$ - the normal safety indices,

$$\tfrac{1}{2} - \tfrac{1}{2}\ \mathrm{erf}\ \frac{t_{\omega/\ell}}{\sqrt{2}} = 1 - \sqrt[l]{1 - \omega} \approx \frac{\omega}{l} \tag{25}$$

$\omega = \omega_P$ or ω_R, $l = n$ or m, respectively.

Obvious conclusions are that the characteristic and design loads, P_k and P_d, depend on the service life of the structure T, and the characteristic and design strengths, R_k and R_d, depend on the structure's size, defined by the number of structural elements m. However, authors of semi-probabilistic design methods sometimes disregard these facts.

PROBABILISTIC MODELS

Intrinsic theoretical models of loading process and the structural response are formulated for the probabilistic analysis of structures. Coupled effects of time and size are observed in many cases. Two such cases will be presented below, according to the previous solutions [4], [5].

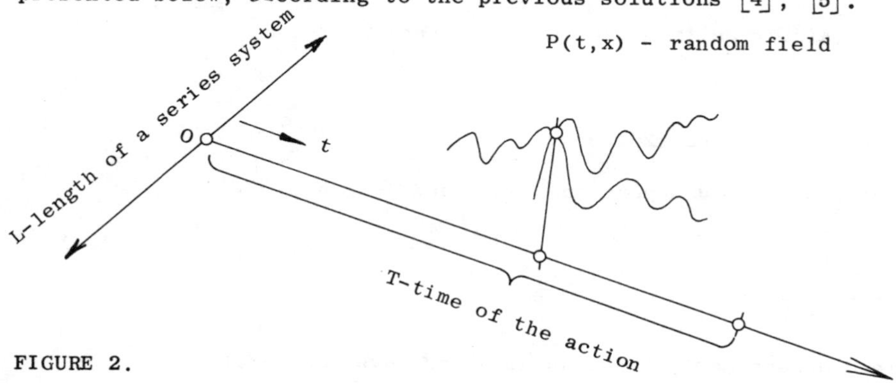

FIGURE 2.

The first case under consideration is a two-dimensional random load field P(t,x). It is defined for long structures such as multi-span bridges, industrial framed halls, slender ship hulls etc. (Fig 2). The point is that a failure of one structural segment means the failure of the entire structural system. A fully determined, but not necessarily constant load-carrying capacity Q(x) is assumed, $-\frac{L}{2} < x < +\frac{L}{2}$. The problem of exceeding of the level Q(x) by the Gaussian stationary field P(t,x) is solved so that a generalized formula for the risk of overloading has been derived, [4],

$$r = \frac{1}{\theta_L \theta_T} \exp\left[-\frac{(\overline{P} - Q)^2}{2 \mu_P^2}\right] , \tag{26}$$

where $\overline{P}$, μ_P - mean value and standard deviation of the instantaneous load,

$$\theta_L = \frac{\pi}{\ddot{\zeta}_L} , \qquad \theta_T = \frac{\pi}{\ddot{\zeta}_T} ,$$

$$\ddot{\zeta}_L = \left.\frac{d^2\zeta_L(x)}{dx^2}\right|_{x=0} , \qquad \ddot{\zeta}_T = \left.\frac{d^2\zeta_T(t)}{dt^2}\right|_{t=0}$$

An assumption was made that the two-dimensional auto-correlation function can be decomposed

$$\zeta(t,x) = \zeta_L(x)\, \zeta_T(t) , \tag{27}$$

for

$$0 \leqslant t \leqslant T, \qquad -\frac{L}{2} < x < +\frac{L}{2}$$

The probability of over-loading at any cross-section of the structure and at any time is calculated from the known exponential formula

$$\omega_P = 1 - \exp(-rLT) \tag{28}$$

where r = const., according (26).

The characteristic load depend on time T and length L

$$P_k = \overline{P} + k_{LT}\, \mu_P, \tag{29}$$

where $k_{LT} = \sqrt{\ln\left(\frac{L\ T}{\theta_L \theta_T}\right)^2}$;

the design load depend on T, L and an accepted level of safety ω_P

$$P_d = \overline{P} + \mu_P \sqrt{k_{LT}^2 + 2t_\omega} \approx \overline{P} + (k_{LT} + \frac{t_\omega}{k_{LT}})\ \mu_P, \qquad (30)$$

where $t_\omega = -\ln\left[-\ln(1-\omega_P)\right]$.

The second, approximate formula (26) may be applied, if $t_\omega << k_{LT}^2$. Then, the maximum load may be characterized by the Gumbel asymptotic distribution function. This has been verified by means of an analytic 'collocation' method, extended from a graphical estimation method, for which coordinate systems with functional scales are used.†

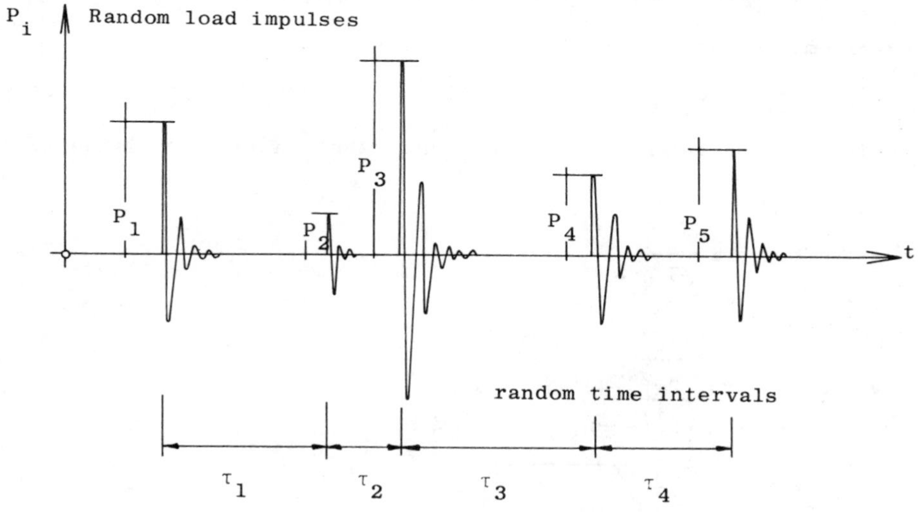

Figure 3

Another case, which has been analysed in paper [5] and also in the book [3] is concerned with a combined fatigue and over-loading problem++. The

---

† The statistical 'collocation' method is developed in the Politechnika Krakowska in co-operation with Dr A. Sowa.

++ Dr A. Winiarz (Politechnika Krakowska) co-operated in solving this problem.

structural member under consideration is subject to random load impulses P_i at random time intervals τ_i,

$$\tau_i = t_{i+1} - t_i, \quad i = 1,2,3 \ldots\ldots \tag{31}$$

Damped vibrations arise after each load impulse (Fig 3).

The cumulative decohesion of the structural material, $\lambda(t)$, must not exceed a specific constant λ_{ult},

$$\lambda = 1 - \exp\left(- \sum_{i-1}^{n} \sqrt[w]{\frac{\phi P_i}{\hat{R}}}\right) < \lambda_{ult} \,, \tag{32}$$

where $\hat{R}$, w - the Weibull parameters of the random micro-strength.

The logarithmic decohesion ratio,

$$\lambda' = \ln \frac{1}{1-\lambda} \tag{33}$$

is taken under consideration and the Gauss distribution function is derived for it

$$F(\lambda') = \tfrac{1}{2} + \tfrac{1}{2} \operatorname{erf} \frac{\lambda' - \overline{\lambda'}}{\sqrt{2}\, \mu_\lambda^{\;2}} \quad , \tag{34}$$

where $$\overline{\lambda'} = \sqrt[w]{\frac{\phi \check{P}}{\hat{R}}} \; \frac{T}{\overline{\tau}} \; \frac{\exp \frac{\nu^2}{2w^2}}{1 - \exp(-\frac{\delta}{w})}$$

$$\mu_\lambda = \sqrt[w]{\frac{\phi \check{P} T}{\hat{R}\, \overline{\tau}}} \; \frac{\exp \frac{\nu^2}{2w^2} \sqrt{\exp \frac{\nu^2}{w^2} - 1}}{1 - \exp(-\frac{\delta}{w})}$$

$\check{P}$, ν - median and logarithmic c.o.v. of load impulses,

$\overline{\tau}$ - mean time interval of the load impulses,

δ - logarithmic decrement of damping,

ϕ - proportionality factor between a load impulse and the stress in the structural member.

A definition of design value of λ' or λ would be rather useless because those hypothetical measures cannot be controlled. A simple safety condition is to check a design service life

$$T_d = \breve{T} + t_{\omega}^2 \, \bar{\tau}' \left(\tfrac{1}{2} - \sqrt{\frac{1}{4} + \frac{\breve{T}}{t_{\omega} \bar{\tau}'}}\right) \tag{35}$$

where $\breve{T} = \bar{\tau} \, \dfrac{1 - \exp(-\delta/w)}{\exp\dfrac{(1-w)\nu^2}{2w^2}} \sqrt[w]{\dfrac{\hat{R}}{P}} \ln \dfrac{1}{1-\lambda_{ult}}$, $\quad \bar{\tau}' = \bar{\tau}\,(\exp \dfrac{\nu^2}{w^2} - 1)$,

t_{ω} - the normal tolerance index for the probability ω.

A disadvantage of this approach is that a catastrophic failure of $100_{\omega}\%$ of cases will happen and nothing can be done in order to reduce this number, unless any control values of loads and material properties are defined.

SAFETY MEASURES

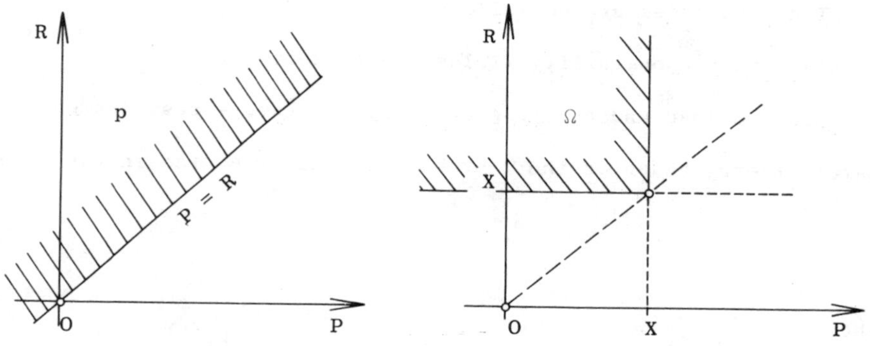

No control! Perfect control!

Figure 4

The reliability of a system depends on the efficiency of control. It could be no random failures, if the control was perfect. The structural reliability may be estimated in realistic conditions as follows

$$R = \psi \, \Omega + (1-\psi) p, \tag{36}$$

where ψ - efficiency of control,

$\omega=1-\Omega$ - probability of disqualification under condition of the perfect control.

$q=1-p$ - probability of failure under condition of no control.

The difference between the probabilities:

$$\Omega = \text{probability } (P < X, \quad R > X) \tag{37}$$

and $$P = \text{probability } (P < R) \tag{38}$$

is sometimes not understood. Figure 4 illustrates the fact that

$$\Omega \neq p. \tag{39}$$

Usually, the probability p is supposed to be an adequate measure of safety. However, if an information about the control efficiency is not available, the probability Ω shall be a better measure of the structural safety. The advantages are as follows:

i) - separate responsibility fields - easy control!

ii)- separate load and carrying capacity analyses - easy design!

A safety theory based on the Ω probability is presented in the reference [3].

REFERENCES

1 V. A. Baldin and others, Design of building structures according to the limit states, Russian edition - Stroizdat, Moscow (1951).

2 ISO/R 2394-1973, General principles for the verification of the safety of structures.

3 J. Murzewski, Safety of building structures, Polish edition-Arkady, Warszawa (1970, German Edition - VEB Verlag für Bauwesen, Berlin (1974).

4 J. Murzewski, Stochastic models of structural loads, Wissenschaftliche Zeitschrift der Hochschule fur Architektur und Bauwesen Weimar, (Vol 22), pp 192-194 (1975).

5 J Murzewski, Cumulative damage of solids for random stress, Engineering Fracture Mechanics (Vol.8), pp 131-146, (1976).

Professor J. Murzewski, Professor of Civil Engineering, Politechnika Krakowska, Kraków, Poland.

Y YAMANOUCHI

Nonlinear response of ships on the sea

NON-LINEAR RESPONSE OF SHIPS ON THE SEA

Even the viscosity and the compressibility of the water being ignored, the fundamental equation of motion of the surface waves is essentially non-linear. Usually however, under proper assumptions, the linear approximation is adopted in general cases for deep water waves, with practically applicable results.

As for the actual sea waves, the Gaussian process approximation has also been accepted generally, and the spectrum expression of waves has been utilised popularly, from wave generation analysis to many engineering purposes. Now, taking account of the angular propagation of energy, so called directional spectrum expression is usually used.

The response characters of the ship's behaviour of waves have also been approximated as linear, for most of the cases, for moderate responses. As the results, the ship's behaviour on the sea have been treated as Gaussian process too, and the stochastic process analysis technique, introduced and developed for these two decades, have been successfully applied with many fruitful results.

The spectrum expression of the ship's various response and their relations with that of the sea waves have come to be the common interest of the naval architects. The extreme values and other expected values predicted from the spectrum are now popularly used for ship's design. Ship's many realistic response characteristics have been analysed, introducing the cross-correlation or the cross spectrum, besides the auto correlation or the auto spectrum, not only for one input one output case but also for multiple input single output systems [1] together with a trial to attack the non-linearity that is included in one of the inputs.

The effect of non-linearity on the spectrum of sea waves was investigated in 1959 by L. J. Tick [2], and was shown to be rather trivial as far as the depth of water is not small. Accordingly, the sea waves, the input to the ship's behaviour are assumed as staying to be linear.

The effects of weak non-linearity of ship's response character on the shape of spectrum of the response were calculated by this author [3] using the perturbation method for the simplified ship's non-linear oscillation, for example the non-linear rolling with non-linear damping or the non-linear restoring term as are shown as follows:

$$\ddot{\theta} + 2\alpha\dot{\theta} \pm \beta\theta^2 + \omega_o^2\theta = m(t) \tag{1}$$

or

$$\ddot{\theta} + 2\alpha\dot{\theta} + \omega_o^2\theta + k\omega_o^2\theta^3 = m(t) \tag{2}$$

The apprximate expression of the spectra $S_{\phi_1\phi_2}(\omega)$ and $S_{\phi_2\phi_2}(\omega)$ were calculated as follows, for the non-linear rolling, expressed by equations (1) and (2) respectively.

$$S_{\phi_1\phi_1}(\omega) = S_{\phi_o\phi_o}(\omega)+2\sqrt{\frac{2}{\pi}}\,\sigma_{\dot{\phi}_o}\beta S_{\phi_o\phi_o}(\omega)\omega Im|H\phi_o m(\omega)| + \beta^2|H\phi_o m(\omega)|^2$$
$$|\frac{8}{\pi}\sigma_{\phi_o}^2\omega^2 S_{\phi_o\phi_o}(\omega)+ \frac{4}{3\pi\sigma_{\phi_o}^2}\int_{-\infty}^{\infty}\int_{-\infty}^{\infty} S_{\dot{\phi}_o\dot{\phi}_o}(\omega_1)S_{\dot{\phi}_o\dot{\phi}_o}(\omega_2)S_{\dot{\phi}_o\dot{\phi}_o}(\omega-\omega_1-\omega_2)d\omega_1 d\omega_2 \tag{3}$$

$$S_{\phi_2\phi_2}(\omega)= S_{\phi_o\phi_o}(\omega)-6\,k_3\sigma_{\phi_o}^3 S_{\phi_o\phi_o}(\omega).R|H\phi_o m(\omega)|+ k_3^2|H\phi_o m(\omega)|^2$$
$$\left\{9\sigma_{\phi_o}^4 S_{\phi_o\phi_o}(\omega)+ 6\sigma\int_{-\infty}^{\infty}\int_{-\infty}^{\infty} S_{\phi_o\phi_o}(\omega_1)S_{\phi_o\phi_o}(\omega_2)S_{\phi_o\phi_o}(\omega-\omega_1-\omega_2)d\omega_1 d\omega_2\right\} \tag{4}$$

here ϕ_o, $H_{\phi_o m}(\omega)$ being the linear solution and the frequency response function derived from Equation (1) or (2) neglecting the non-linear terms, R, I_m, show the real and imaginary part and $S_{\dot{\phi}_o\dot{\phi}_o}$, $\sigma_{\dot{\phi}_o}$ being the spectrum of $\dot{\phi}_o$, and the standard deviation derived from it.

These ways of approach might help us to guess the modification of the spectrum by the existence of weak non-linear terms in the response characteristic equation. However it is also very clear that it is only when the process is linear, that the spectrum includes all information on the statistical characters of the original process. When the process is not linear, it is not true any more. The 2nd order moment spectrum namely the ordinary spectrum does not include all information of the statistic characters of the process. Besides the 2nd order moment spectrum, namely the ordinary spectrum, the higher order moment spectrum [4] as the Bispectrum, the third order moment spectrum, or the Trispectrum, the fourth order moment spectrum, and so on come to be necessary to express the statistic characters of the process completely. This corresponds to the functional expression of the response process y(t) with the input x(t) as follows:

$$y(t)= y_0+y_1(t) +y_2(t)+y_3(t) + \dots\dots\dots\dots$$

$$= y_0+\int_{-\infty}^{\infty} h(\tau).x(t-\tau)d\tau+ \int_{-\infty}^{\infty}\int_{-\infty}^{\infty} h(\tau_1,\tau_2).x(t-\tau_1)x(t-\tau_2)d\tau_1 d\tau_2$$

$$+ \int_{-\infty}^{\infty}\int_{-\infty}^{\infty}\int_{-\infty}^{\infty} h\ (\tau_1\tau_2\tau_3)x(t-\tau_1)x(t-\tau_2)x(t-\tau_3)d\tau_1.d\tau_2.d\tau_3 +\dots\dots \quad (5)$$

Before getting into the analysis of the higher order impulse response function $h(\tau_1,\tau_2)$, $h(\tau_1,\tau_2,\tau_3)$,--- or their double or triple Fourier integral, the higher order frequency response function $(H\omega_1,\omega_2)$, $H(\omega_1\omega_2\omega_3)$,- the 3rd order moment spectrum namely the Bispectrum $B(\omega_1,\omega_2)$of the rolling of a ship on the sea was investigated as the first step as in Fig 1 [5]. This is the Bispectrum of the rolling of a ship named Seiun-Maru on the sea, when the skewness of the original time series was 0.22 and the peakedness was 2.79, namely of a rolling that looked like rather non-linear, non-Gaussian. Looking at this figure, we find the typical trace of ridges that

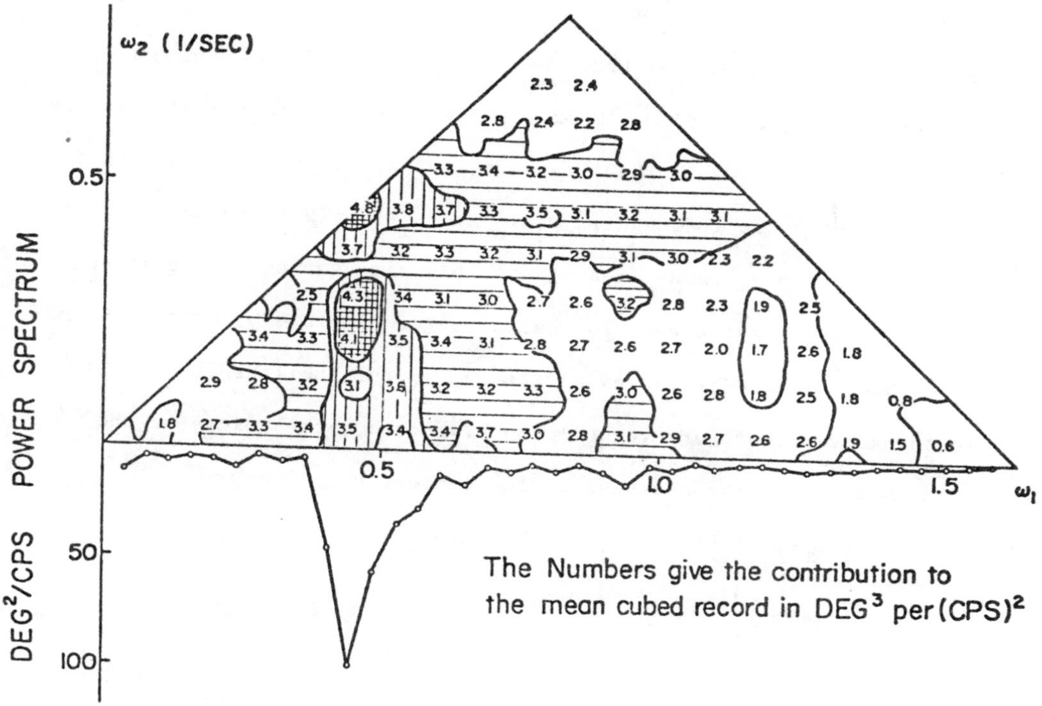

FIGURE 1. Bispectrum of a Rolling of Seiun-Maru on the Sea.

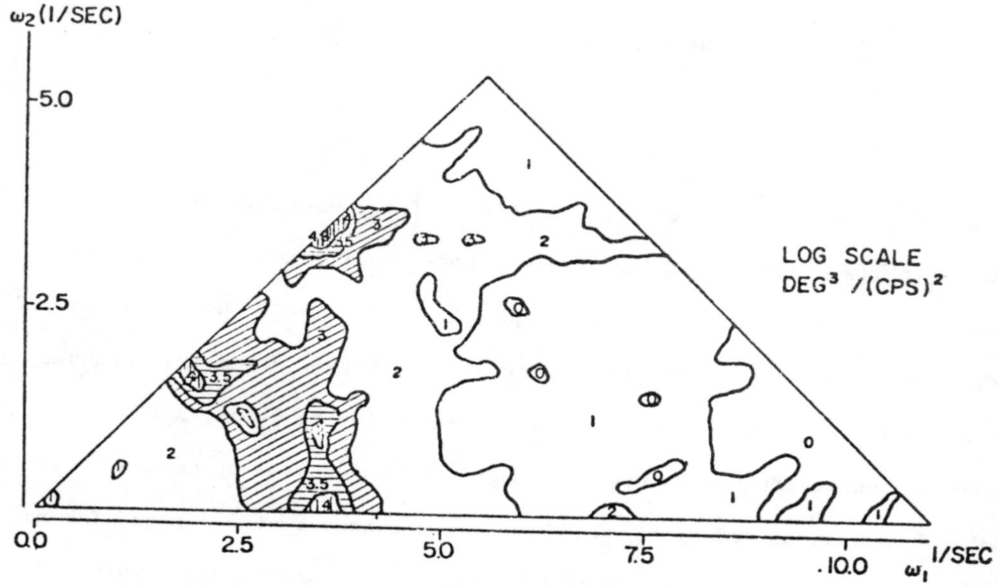

FIGURE 3. Bispectrum of a Non-linear Rolling by Simulation.

appeared on the line of $\omega_1 + \omega_2 = \omega_0$, $\omega_1 = \omega_0$, $\omega_2 = \omega_0$, ω_0 being the peak frequency of the 2nd moment spectrum that is drawn under the abscissa ω_1 axis, as the marginal case of $\omega_2 = 0$ of the Bispectrum. The highest peak is also recognised at $\omega_1 = \omega_2 = \omega_0$. These facts show that there are significant dependencies between the frequencies $\omega_1 = \omega_0$ and $\omega_1 + \omega_2 = 2\omega_0$ and so on, that show the non-Gaussianity, non-linearity of this time series. Here we can refer a Bispectrum of sea waves in shallow water obtained by K Hasselmann et al [6], that is one of very few Bispectra, so far published.

If we think over the characters of the Bispectrum, the volume of the Bispectrum should show the skewness of the process as,

$$\text{Volume of the Bispectrum} = \int_{-\infty}^{\infty}\int_{-\infty}^{\infty} B(\omega_1,\omega_2)\,d\omega_1 d\omega_2$$

$$= \left|\int_{-\infty}^{\infty}\int_{-\infty}^{\infty} B(\omega_1,\omega_2)e^{j(\omega_1\tau_1+\omega_2\tau_2)}d\omega_1.d\omega_2\right|_{\tau_1=0,\,\tau_2=0} \equiv \left|M(\tau_1,\tau_2)\right|_{\tau_1=0,\,\tau_2=0}$$

$$= \left|E\{x(t).x(t+\tau_1).x(t+\tau_2)\}\right|_{\tau_1=0,\,\tau_2=0}$$

$$= \left|\lim_{T\to\infty}\frac{1}{T}\int_{-\frac{1}{2}}^{\frac{1}{2}} x(t).x(t+\tau_1).x(t+\tau_2)\,dt\right|_{\tau_1=0,\,\tau_2=0} \qquad (6)$$

Skewness should stay to be zero, however, if the oscillation is purely symmetry to its neutral position, even if non-linear damping or non linear restoring term exist, as in eqns. (1) and (2). Accordingly the skewness 0.22 of this run should show that the rolling must have been asymmetry, because of the non upright neutral position by the wind blowing abeam, and also because of the different values of damping and the restoring moment by the sides. This was checked by Fig 2, where the skewness of the series of runs, in which the direction of the winds relative to the ship's course were

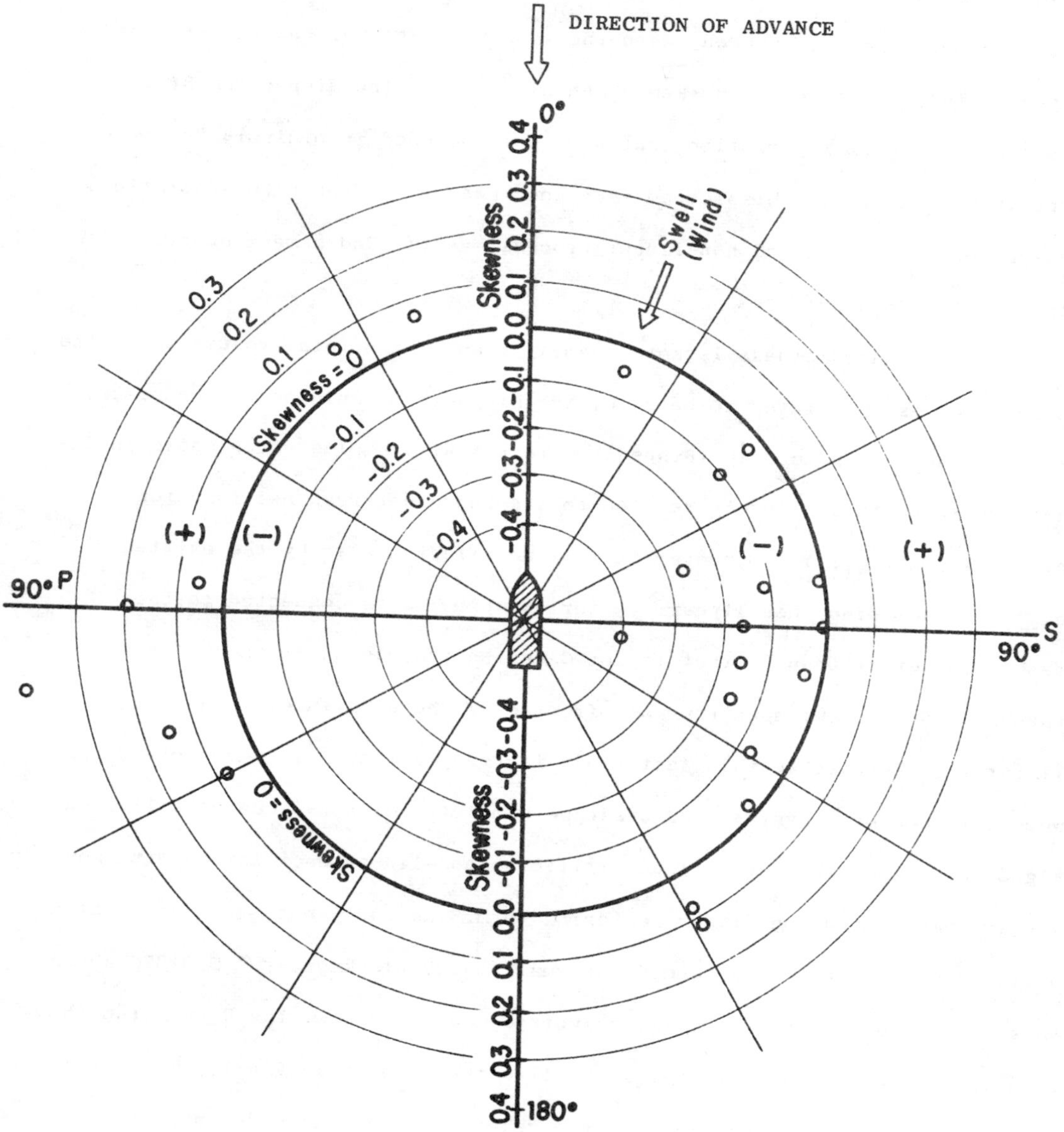

FIGURE 2. Variation of the Skewness of Rolling by the direction of the Wind

changed by runs, were plotted. The value of skewness deviated from zero to plus or minus values, depending upon the direction of the wind. The rolling becomes more and more skewed, when the wind blows closer and closer from abeam. When the process is skewed and non-linear, the Bispectrum $B(\omega_1,\omega_2)$ includes a lot of information that was not included in ordinary 2nd order spectrum. Namely the Bispectrum is a good measure to show the statistic characters of the skewed non-linear process, as the 2nd moment spectrum was for the linear process.

Even when the skewness is zero, namely even if the total volume under the Bispectrum is integrated to be zero, the Bispectrum could show some peaks, ridges and valleys locally,(Bispectrum is not always plus as is shown in its Argument), as it is closely related to the 2nd order response function $h(\tau_1,\tau_2)$ or $H(\omega_1,\omega_2)$. Moreover it will be affected also by the existence of the higher order non-linearity, for example, as is connected to the peakedness other than 3, that is for Gaussian process, just as the ordinary, the 2nd order moment spectrum was affected by the existence of the non-linearity. We checked this fact analysing many artificial time series, that were obtained by a synthesized analogue simulator of non-linear oscillation. Fig 3 shows the Bispectrum of an artificial non-linear oscillation with non-linear damping and non-linear restoring term. As the input, the output of a white noise generator was used. The small value of skewness - 0.01276 shows that this time series was really symmetry. Nevertheless, the Bispectrum shows a pattern of non-linearity that looks like the same as in Fig 1. This comes mainly from the non-linearity that are related to the 2nd order non-linear response character and also to the other kind of non-linear response character and also to the other kind of non-linearity as to give the peakedness of 2.388.

We know the peakedness is closely related to the fourth moment, namely to the fourth moment spectrum, or the Trispectrum,

$$T(\omega_1\omega_2\omega_3) = (\frac{1}{2\pi})^3\int_{-\infty}^{\infty}\int_{-\infty}^{\infty}\int_{-\infty}^{\infty} M(\tau_1\tau_2\tau_3)e^{-j(\omega_1\tau_1+\omega_2\tau_2+\omega_3\tau_3)} d\tau_1 d\tau_2 d\tau_3$$

where

$$M(\tau_1\tau_2\tau_3) = E\left[x(t).x(t+\tau_1).x(t+\tau_2) .x(t-\tau_3)\right] \tag{7}$$

However, it is not realistic nor practical at present to treat the Trispectrum, even if the peakedness μ_4 is another important parameter besides variance σ^2, skewness μ_3 to express the statistical characters of a non-linear process.

As another example of non-linear process, skewed this time, an analogue simulation was also tried for another non-linear oscillation, that is with non-linear asymmetry damping $\beta\dot{x}^2$(not $\beta|\dot{x}|\dot{x}$) that resulted with the skewness of 0.3122. The Bispectrum of the time series of this system is shown in Fig 4, where again the typical pattern as was in Fig 1 can be recognised clearly. The Bispectrum of the white noise that was used as the input for this simulation is shown in Fig 5. Almost nothing appeared as was expected, and checks that this white noise was quite Gaussian.

Thus we can get information on non-linear process from the Bispectrum, and may be we can proceed to the cross-Bispectrum analysis [7] [9], from which we expect to get the two dimensional transfer function $H(\omega_1\omega_2)$ or $h(\tau_1\tau_2)$. However, we are not familiar even to the two dimensional expression of frequency response or impulse response function $H(\omega_1\omega_2)$ $h(\tau_1\tau_2)$ to show the non-linear response characters. We have to study a lot more on the non-linear response system.

Besides, even after getting the Bispectrum or the non-linear response characters, we have to know a lot more to express the statistical characters of the original process. We have no probability theory to predict the extreme

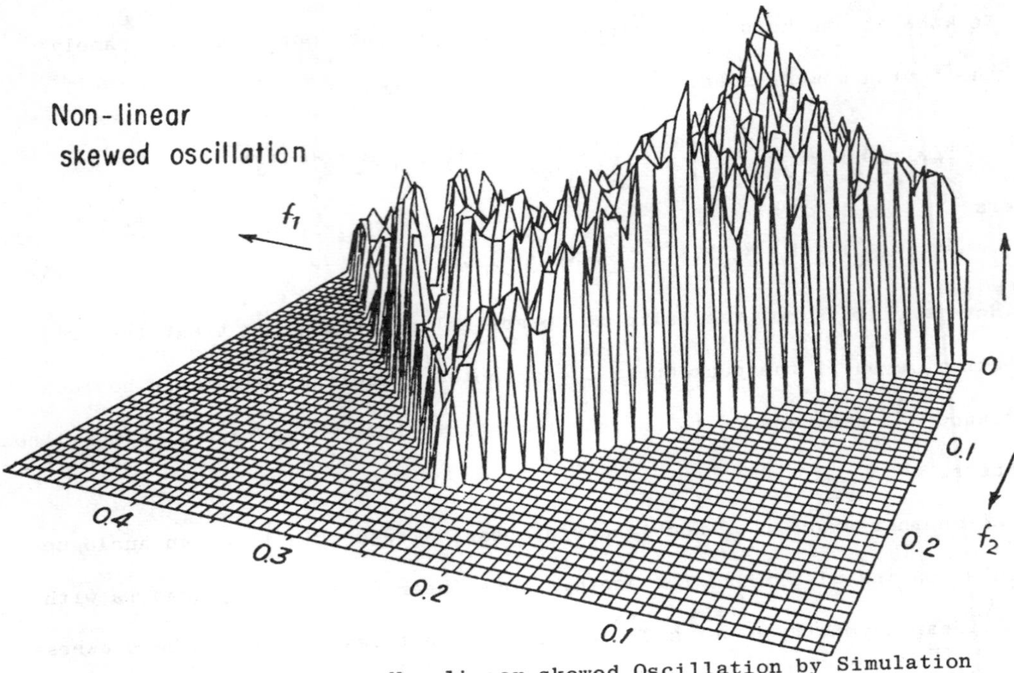

FIGURE 4. Bispectrum of a Non-linear skewed Oscillation by Simulation

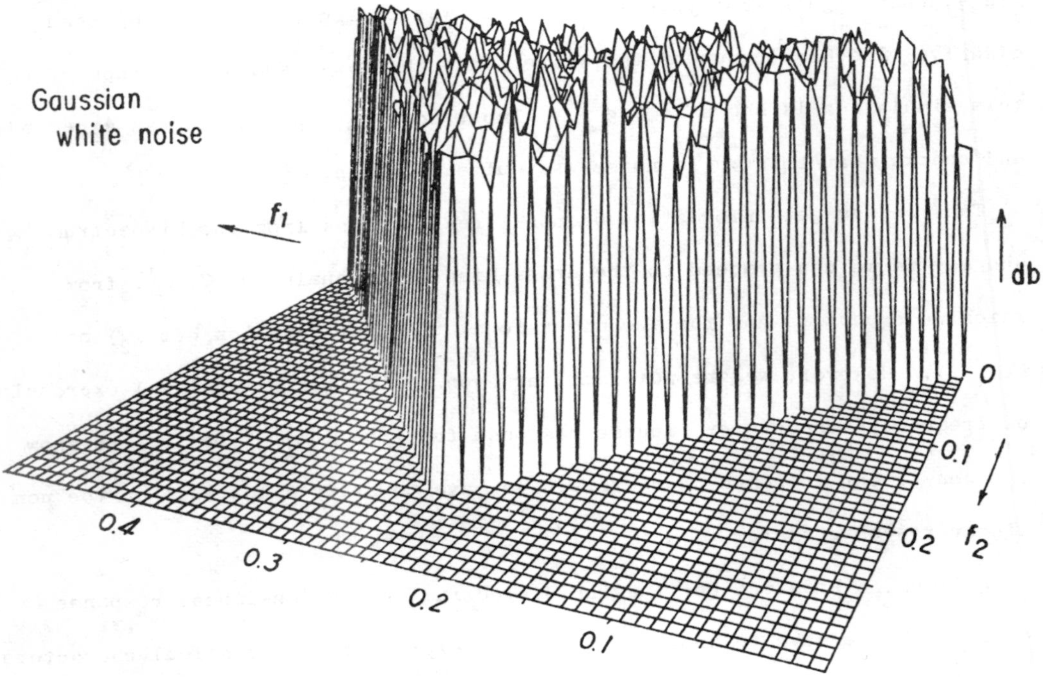

FIGURE 5. Bispectrum of Input White Noise

values or other expected values from them, in case of non-linear process, as we had Cartwright and Longuet-Higgins' [8] theory for Gaussian process. We wish we can have this kind of theory developed in near future, to enable us to utilise the results of Bispectrum analysis, and encourage us to do more in this kind of higher order spectrum analysis.

At the end of this short note, this author wants to express his sincere thanks to Prof K. Ohtsu and Mr H Oda on their co-operation in pursuing these studies.

REFERENCES

1 Y. Yamanouchi, On the application of the multiple input analysis to the study of ship's behaviour and an approach to the non-linearity of responses. Selected Papers from Journal SNA Japan, 7, 92-111 (1971).

2 L. J. Tick, A non-linear random model of gravity waves -1. J. Math. Mech.8,(No 5) 643-652 (1960).

3 Y. Yamanouchi, On the non-linearity of the ship's response on the sea. Proceedings on the symposium of Inst of Stat Math, Part 1, Analysis of Time Series, 60-82 (1971).

4 D.R. Brillinger, Asymptotic theory of estimates of K-th order spectra, Proc. Advanced Seminar on Spectral Analysis of Time Series, Ed. B Harris 153-188 (1967).

5 Y. Yamanouchi and K. Ohtsu, On the linearity of ship's response and the higher order spectrum - Application of the Bispectrum Jour SNAJ, 131, 115-135, (1972).

6 K. Hasselmann, W. Munk and G. MacDonald, Bispectra of Ocean Waves. Chapt. 8, Time Series Analysis, Ed M Rosenblatt, 125-139 John Wiley & Sons, New York, (1963).

7 L. J. Tick, The estimation of Transfer Function of quadratic systems, Technometrics, 3 (no 4) 563-567 (1961).

8 D. E. Cartwright and M. S. Longuet-Higgins, The Statistical distribution of the maxima of a random function, Proc R Soc, 237A, 212-232 (1956).

9 J.F. Dalzell, Cross-Bispectral analysis: Application to ship resistance in waves, Jour. of Ship Research, 18, 1, 62-72, (1974).

Dr. Y. Yamanouchi, Mitsui Engineering and Shipbuilding Co. Ltd.,
6-4 Tsukiji 5-Chome, Chuo-Ku, Tokyo 104, Japan.

K OTZU and G KITAGAWA

The stochastic control of ship's course keeping motion

INTRODUCTION

The essential aim of the conventional auto pilot system of a ship has been laid on merely to keep the desired course. However, we think that the up-to-date auto pilot system should furnish a faculty not only to keep the course faithfully but also to accomplish it through as little energy consumption as possible.

In this paper, we propose a way of designing a new auto pilot system which also take account of the energy consumption of a control device (i.e. a steering gear). In our approach, a discrete-time stochastic model of a ship is identified by the minimum AIC procedure proposed by Akaike [2], [3]. A new criterion function which is proper to our aim is proposed and the optimal control law is derived by the technique of Dynamic Programming.

The feasibility of our controller is examined by the simulation study.

AR MODEL AND ITS IDENTIFICATION

Hereafter, r-dimensional controlled variable (yawing, rolling, pitching, etc) and s-dimensional control variable (rudder angle, s=1) are represented by x_n and y_n, respectively.

Assuming that k-dimensional (k=r+s) process $X_n^t = (x_n^t, y_n^t)$ is a Gaussian stationary stochastic process with a finite second order moment, we represent the behaviour by an autoregressive model

$$X_n = \sum_{m=1}^{M} A_m X_{n-m} + U_n, \tag{1}$$

where A_m is a k x k matrix and U_n is a k-dimensional white noise which is independent of X_{n-m}, m = 1,2, ...

Then, partitioning A_m and U_n as

$$A_m = \begin{bmatrix} a_m & b_m \\ * & * \end{bmatrix}, \quad U_n = \begin{bmatrix} u_n \\ v_n \end{bmatrix},$$

where a_m and b_m are r x r and s x s matrices and u_n and v_n are r and s-dimensional white noise, respectively, we get an autoregressive type representation of controlled variable x_n,

$$x_n = \sum_{m=1}^{M} a_m x_{n-m} + \sum_{m=1}^{M} b_m y_{n-m} + u_n \tag{2}$$

In fitting the model, the determination of the order M is the most difficult problem. But we can determine the order reasonably by the minimum AIC procedure. In the present case, the AIC is given by

$$AIC(M) = N \, Log(det(d_M)) + 2r(r+s)M, \tag{3}$$

where N denotes the data length and $det(d_m)$ the determinant of the maximum likelihood estimate of the residual covariance matrix of u_n in equation (2).

By the minimum AIC procedure, the order M is given as the one which gives the minimum of AIC(M) and the parameters a_m and b_m are obtained by the maximum likelihood method.

STATISTICAL CONTROLLER DESIGN FOR SHIP'S COURSE KEEPING MOTION

Now we can proceed to design an optimal controller for the ship's course keeping motion (see Figure 1).

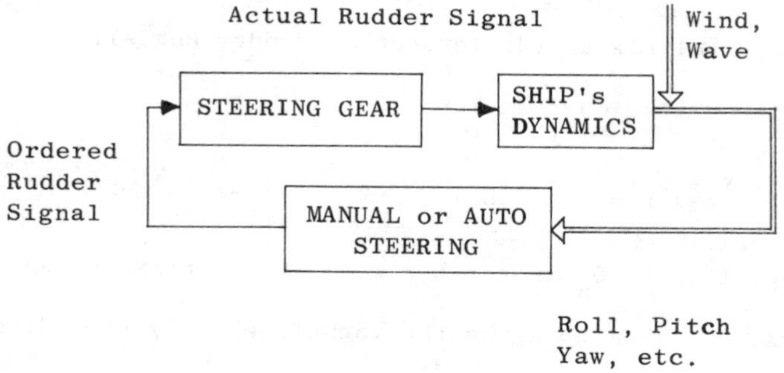

Figure 1 Steering System of Ship's Course Keeping Motion

Putting

$$\Phi = \begin{bmatrix} a_1 & I & 0 & \dots & 0 \\ a_2 & 0 & I & \dots & 0 \\ \vdots & \vdots & \vdots & \ddots & \vdots \\ a_{M-1} & 0 & 0 & \dots & I \\ a_M & 0 & 0 & \dots & 0 \end{bmatrix}, \quad \Gamma = \begin{bmatrix} b_1 \\ b_2 \\ \vdots \\ b_M \end{bmatrix}, \quad W_n = \begin{bmatrix} u_1 \\ 0 \\ \vdots \\ 0 \end{bmatrix}, \tag{4}$$

$Y_n = y_n$ and $H = [I\ 0\ \dots\ 0]$, we have a state space representation of the ship's dynamics

$$\begin{cases} Z_n = \Phi Z_{n-1} + \Gamma Y_{n-1} + W_n \\ X_n = H Z_n, \end{cases} \tag{5}$$

with (M x r)-dimensional state vector Z_n, which is equivalent to the autoregressive type representation (2).

In designing the auto pilot system, we must select a criterion function which is suitable for our aim.

The criterion function must penalize;

(i) the deviation of the controlled variables from a desired value,

(ii) the amount of control (rudder angle),

(iii) the rate of control signal (difference of rudder angle).

Thus, we adopt the criterion function

$$E\left[\sum_{n=1}^{J} \{X_n^t Q_n^t X_n + Y_{n-1}^t R_n Y_{n-1} + (Y_{n-1} - Y_{n-2})^t T_n (Y_{n-1} - Y_{n-2})\} \right], \tag{6}$$

where E denotes expectation, Q_n is a r x r weighting matrix, R_n and T_n are s x s weighting matrices. Then applying the technique of Dynamic Programming, we get the control law,

$$Y_n = G_n Z_n + H_n Y_{n-1}, \tag{7}$$

where the feedback gains G_n and H_n are given in [6]. Usually G_n and H_n converge to some gains G_∞ and H_∞ as n increase to infinity. Thus we can make use of the control law

$$Y_n = GZ_n + HY_{n-1},$$

where $G = G_1$ and $H = H_1$ for sufficiently large J.

SIMULATION STUDY

In this section, we proceed to discuss about the feasibility of our new auto pilot system. A set of data obtained by an experiment was considered, which was carried out on a standard container ship (about 16,400 gross tons and 175 meters in length) under manual steering at the speed of 10 knots an hour in the Pacific Ocean.

A ship's motion has six degrees of freedom. But we selected the yawing (Yaw), rolling (Roll) and lateral acceleration at the forepeak (Yacc) to represent the course keeping motion. They are considered as controlled variables and rudder angle (Rud) as control variable. A sixth order 4-dimensional autoregressive model was fitted by the minimum AIC procedure. The power spectra of the motion (Figure 2) are derived through the fitted model. It is noted that the power spectra of Yaw and Rud take their significant peaks around at D.C. .

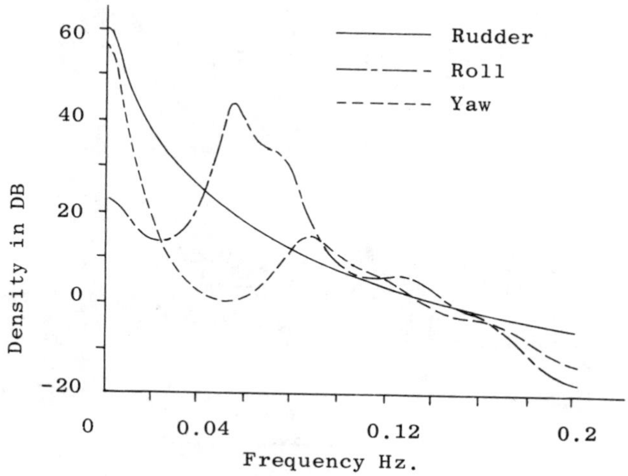

Figure 2 Power Spectra - by Manual Steering

Basing upon the result of identification of the model, we proceeded to design an optimal control system for the ship's course keeping motion. Effective weighting matrices Q, R and T were sought by the method of trial and error or a fairly systematic method [6].

To evaluate the performance of the controllers digital simulations were carried out using the equation of dynamics and the control law

$$Z_n = \Phi Z_{n-1} + \Gamma Y_{n-1} + W_n,$$

$$Y_n = GZ_n + HY_{n-1}$$

The system was driven by the pseudo white noise having the same covariance matrix as that of the fitted model.

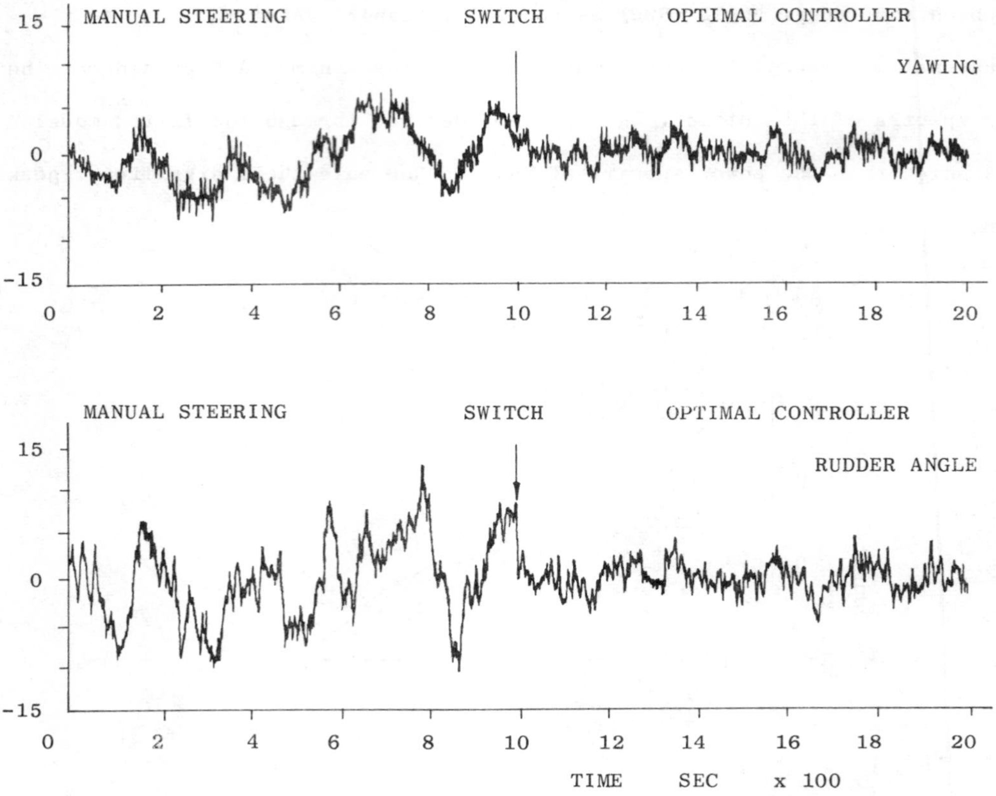

Figure 3 Comparison of Manual Steering and Simulation of Optimal Controller Based on the Fitted Model
Weighting Matrix; $Q = \begin{bmatrix} 7 & 0 & 0 \\ 0 & 35 & 0 \\ 0 & 0 & 1 \end{bmatrix}$, $R = 5$, $T = 3$.

Figure 3 illustrates a result of simulation where the left half of each record shows a simulation of manual steering and the right half that of steered by one of our optimal controllers.

From the results of simulation it is shown that by our controller

(i) the amount of yawing will be reduced compared with that of by manual steering,

(ii) the ship's course keeping motion will be able to be controlled through less and smoother rudder motion than by present auto pilot system [6],

(iii) the amount of rolling might be reduced [6].

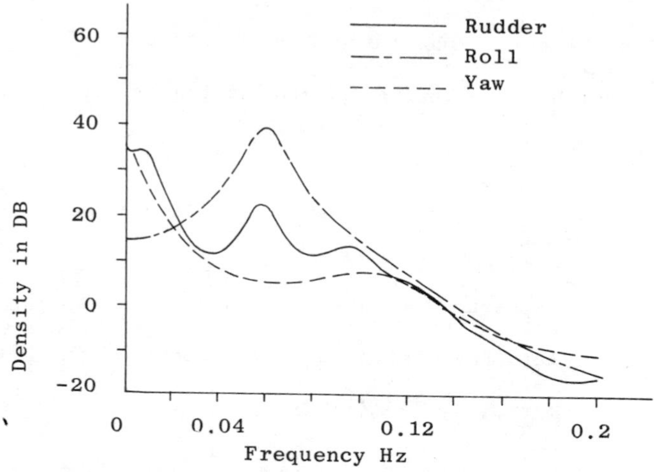

Figure 3 Power Spectra - by Optimal Controller

$$Q = \begin{bmatrix} 7 & 0 & 0 \\ 0 & 35 & 0 \\ 0 & 0 & 1 \end{bmatrix}, \quad R = 5, \quad T = 3.$$

Figure 4 illustrates the power spectra of Yaw, Roll and Rud when the ship is controlled by our controller, from which we can reconfirm the above insight.

CONCLUSIONS

A new method of designing a ship's auto pilot system is presented. The method is based on the statistical identification technique, in particular, the minimum AIC procedure. The simulation results show the possibility of developing an entirely new direct digital controller.

ACKNOWLEDGEMENT

The Authors express their hearty thanks to Dr Akaike of the Institute of Statistical Mathematics for his helpful suggestion. Thanks are also devoted to Dr Y. Yamanouchi of Mitsui Shipbuilding Co., and to Dr A. Ogawa of the Institute of Ship Research for allowing the use of the original record of the ship.

REFERENCES

1 H. Akaike Spectrum Estimation Through Model Fitting (q.v.)

2 H. Akaike Auto Regressive Model for control, Ann. Inst. Statis. Math., Vol.23 (1971).

3 H. Akaike Information Theory and an extension of the maximum likelihood principle, Problem of control and Information Theory, AKADEMIAI KIADO (1973).

4 T. Ootomo, T. Nakagawa, and H. Akaike Statistical Approach to Computer Control of Cement Rotary Kilns, Automatica Vol 8. (1972).

5 H. Akaike A New Look at the Statistical Model Identification, IEEE Trans. Automat. Control AC-19 (1974).

6 K. Otsu, G. Kitagawa and M. Horigome, The Prediction of Ship's Motion and Stochastic Control, IFAC Symposium on Ship Operation and Automatiion (1976).

K. Otsu, University of Mercantile Marine, Tokyo, Japan.

G. Kitagawa, The Institute of Statistical Mathematics, 4-6-7 Minami-Azabu, Minato-Ku, Tokyo, Japan.

G KITAGAWA

On the identification of ship's steering dynamics

INTRODUCTION

In designing an auto pilot system or analysing manoeuvrability of a ship, the dynamics of a ship's steering motion is of interest. The dynamics of a ship is usually represented by some deterministic models using differential equations. However, these models are not satisfactory, since the ships are usually under strong influence of disturbance such as wave and wind. The models also have a disadvantage that they are difficult to be identified from data obtained by experiments.

In this paper, the motion of ships are regarded as stochastic processes and behaviour of yawing (heading angle) and steering (rudder angle) is described by linear discrete-time models. The models have an advantage that they can be identified quite reasonably by the minimum AIC procedure.

Impulse response function and frequency response function are derived easily through the identified models.

Thus we established a procedure to obtain the characteristics of the manoeuvrability of a ship.

INPUT-OUTPUT RELATION MODELS

Let the rudder angle x_n be the input and the yawing y_n be the output. We assume that the behaviour of a ship is represented by an Input-Output Relation Model

$$y_n + \sum_{m=1}^{M} a_m y_{n-m} = \sum_{m=1}^{L} b_m x_{n-m} + u_n ,$$

where u_n is a disturbance to the system, which is not generally a white noise.

Since, for sufficiently large K, the disturbance would be represented by a moving average model

$$u_n = \varepsilon_n + \sum_{m=1}^{K} c_m \varepsilon_{n-m} ,$$

with a white noise ε_n, the equation becomes

$$y_n + \sum_{m=1}^{M} a_m y_{n-m} = \sum_{m=1}^{L} b_m x_{n-m} + \varepsilon_n + \sum_{m=1}^{K} c_m \varepsilon_{n-m} .$$

For the identification of the model, we must determine the order M, L and K of the model. By the minimum AIC procedure, the orders of the model are selected as the ones which attain the minimum of

$$\begin{aligned} \text{AIC} &= -2 \text{ (maximum likelihood)} \\ &\quad +2 \text{ (number of independently adjusted parameters)}. \end{aligned}$$

Under the Gaussian assumption of ε_n, an estimate of the AIC of the model is given by

$$\text{AIC}(M,L,K) = N \log (\sigma^2(M,L,K)) + 2(M+L+K),$$

where N denotes the data length and $\sigma^2(M,L,K)$ is the residual variance of ε_n after fitting the (M,L,K)-th order model by the maximum likelihood method.

A set of data (Figure 1) obtained by an experiment was considered. The experiment was carried out on a container ship, and yawing, rudder angle and other related variables are sampled at every four second.

The maximum likelihood estimates of the model were obtained using the Powell's least squares method, which minimizes the residual variance

$$\sigma^2(M,L,K) = \frac{1}{N} \sum_{n=1}^{N} \varepsilon_n^2 ,$$

without using derivative of σ^2. The residuals $\varepsilon_n (n=1,\ldots,N)$ are calculated easily when the initial values ε_n $(n=0,-1,\ldots,-K+1)$ are given. They are also considered as unknown parameters of the model. But in practical computation, they might be set to zero.

In fitting the models the orders M and L are restricted at most to three. For various combinations of M and L, (M,L,K)-th order models (K=0,1,...,9) are fitted and the value of AIC were compared.

* Value of AIC (y_n: yawing, x_n: rudder) *

K	M=1,L=1	M=2,L=1	M=2,L=2	M=3,L=3
0	164.8	107.3	108.3	80.1
1	109.2	100.2	91.2	81.1
2	85.7	70.3	70.7	71.7
3	85.5	66.8*	68.8	72.6
4	87.5	68.8	70.8	74.6
5	85.1	69.5	71.4	74.1
6	86.8	70.3	72.2	76.1
7	87.4	72.3	74.2	77.9
8	88.8	74.1	76.1	79.8
9	90.6	76.1	78.1	81.8

* shows the minimum

The AIC gives a (2,1,3)-th model

$$y_n - 1.822\, y_{n-1} + 0.821\, y_{n-2}$$

$$= -\ 0.021\, x_{n-1}$$

$$+ \varepsilon_n - 1.631\, \varepsilon_{n-1} + 0.952\, \varepsilon_{n-2} - 0.191\, \varepsilon_{n-3}$$

$$\sigma^2 = 1.308, \qquad \mathrm{AIC} = 66.84$$

Since the yawing is ideally given by integrating the yaw rate, the difference $Y_n = y_n - y_{n-1}$ of yawing y_n are also considered and the models

$$Y_n + \sum_{m=1}^{M} a_m Y_{n-m} = \sum_{m=1}^{L} b_m x_{n-m} + \varepsilon_n + \sum_{m=1}^{K} c_m \varepsilon_{n-m}$$

are fitted.

* Value of AIC (Y_n: yaw rate, x_n: rudder) *

K	M=1,L=1	M=1,L=2	M=2,L=1	M=2,L=2
0	105.9	106.8	97.1	91.6
1	99.9	91.5	98.2	93.3
2	69.5	69.7	67.0	68.6
3	65.7*	67.6	67.7	69.7
4	67.7	69.6	69.7	70.0
5	68.2	70.2	69.4	71.0
6	69.2	71.1	71.2	73.0
7	71.2	73.1	73.2	74.8
8	73.1	75.0	75.2	76.7
9	75.1	77.0	77.2	78.7

From the above table, we choose the model

$$Y_n - 0.828\, Y_{n-1} = -\, 0.020\, x_{n-1} + \varepsilon_n - 1.640\, \varepsilon_{n-1} + 0.962\, \varepsilon_{n-2}$$
$$-\, 0.196\, \varepsilon_{n-3}$$

or equivalently,

$$y_n - 1.828\, y_{n-1} + 0.828\, y_{n-2} = -\, 0.020\, x_{n-1}$$
$$+\, \varepsilon_n - 1.640\, \varepsilon_{n-1} + 0.962\, \varepsilon_{n-2} - 0.196\, \varepsilon_{n-3},$$
$$\sigma^2 = 1.315, \quad \text{AIC} = 65.67.$$

The value of AIC is shown to be reduced, from which this model is considered to be better fitted than the original model.

Through the fitted model, impulse response function from rudder to yawing is given by

$$W_i = b_i - \sum_{j=1}^{i-1} a_j w_{i-j},$$

where $a_m = 0$ for $m > M$ and $b_m = 0$ for $m > L$. Frequency response function $G(f)$ $(0 \leqq f \leqq \frac{1}{2\Delta T})$ are also given by

$$G(f) = \frac{\sum_{m=1}^{L} b_m \exp(-2\pi imf)}{1 + \sum_{m=1}^{M} a_m \exp(-2\pi imf)}$$

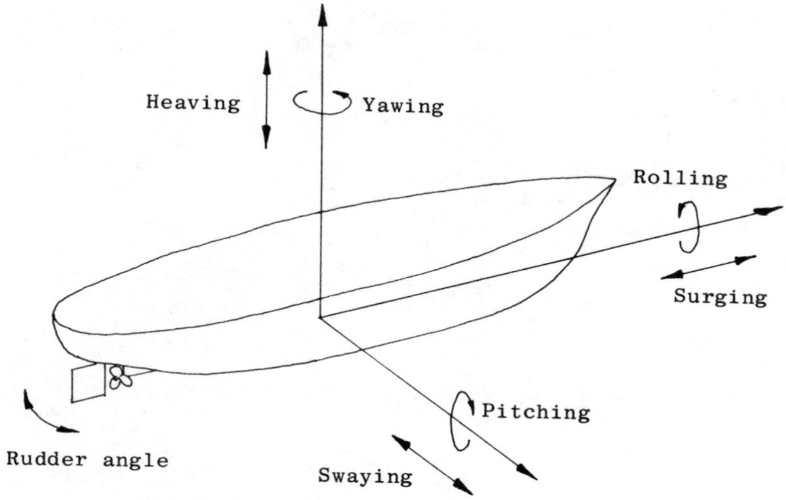

Figure 1

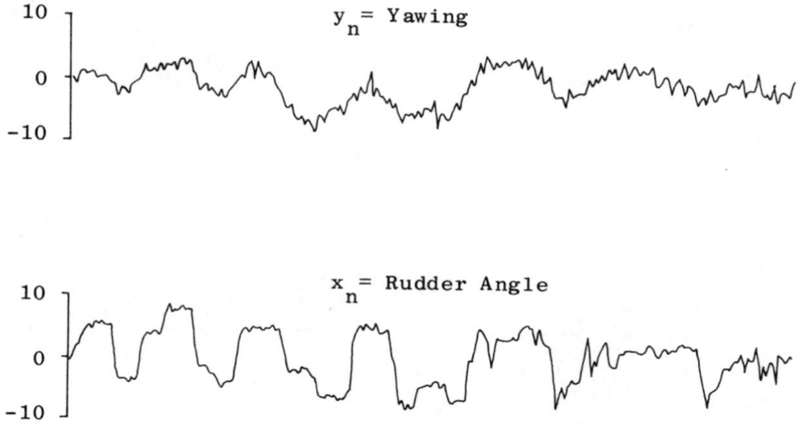

Figure 2 Data

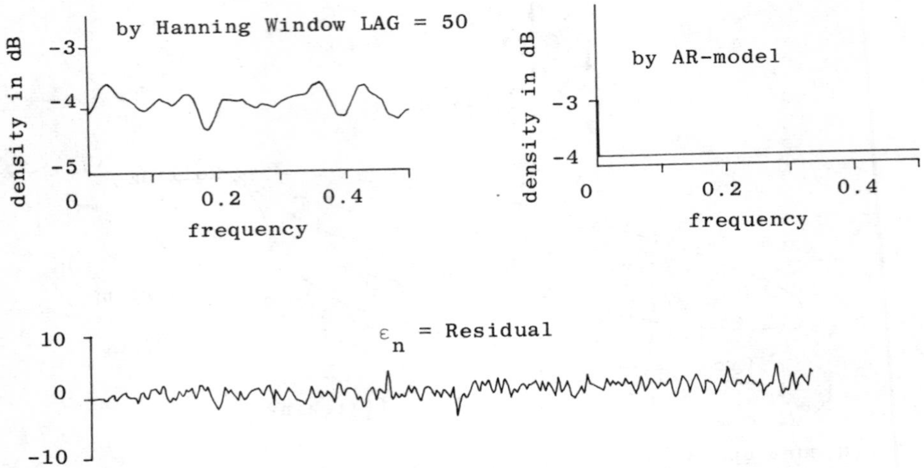

Figure 3 Residual ε_n and its power spectra

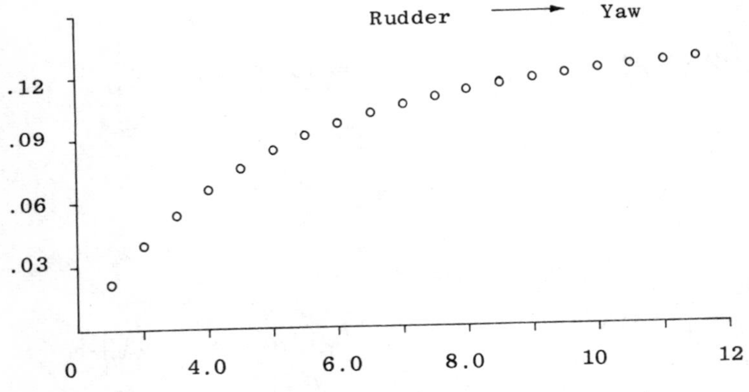

Figure 4 Impulse response function

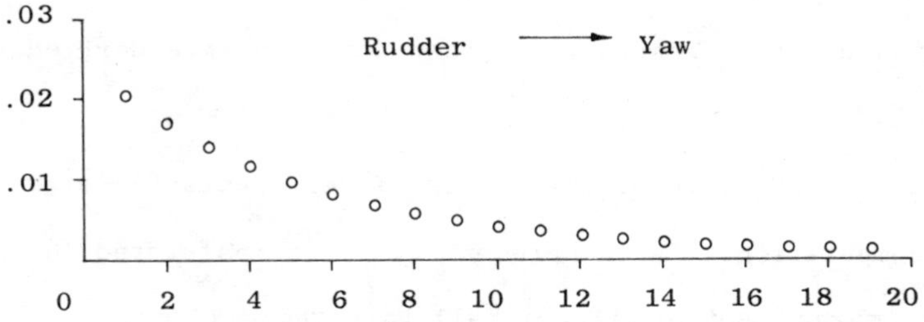

Figure 5 Impulse response function

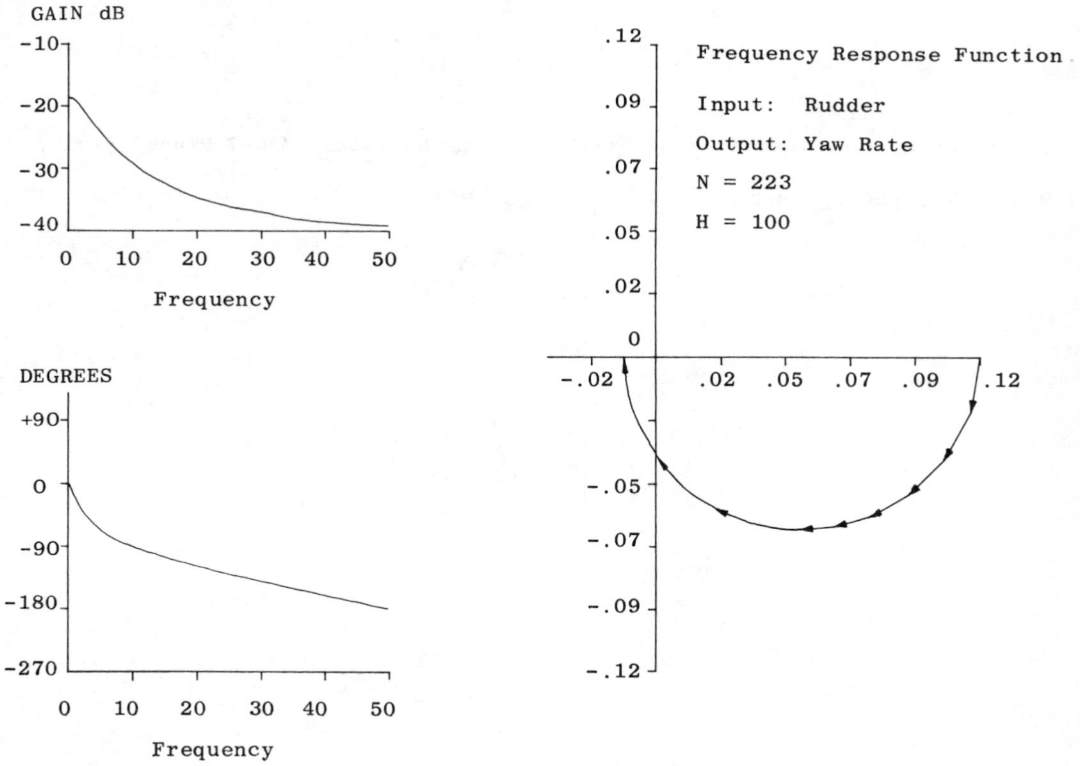

Figure 6 Frequency response function

CONCLUSION AND DISCUSSION

Ship's steering dynamics was represented by some linear discrete-time models. The models were identified by the minimum AIC procedure and the impulse response function and the frequency response function were derived through the fitted models.

In fitting the model, lack of uniqueness of representation might cause a difficulty in computation. In the present case, we restricted the orders M and L to at most three, and we did not fall into the difficulty.

G. Kitagawa, The Institute of Statistical Mathematics, 4-6-7 Minami-Azabu, Minato-Ku, Tokyo, Japan.